**IEE CONTROL ENGINEERING SERIES 62**

Series Editors: Professor D. P. Atherton
Professor G. W. Irwin
Professor S. Spurgeon

# Active Sound
# and Vibration Control
*theory and applications*

**Other volumes in print in this series:**

# Active Sound and Vibration Control
## *theory and applications*

**Edited by**
**Osman Tokhi** *and*
**Sandor Veres**

The Institution of Electrical Engineers

Published by: The Institution of Electrical Engineers, London,
United Kingdom

620·23 ACT

The Institution of Electrical Engineers,
Michael Faraday House,
Six Hills Way, Stevenage,
Herts. SG1 2AY, United Kingdom
www.iee.org

**British Library Cataloguing in Publication Data**

Active sound and vibration control.—(IEE control engineering
series; no.62)
1. Vibration  2. Acoustical engineering
I. Tokhi, M. O.  II. Veres, Sandor M.,1956-  III Institution of Electrical Engineers
620.3′ 7

**ISBN 0 85296 038 7**

Printed from authors' CRC
Printed in the UK by Bookcraft Ltd, Bath

# Contents

# Foreword

Active control of sound and vibration has emerged as an important area of scientific and technological development in recent years. The general theory of wave interference, which is used as the basis of active control of sound and vibration, reportedly emerged in the 17th century. Initial principles of active control, on the other hand, have been reported in the late 1870s and the early 1930s. Since then, considerable effort has been devoted to the development and realisation of methodologies for the control of sound and vibration in various application areas. These have included analyses and reformulations of Huygen's principle of wave interference, parametric and nonparametric interpretations of the process of cancellation, performance assessment methods and fixed and adaptive/intelligent techniques within signal processing and control frameworks. These efforts have also identified and, to some extent, have addressed important issues of theoretical and practical significance, such as the nature of the sound/vibration (disturbance), system complexity, stability, causality, and implementation-related factors.

The type and nature of the source of disturbance has provided the main motivation for exploring different control structures and design criteria. The type of source, whether considered as compact or distributed, provides an opportunity to investigate single-input and multi-input types of control structure, accordingly. Similarly, the performance requirements at the system output level, whether cancellation is required at a specific point or over an extended region within the propagating medium, has stimulated investigations into single-output and multi-output types of control structure. Permutations of such options at the input and output levels thus provide considerable room for design flexibility. In either case, two distinct categories of control structure, namely, feedback and feedforward, have emerged over the years. Among these, the feedback control structure corresponding to a standard classic feedback control system has attracted considerable attention. However, the main difficulty arising in the design and realisation of active control systems based on such a control structure has been due to large controller gain requirements and the resulting implications on the stability of the system. This control structure does offer a great deal of potential in that a vast number of standard control system design methods and criteria can be utilised without too much difficulty. However, the design of active control systems based on this type of control structure will inevitably require a compromise between system performance and system stability. Feedforward control structures, on the other hand, offer a great deal of flexibility because they include feedback control structures as special cases. Moreover the problem of controller realisability due to large gains is resolved with this control structure. However, the issue of nonminimum phase characteristics, commonly exhibited by sound/vibration systems, is of relevance and forms an important design consideration when maintaining controller and system stability.

The nature of the source of disturbance is an influential factor in the choice of control strategy. A simple control mechanism may suffice to deal effectively with tonal disturbances. However, if the disturbance is of a broadband nature, the control mechanism has to produce a required set of characteristics over a broad frequency range. Further levels of complexity become necessary, depending on whether the disturbance is stationary, time varying or random in nature. Previous work has established that with time-varying or random phenomena in the system, the control mechanism is required to incorporate an adaptive capability to track variations in system characteristics and modify the control signal accordingly so that the desired performance is achieved and maintained. This has formed the motivation for much of the research into adaptive methods such as adaptive filtering techniques, self-tuning control and model-reference adaptive control. These have mainly incorporated traditional parameter estimation techniques such as least mean squares (LMS), recursive least squares (RLS) and their variants. Moreover, these methods have commonly used linear parametric models. This is largely due to the simplicity which such models provide in the analysis and design process. However, new methodologies, which make use of the optimisation capabilities of genetic algorithms (GAs) and the learning capabilities of neural networks, are finding their way into the area of active control. In parallel with these, investigations and development of non-linear modelling and control techniques for active control are currently in progress. The design of active control systems incorporating soft computing methods such as neural networks, GAs and fuzzy logic techniques provide a further level of development in the area of active control, and some of the recent developments are reported in this book.

Intelligent control methods provide opportunities for developing new control strategies. Neural networks, for example, have successfully been used in a number of engineering applications, including modelling and control of linear and nonlinear dynamic systems. The potential of neural networks can be exploited in devising suitable active control techniques. Similarly, GAs provide efficient nonlinear search methods resulting in accurate solutions. Accordingly, an opportunity is provided with GAs to explore, establish and optimise the geometrical arrangement of components in an active control system for enhanced performance as well as to identify the system model. The exploitation of the potentials of neural networks and GAs is evidenced in the literature, to some extent, in relation to the design of active control systems, and some of the recent developments are reported in this book. However, further developments in the analysis and design of active control systems based on soft computing methods are still emerging.

Stability has formed an important design issue in active control systems. This issue has been addressed in the case of airborne noise in ducts, for example, by developing multi sensor/multi source systems, so as to isolate the detector sensor from the secondary source radiation. Moreover, analyses based on relative stability measures have been provided for active noise control systems in three-dimensional sound fields. Design methods such as H-infinity and internal model control have been used to further address this issue, and some of these developments are reported in this book.

Owing to the advantages that digital implementation of controllers provide over corresponding analogue implementations, active control algorithms are commonly implemented using digital computing techniques. As the performance demands and hence the design complexity of the system increase, so do the computing power requirements of the computation domain utilised for the application. Such a trend has motivated the utilisation of special-purpose digital signal processing (DSP) devices, in either sequential or parallel forms, in active control applications. In either case, owing to a mismatch between the computation requirements of the algorithm and the computing resources of the processors, efficient real-time performance may not be achieved. This has economic implications on the design and development of such systems. Thus, investigations into efficient computing methods involving a variety of processor types to allow close matching of algorithms to processors, using DSP and parallel processing methods, can lead to favourable developments in the realisation of active control systems. Some of the recent developments in high-performance computing for real-time implementation of active control algorithms are reported in this book.

Developments in active control have allowed successful application of the concept in numerous industrial areas. Some of these include control of noise in ventilation systems, aircraft, automobiles and refrigerators and vibration control in vehicle driver/passenger seats and helicopters. Several companies sell headsets for the cancellation of noise in a variety of environments. Other applications include the control of flexible structures such as flexible robot manipulators and rotating machines, and in building engineering.

This book is an attempt to address some of the issues highlighted above. Accordingly, the purpose of this book is to report on established fundamental and new knowledge in the area of active sound and vibration control. This includes theoretical developments, algorithmic developments and practical applications. The book is divided into 15 chapters. The opening Chapter 1 provides a comprehensive review of previous developments in active control techniques, from the initial principles to the current state. The process of noise cancellation in a three-dimensional propagation medium, design assessment of active noise control systems using to the zones of cancellation and reinforcement, and additionally geometrical design considerations to achieve robust and stable performance are presented in Chapter 2. Methodologies and issues surrounding adaptive and iterative control techniques are addressed in Chapters 3–7. Among these, Chapter 3 reviews a number of adaptive control techniques, followed by specific methodologies and treatments in Chapters 4–7. Chapter 4 proposes stability-assured adaptive algorithms to update adaptive feedforward controllers. Chapter 5 discusses adaptive control based on control at individual harmonics for cancellation of periodic disturbances. In Chapter 6 a model-free time-domain iterative controller tuning method is introduced and extended for periodic noise cancellation using a two-degree-of-freedom controller. Chapter 7 summarises the main ideas of controller tuning via iterative refinement of models and controller redesign. Some of the new and recent developments in the areas of neuro and genetic modelling and control adopted for active control are presented in Chapters 8 and 9. In these chapters, the optimisation capabilities of GAs and the learning capabilities of neural networks are demostrated in terms of their exploitation and utilisation for active control solutions. Chapters 10–14 are essentially addressing the applications domain of active control, where specific methodologies are adopted and applied to relevant applications. Chapter 15 discusses digital computing techniques and demonstrates how these techniques can be utilised and adopted for practical realisation of active control methodologies.

The editors would like to thank Thomas Meurers, Galina Veres, Alfred Tan at Southampton University and Takatoshi Okuno at Sheffield University for their assistance in correcting the proofs and to Roland Harwood and Diana Levy at the IEE for their encouragement throughout the preparation of this book.

M O Tokhi
*University of Sheffield*
S M Veres
*University of Southampton*
March 2002

# Authors

## S. J. Elliott
*Institute of Sound and Vibration Research, University of Southampton, Southampton, UK, Email: sje@isvr.soton.ac.uk[1]*

## M. Fontana
*University of Naples, Federico II, Italy, Email: mfontana@unina.it*

## D. Guicking
*Drittes Physikalisches Institut, University of Göttingen, Göttingen, Germany, Email: guicking@physik3.gwdg.de*

## S. Honda and H. Hamada
*Department of Information and Communication Engineering, Tokio Denki University, Tokio, Japan, Email: honda@acl.c.dendai.ac.jp*

## M. A. Hossain
*School of Engineering Sheffield Hallam University, Sheffield, UK*

---

[1] email addresses of corresponding authors are indicated

## C. H. Hansen, M. T. Simpson and B. S. Cazzolato

*Department of Mechanical Engineering, University of Adelaide,*
*Adelaide, South Australia, Email: chansen@watt.mecheng.adelaide.edu.au*


## Y. Park, H. S. Kim and S.H. Oh

*Department of Mechanical Engineering*
*Korea Advanced Institute of Science and Technology*
*Science Town, Taejon, Email: osh@cais.kaist.ac.kr*


## R. S. Langley

*Department of Engineering, University of Cambridge, UK*


## B. Mocerino

*Ansaldo Transporti, Napoli, Italy*


## E. Rogers

*Department of Electronics and Computer Science, University of Southampton,*
*UK, Email: etar@ecs.soton.ac.uk*


## A. Sano, T. Shimizu, T. Kohno, H. Ohmori

*Department of System Design Engineering, Keio University,*
*Keio, Japan, Email: sano@sano.elec.keio.ac.jp*


## J. Stoustrup

*Department of Control Engineering, Aalborg University, Denmark*


## M. O. Tokhi, R. Wood, K. Mamour, Wen-Jun Cao, Jian-Xin Xu, H. Poerwanto

*Department of Automatic Control and Systems Engineering,*
*University of Sheffield, UK, Email:o.tokhi@sheffield.ac.uk*

## S. M. Veres, G. S. Aglietti, T. Meurers, S. B. Gabriel
*School of Engineering Sciences, University of Southampton, Southampton, UK, Email: s.m.veres@soton.ac.uk*


## A. Vecchio, L. Lecce
*Universita di Napoli, Napoli, Italy*


## M. Viscardi
*Active S.r.l., Napoli, Italy, Email: massimo.viscardi@tin.it*

# Part I

# Review of fundamentals

*Chapter 1*

# An overview of ASVC: from laboratory curiosity to commercial products

## D. Guicking

*Drittes Physikalisches Institut, University of Göttingen, Göttingen, Germany,*
*Email: guicking@physik3.gwdg.de*

*A short historical review of active noise control is followed by brief discussions of the physical principles and basis concepts of technical realisations, grouped by the geometrical system topology (one-dimensional, local and three-dimensional control). The Section on active vibration control starts, again, with historical remarks, followed by examples of technical applications to active mounts, buildings, active and adaptive optics and sound radiation control by structural inputs. The Chapter ends with a discussion of active flow control and some statistical data on publications in the field of active controls.*

## 1.1 Introduction

The concept of cancelling unwanted sound or vibrations by superimposing a compensation signal exactly in antiphase is not new. In acoustics, most of the early publications in this field are patent applications, showing that technical applications must have been intended. However, for a long time experiments were nothing more than laboratory demonstrations which were smiled at as curiosities, far from reality. Only modern electronics made technical applications feasible. The situation was different with the compensation of low-frequency mechanical vibrations; these techniques were used in practice at a very early stage. In the following, an overview is given of the historical development, technical realisations and present research activities.

## 1.2    Active Noise Control

### 1.2.1    Early investigations

The first experiments on the superposition of sound fields were presumably made in 1878 by Lord Rayleigh [246]. He describes under the heading 'Points of Silence' how he scanned, with his ear, the interference field produced by two electromagnetically synchronised tuning forks, and that he found maxima and minima of loudness. Although it can be assumed that these experiments should only prove that coherent sound fields can interfere in the same way as do optical fields (which was known since the days of Thomas Young), patent applications by Coanda [55, 56] and Lueg [181] aimed at possible noise reduction, however only in Lueg's proposal in a physically realistic way. Lueg's German and especially the related US application [179] with an additional sheet of drawings are therefore considered rightly as the first written documents on active control of sound.

Lueg had already proposed the usage of electroacoustic components, but the first laboratory experiments were documented by Olson [227, 228] who also listed far-sighted prospective applications. Technical applications were not possible at that time because of the clumsy electronic vacuum tube equipment, lacking sufficient versatility. Also, our ears present a problem, namely the nearly logarithmic dependence of the perceived loudness on the sound pressure. For example, a sound level reduction by 20 dB requires an amplitude precision of the compensation signal within 1 dB and a phase precision within six degrees of the nominal values – for all frequency components of the noise signal. These demands, together with the requirement of temporal stability, have impeded for a long time the technical use of coherent active compensation of acoustical noise, also termed anti-sound, until in recent years digital adaptive filters proved to be the appropriate tool.

### 1.2.2    The energy objection

In the context of the active cancellation of sound fields a question often posed is 'Where does the energy go?'. With the seemingly convincing argument that the primary field energy can only be enhanced by adding secondary sound sources, the concept of active noise control (ANC) is principally questionable. The objection is correct if the cancellation is achieved by interference only; a local cancellation leads to doubling of the sound pressure elsewhere. But a more detailed consideration reveals that the secondary sources can, properly placed and driven, absorb the primary energy. In other situations, the sources interact such that the radiation impedance is influenced and thereby the sound production reduced. This will be elucidated in the following sections.

### 1.2.3   The JMC theory

M. Jessel and his coworkers G. Mangiante and G. Canevet developed a theory which has become known, after their initials, as the JMC theory. They considered the problem sketched in Figure 1.1 (e. g. [149]). Sound sources $Q$ are located within a volume $V$ with surface $S$. Along $S$, secondary sources shall be arranged such that they compensate the sound field radiated to the outside, but do not alter the field within $V$. This is possible according to Huygens's principle: substitute sources $q$ continuously distributed along $S$ can create the same sound field in the outside as do the primary sources $Q$. With reversed poling, they produce a field which is in antiphase to the original one. Assuming that such reversed (and acoustically transparent) substitute sources operate together with $Q$, the sound fields in the outside cancel each other.

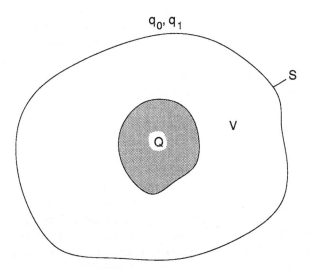

Figure 1.1    *JMC theory: primary sources $Q$ in a volume $V$ enclosed by a Huygens surface $S$ with monopole distribution $q_0$ and dipole distribution $q_1$*

If the cancellation sources are acoustic monopoles they radiate not only into the outside but also into $V$, creating standing waves and enhancing the sound energy in $V$. The inward radiation can be prevented by combining monopoles $q_0$ along $S$ with dipoles $q_1$ so that the primary field in $V$ is not altered. As to the energy, the tripoles formed by the $q_0$ and $q_1$ (directional radiators with cardioid characteristic) absorb, along $S$, the sound coming from $Q$. They serve as perfectly matched absorbers with an acoustic input impedance equal to the characteristic impedance of the medium.

With the same argument it follows that a source-free region $V$ can actively be shielded against sound influx from outside by arranging appropriate compensation sources along the surface $S$ of $V$. Monopole distributions along $S$ reflect, tripoles absorb the incident sound.

For a given surface $S$ and primary source distribution $Q(r)$, where $r$ is the position vector, the substitution sources $q_0(r)$ and $q_1(r)$ can be calculated from the Helmholtz-Huygens integral equation which links the sound field in a region to the sound pressure and its gradient along the surface [185].

For practical applications, the theoretically required continuous source distribution has to be replaced by discrete sources. Their minimal surface density follows from their absorption crosssection $A = \lambda^2/4\pi$ [235] and the smallest sound wavelength $\lambda$ for which the system shall be effective. This concept has been verified in computer simulations [186] and experimentally in an anechoic room [237]. A practical application is noise shielding of large open-air power transformers by an array of loudspeakers to save the people living in the surroundings from the annoying hum [168].

### 1.2.4  One-dimensional sound propagation, algorithms

Primary and cancellation sound must have the same direction of propagation. It is therefore easier to cancel plane, guided waves in ducts (below the onset frequency of the first lateral mode) than, for example, three-dimensional sound fields in rooms with omnidirectional propagation. In a setup as sketched in Figure 1.2 (which has in principle already been proposed by Lueg in 1934) the sound incident from the left is received by the microphone and, after some signal processing, fed to the loudspeaker such that to the right-hand side the primary (solid wavy line) and the additional signal (dashed) cancel each other.

After Lueg's idea, the signal processing should comprise amplitude adjustment, sign reversal and time delay in accordance with to the acoustic path length. However, an Active Noise Control (ANC) system is not practically applicable in this simple form. First, the acoustic feedback from the loudspeaker to the microphone has to be avoided and, second, in most cases it is necessary

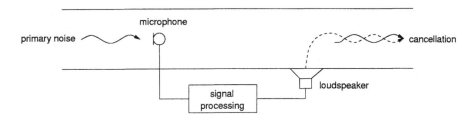

*Figure 1.2    Principle of active feedforward cancellation of sound in a duct*

to follow up the transfer function adaptively since the time delay and spectral decomposition can change as a result of temperature drift, superimposed flow and other environmental conditions. It is therefore common practice today to apply adaptive digital filters which are implemented on fast signal processors to enable online updating. Figure 1.3 shows a typical block diagram (amplifiers, A/D and D/A converters, as well as antialiasing lowpass filters being omitted).

The transfer function of the acoustic feedback path from the loudspeaker L to the reference microphone R is modelled by the feedback compensation filter FCF so that the input signal $x(t)$ of the main filter does not contain contributions from L. The error microphone E receives, in the case of incomplete cancellation, an error signal $e(t)$ which serves for the adaptation of the main filter A. The filter A adapts such that it models the acoustic transfer function from R to L, including the (complex) frequency responses of R and L. The filters A and FCF are often realised as transversal filters (finite impulse response or FIR filters), and the most common adaptation algorithm is the filtered-x LMS algorithm after Widrow and Hoff [337] where LMS stands for least mean squares. The algorithm is controlled by the product $e(t)x(t)$ and adjusts the filter coefficients by a stochastic gradient method so that $x(t)$ and $e(t)$ are decorrelated as far as possible. If the primary sound is broadband, the propagation delay from L to E decorrelates $x(t)$ and $e(t)$ to a certain degree which impairs the performance of the ANC system. In order to compensate for this effect, $x(t)$ is prefiltered in the update path (lower left) with a model $\tilde{H}_{LE}$ of the error path $H_{LE}$. The necessary error path identification is performed with an auxiliary broadband signal of the noise generator NG in the adaptation unit shown at the lower right of Figure 1.3. The coefficients of $\tilde{H}_{LE}$ (and also of FCF) are either determined once at startup and then kept constant or, if the transfer functions vary too much with time, permanently;

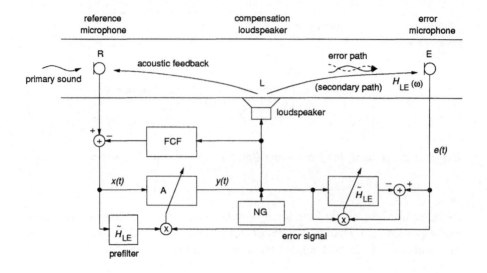

Figure 1.3    ANC in a duct by adaptive feedforward control with feedback cancellation and error path identification for the filtered-x LMS algorithm

in the latter case, however, the (weak) auxiliary signal remains audible at the duct end since it is not compensated for by the loudspeaker signal $y(t)$.

After adaptation the loudspeaker acts as a sound-soft reflector for the wave incident from the left which is, hence, not absorbed but reflected to the left. With a different control strategy the loudspeaker could be operated as an active absorber, but the maximum possible absorption is half of the incident sound power; either one quarter are reflected and transmitted. The reason is that it is not possible to achieve perfect impedance matching with a single loudspeaker mounted at the duct wall. The incident wave 'sees' the parallel connection of the loudspeaker input impedance and the characteristic impedance of the ongoing part of the duct. (But a loudspeaker at the end of a duct can be driven to perfectly absorb the incident sound [119].)

If the standing waves or the stronger sound propagation to the left in arrangements such as those in Figures 1.2 and 1.3 cannot be tolerated, a true active absorber can be realised with loudspeaker pairs or linear arrays [118, 285, 339].

A series of commercial ANC systems working on the principle of sound-soft reflection have been developed by the US company Digisonix and successfully installed mainly in industrial exhaust stacks since 1987 [84]. The filters A and FCF are combined to one recursive, infinite-impulse response (IIR) filter, often employing the Feintuch algorithm [91]. The signal processors allow on-line operation at least up to 500 Hz, suppress tonal noise by up to 40 dB and broadband noise typically by 15 dB. Similar systems have also been installed in Germany [66, 123, 129] and elsewhere. The lower frequency limit is given by pressure fluctuations of the turbulent flow, the upper limit by the computational speed of the signal processor and the lateral dimensions of the duct. The higher modes occurring at higher frequencies can only be cancelled by using a greater amount of hardware [88]; few such systems have therefore been installed so far.

The filtered-x LMS algorithm is very popular because of its moderate signal processing power requirement (the numerical complexity is $O(2N)$ if $N$ is the filter length), but its convergence is very slow for spectrally coloured random noise. Fan noise spectra have typically a steep roll-off with increasing frequency so that the convergence behaviour of the algorithm is often insufficient. Efforts have therefore been made in recent years to develop faster algorithms which are still capable of being updated in real time. One example is the SFAEST algorithm [214] which is independent of the signal statistics and has a complexity of $O(8N)$. Since it further calculates the optimal filter coefficients in one single cycle, it is particularly useful for nonstationary signals and nonstationary transfer functions. Stability problems in the initialisation period could be solved by the FASPIS configuration [240, 256, 257]. More on algorithms can be found in References [162, 304].

An important concept in many fields of ANC is adaptive noise cancelling which has become widely known since B. Widrow *et al.*'s 1975 seminal paper [338] (see Figure 1.4). A primary sensor picks up a desired signal which is corrupted by additive noise, its output being $s_p$. One or more reference sensors are placed such that their output $s_r$ is correlated (in some unknown way) with the primary noise, but does not essentially contain the desired signal. Then, $s_r$ is adaptively filtered and subtracted from $s_p$ to obtain a signal estimate with improved signal-to-noise ratio (SNR) since the adaptive filter decorrelates the output and $s_r$. This concept, realised by a linear predictive filter employing the least mean squares (LMS) algorithm, has been patented by McCool *et al.* [194] and found wide applications especially in speech transmission from a noisy environment [264], but also in seismic exploration [336], medical ECG diagnostics [335], a noise cancelling stethoscope [125] etc.

In adaptive feedforward control systems as shown in Figure 1.3 the sound propagation path from microphone R to loudspeaker L must be long enough

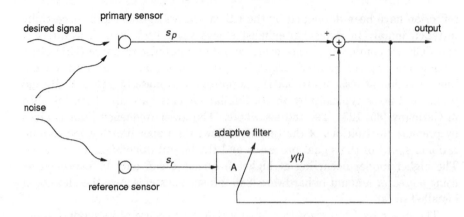

*Figure 1.4    Adaptive noise cancelling*

to provide the time required for calculating the signal to be fed to L (causality condition). The limiting factor is mostly not the computation time in the signal processor but the group delay in the antialiasing lowpass filters which are necessary in digital signal processing.

Problems in technical ANC applications are often posed by the loudspeakers. Very high low-frequency noise levels are typically encountered in exhaust stacks or pipes, demanding high membrane excursions without nonlinear distortion and, often, robustness against aggressive gases and high temperatures. On the other hand, a smooth frequency response function (as for hi-fi boxes) is not an issue because frequency irregularities can be compensated for by the adaptive filter. Special loudspeakers for ANC systems are being developed [44, 62, 72].

### 1.2.5   Interaction of primary and secondary sources

The ANC systems discussed in the preceding sections aimed at absorption or at least reflection of the primary sound power, tacitly assuming that primary power radiation is not influenced by the cancellation sources. However, if it is possible to reduce the primary sound production by the operation of the secondary sources, this will be a particularly effective method of noise reduction.

A monopole radiator of radius $a$ with a surface particle velocity $v$ produces a volume velocity $q_0 = 4\pi a^2 v$. The sound power radiated into a medium of density $\rho$ and sound velocity $c$ at a frequency $\omega$ (wavelength $\lambda$, wave number $k = 2\pi/\lambda$) is $P_0 = \rho\omega^2 q_0^2/(4\pi c)$. Adding an equal but antiphase monopole at a distance $d \ll \lambda$, produces a dipole which radiates the power $P_1 = P_0(kd)^2/3$. Supplementing this dipole by another one to form a quadrupole, the radiated power is further reduced to $P_2 = P_0(kd)^4/15$, assuming $kd \ll 1$ [211].

These conditions are correct if the volume velocity $q$ is the same in all three cases, but this is not necessarily so because ANC, by adding a compensation source in close proximity, does not only raise the multipole order, but can also alter the radiation impedance $Z_r = R_r + j\omega M_r$. The surrounding medium acts upon a monopole with the radiation resistance $R_{r0} = 4\pi a^2 \rho c$ and the mass load $M_{r0} = 4\pi a^3 \rho$ (three times the replaced fluid mass), upon a dipole with $R_{r1} = R_{r0} \cdot (ka)^2/6$ and $M_{r1} = M_{r0}/6$, and on a quadrupole with $R_{r2} = R_{r0} \cdot (ka)^4/45$ and $M_{r2} = M_{r0}/45$ [211]. The mass load leads to a reactive power, an oscillation of kinetic energy between primary and secondary source (acoustical short-circuit). The product $v^2 R_s$ determines the radiated (active) power. The particle velocity $v$ of the primary source depends on its source impedance and the radiation resistance. A low-impedance source (sound pressure nearly load independent) reacts on a reduced radiation resistance $R_s$ with enhanced particle velocity $v$ so that the reduction of radiated power by the higher multipole order is partially counteracted. But the sound radiation of impedance-matched and of high-impedance (velocity) sources is reduced in the expected way by an antiphase source in the nearfield.

These relationships can be utilised, e. g. for the active reduction of noise from exhaust pipes (ships, industrial plants, automobiles with internal combustion engines). A large demonstration project was implemented as early as 1980: the low-frequency hum (20 to 50 Hz) from a gas turbine chimney stack has been cancelled actively by a ring of antisound sources [287]. Each loudspeaker has been fed from one microphone pair through amplifiers with fixed gain and phase settings. Such a simple open-loop control system was sufficient in this case owing to the highly stationary noise and its narrow frequency band.

The pulsating gas flow emanating from a narrow exhaust pipe is a very efficient high-impedance monopole sound radiator; an adjacent antiphase source turns it into a dipole or, in the case of a concentric annular gap around the exhaust mouth, into a rotationally symmetric quadrupole. Active mufflers for cars using the principle of source interaction have often been proposed (e. g. Reference [47]), but practical installations are still lacking, for technical and economical reasons: the strong vibrational input beneath the car body, splash water, thrown-up gravel and the hot, aggressive exhaust gas form a hostile environment for loudspeakers and microphones; furthermore, active systems have to compete with the highly efficient and comparatively cheap conventional mufflers of sheet metal.

### 1.2.6   Waveform synthesis for (quasi)periodic noise

A conceptually simple adaptive algorithm has been developed by a British research team [47]. It assumes (quasi)periodic noise, the source of which is accessible for obtaining synchronisation pulses (e.g. vehicle engine noise). The principle is explained in Figure 1.5. A loudspeaker is mounted next to the exhaust pipe end, and is fed from a waveform synthesiser, realised with digital electronics. An error microphone is placed in the superposition zone and yields a control signal by which the loudspeaker output is optimised. The sync pulses (obtained, e.g., by a toothed wheel and an inductive probe) guarantee that the compensation signal tracks changing engine rotation speed automatically. The waveform is adapted using a trial-and-error strategy either in the time domain or, more quickly, in the frequency domain. In the latter case, the amplitudes and phases of the (low-order) harmonics of the engine noise are adapted. The prominent feature of this active system is that no microphone is required to receive the primary noise because the signal processor performs the waveform synthesis by itself. The loudspeaker must only provide the necessary acoustic power; resonances, nonlinearities and ageing are automatically compensated for. A disadvantage is the slower convergence as compared with true adaptive algorithms.

An example for a technical application of ANC with waveform synthesis in medicine is a noise canceller for patients undergoing a magnetic resonance imaging (MRI) inspection. The annoying impulsive noise is cancelled with the help of a pneumatic headset, made entirely of plastic material since no ferromagnetics and preferably no metal at all should be brought into the MRI tube [76].

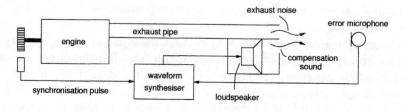

Figure 1.5    *Active cancellation of (quasi)periodic noise by tracking control with sync input and waveform synthesis*

### 1.2.7   Small volumes - personal noise protection

An acoustically simple ANC problem is presented by an enclosure the dimensions of which are small compared with the wavelength even at the highest frequencies of interest. The sound pressure is then spatially almost constant, and the cancelling source can be placed anywhere in the enclosure. Correctly fed, it acts as an active absorber.

One such small enclosure is the space between headphones and ear drum. The idea of personal noise protection by actively controlled headphones was originally documented in a Russian patent application [41] in 1960, but reliable signal processing was, in spite of intense research work in many countries, possible only very much later. Nearly simultaneously the US company Bose [196] and Sennheiser in Germany [320] presented active headsets for aircraft pilots; these have also been produced by other companies, as general purpose hearing protectors for low-frequency noise.

### 1.2.8   Local cancellation

Placing an anti-source in the immediate near-field of a primary noise source gives a global effect, as explained in Section 1.2.5, but if the distance between the two sources gets wider, then only local cancellation by interference remains [235]. Such systems did not, however, receive general attention because of their very limited spatial range of efficiency (in the order of $\lambda/10$). Testers also disliked the strong sound level fluctuations which occurred when they moved their heads. But local cancellation can be very useful for acoustic laboratory experiments, such as head-related stereophony when dummy head recordings are reproduced by two loudspeakers [259]. As the sound radiated from the left loudspeaker should be received by the left ear only, a compensation signal is superimposed onto the right channel which cancels the sound coming from the left loudspeaker to the right ear, and vice versa, see Figure 1.6. As compared to the familiar source localisation between the loudspeakers of a conventional stereo set, this procedure provides true three-dimensional sound field reproduction with source localisation in any direction, including elevation, and also gives a reliable distance impression.

Of great practical relevance is local active sound field cancellation for teleconferencing and hands-free telephones (speakerphones) in order to compensate, at the microphone location, for acoustic room echoes which degrade the speech quality and tend to cause howling by self-excitation (dereverberation of the room response) [124, 160].

A related older problem is the removal of electric line echoes in long-distance telephony with satellite communication links where the long trans-

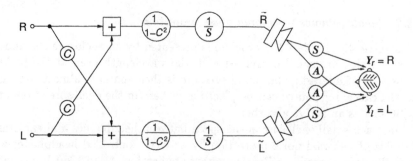

Figure 1.6    *Crosstalk cancellation in stereophonic sound field reproduction with loudspeakers by prefiltering. $S = S(\omega)$ and $A = A(\omega)$ are nearside and farside transfer functions, respectively, between loudspeakers and eardrums; circles to the left from the loudspeakers indicate filters with inscribed transfer functions, $C = C(\omega) = -A(\omega)/S(\omega)$*

mission path also leads to speech degradation by audible echoes [274]. The signals are reflected from an impedance mismatch at the so-called hybrid where the two-wire line branches into the four-wire local subscriber cable. The geostationary satellites are positioned at 36 000 km height so that the echo return path (transmitter → satellite → receiver → satellite → transmitter) is $4 \times 36\,000$ km which yields an echo delay time of nearly 0.5 s.

Locally effective ANC systems with compact microphone/loudspeaker systems in feedback configuration have been described by Olson as early as 1956 [227]; they absorb low-frequency sound in a narrow space around the microphone and have been proposed for aircraft passengers and machine workers [71].

The application of acoustic echo cancellation has also been proposed for ultrasonic testing where flaw echoes can be masked by strong surface echoes. It is possible to subtract the latter from the received signal and so improve the detectability of flaws [110, 137]. Similarly, the ANC technique can be applied to cancel the reflection of the ultrasonic echo from the receiver [127].

### 1.2.9    *Three-dimensional sound fields in enclosures*

The active cancellation of complex sound fields in large rooms, possibly with nonstationary sources and time-varying boundary conditions, is far beyond the scope of present ANC technology. More realistic is the concept of reducing room reverberation by placing active absorbers along the walls. The incident sound is received by microphones which feed the loudspeakers so that their acoustic input impedance is matched to the sound field. The situation is the same as that in Figure 1.1 if the enclosure walls are considered as a Huygens surface. The loudspeakers can also be driven so that their reflectivity takes arbitrary values in a wide range (experimentally, reflection coefficients between 0.1 and 3 have been realised). This would facilitate the construction of a room with adjustable reverberation time [331], but at present still with a prohibitive amount of hardware.

Intense research has been devoted to the active cancellation of sound in small enclosures such as vehicle cabins. Four-stroke internal combustion engines have an inherent unbalance at twice the rotational speed (the second engine order) which often coincides with the frequency of the fundamental cabin resonance of cars, so exciting the highly annoying boom. Since this noise is strongly synchronised with the engine speed, its active cancellation is possible with a relatively small amount of hardware and software [79]. It has, however, only been offered in a production car for some time by Nissan for their model *Bluebird* in Japan. Many other car manufacturers have developed their own systems, and some of them have successfully built prototypes, but all of them are hesitating for several reasons to install the ANC systems in series production.

More involved is the cancellation of broadband rolling noise, both inside and outside the car. Laboratory experiments and driving tests have led to preliminary solutions; the nonstationarity of the noise input and of the acoustic transfer functions demand fast adapting algorithms, also for error path identification [32, 37, 256, 257]. The noise and vibration problems are becoming more severe with the small low-consumption cars now under development; they will possibly be equipped with both ANC for the interior space and active vibration control (AVC) for the engine and wheel suspensions. For more luxurious cars the trend in the automobile industry goes to combining ANC technology with "sound quality design" for the car interior so that the driver has the choice, e.g. of a more silent car or a more sportive sound [97, 255].

Also, for economic reasons, the aircraft industry tends to replace, for short and medium distances, jet engines by propeller (or turboprop) aircraft which are, however, much louder in the cabin. Relatively little effort is necessary to employ a technology known as synchrophasing. The eddy strings separating from the propeller blade tips hit the fuselage and excite flexural vibrations of the hull which radiate sound into the cabin. If the right and left propeller are synchronised so that their "hits" meet the fuselage out of phase instead of simultaneously, then higher-order shell vibrations are excited which radiate less and so reduce the noise level inside [99]. Better results, however, with more involved installations are obtained with multichannel adaptive systems. A European research project with the acronym ASANCA has resulted in a technical application [152].

An important issue in ANC applications to three-dimensional sound fields is the placement of microphones and loudspeakers. Attention has to be paid not only to causality, but also to observability and controllability, in particular in rooms with distinct resonances and standing waves (modal control). If, for some frequency, the error microphone of an adaptive system is positioned in a sound pressure node it does not receive the respective frequency component or room mode so that no cancelling signal will be generated and no adaptation is possible. If the loudspeaker is placed in a node, then a compensation signal calculated by the processor cannot effectively be radiated into the room, which usually leads to higher and higher signal amplitudes and finally to an overload error of the digital electronics.

## 1.2.10 *Free-field active noise control*

Technical applications of ANC to three-dimensional exterior noise problems are still quite rare, but many research projects have been reported and a number of patents exist. The problems with active mufflers for cars with internal combustion engines were discussed in Section 1.2.5. A technically similar problem is the fly-over noise of propeller aircraft which mainly consists of two components: the propeller blade tip vortex threads and the equally impulsive exhaust noise. If the exhaust tail pipe is shifted to a position near to the propeller plane, and if the angular position of the propeller on its shaft is adjusted so that in a downward direction the pressure nodes of one source coincide with the antinodes of the other one, then the destructive interference reduces the fly-over noise by several dB [154].

A method for reducing traffic noise by cancelling the tyre vibrations of an automobile is disclosed in Reference[321], proposing electromagnetic actuation of the steel reinforcement embedded into the tyres.

A frequently investigated problem is the cancellation of power transformer noise, either by loudspeakers arranged around the site [60], by force input to the oil in which the transformer is immersed [49] or to the surrounding tank walls [135], or by sound insulating active panels enclosing the transformer [113, 133]. Experimental results are discussed in Reference [13].

## 1.3  Active Control of Vibrations

### 1.3.1  Early applications

In contrast to ANC, AVC has long been applied, in particular to ships. Mallock reported in 1905 on vibration reduction on a steam ship by synchronisation of the two engines in opposite phase [184], Hort in 1934 on the reduction of roll motion by an actively driven Frahm tank [136] (water is pumped between tanks located on the two sides of the ship) and Allen in 1945 on roll stabilisation by buoyancy control with activated fins, auxiliary rudders with variable angle of attack protruding laterally from the ship hull into the water [9]. The latter technology is still practiced today.

Active damping of aircraft skin vibrations was proposed in 1942 [319], providing multichannel feedback control with displacement sensors and electromagnetic actuators, mainly in order to prevent fatigue damage.

Early publications can also be found on the active control of vibrations in beams, plates and composite structures. In mechanical wave filters where a desired longitudinal wave mode in a bar is superimposed by an interfering detrimental flexural wave mode, the latter can be damped by pairs of piezoelectric patches on either side of the bar which are connected through an electrical resistor [192].

In special environments, e. g. ultrahigh vacuum, magnetic bearings without lubricants are preferred for rotating machinery, but their inherent instability requires feedback control which equally reduces vibrations [14].

An early NASA patent [166] provides an active mass damper (see Section 1.3.4) to cancel structural vibration.

In the 1980s, longitudinal vibrations of the ship superstructure caused by nonuniform propulsion have been compensated for with a type of dynamic absorber, realised by a centrifugal pendulum. This is a pendulum swinging along the length of the ship and rotating about an axis pointing also in a lengthwise direction. The swinging of the pendulum is synchronised to the ship's vibration by controlling the rotational frequency, and hence the centrifugal force, which together with gravity determines its natural frequency [188].

### 1.3.2  AVC for beams, plates and structures

Air and spacecraft have had a great impact on investigations in active control of structural vibration. Other than the abovementioned whole-body vibrations of ships which are comparatively easy to control due to their very low frequencies, elastic structures, i. e. continuous media with an infinite number of degrees of freedom the control of which presents fundamentally different problems, must also be considered. First, there are the different wave types in solids (of which longitudinal, torsional and transversal waves are the most important), and their control demands various types of actuator and sensor.

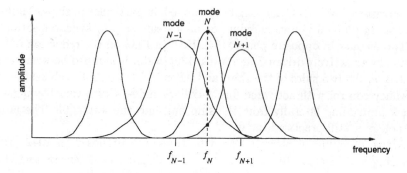

Figure 1.7    Spillover: attempting to control mode $N$ at its resonance frequency $f_N$ also excites neighbouring modes $N-1$ and $N+1$ with lower, but finite amplitudes

Furthermore, the propagation speed is generally higher in water than in air so that causality problems occur with broadband adaptive feedforward controllers. As a consequence, many problems are treated with modal control where, especially in the case of overlapping modes, the spillover problem has to be considered: the unwanted excitation of additional modes the resonance curves of which extend to the controlled frequency. In Figure 1.7 it is assumed that the $N$th mode resonant at frequency $f_N$ shall be controlled; the tails of the neighbouring resonance curves have nonnegligible amplitudes at $f_N$ (the dots on the dashed line) and are therefore also excited by the control signal at $f_N$, to some extent. Owing to the phase slope around a resonance, the neighbouring modes are usually enhanced rather than damped when the $N$th mode is suppressed.

For satellites, the damping of modal vibrations is important after pointing position manoeuvres etc. since they are built from low-loss materials and air friction is not present in space. The optimisation of number and placement of sensors and actuators for the mostly applied adaptive feedback controllers requires precise knowledge of the structural dynamics so that reliable modelling in state-space coordinates and a realistic estimation of discretisation errors are possible. An introduction to this field is given by Meirovitch [203].

Sophisticated controllers have been designed to actively isolate satellite antennas and installations for, e. g., microgravity experiments from structural vibrations caused by the position controllers and other onboard machinery [73].

Damping and stiffness control in mechanical junctions can be achieved by dry friction control where the pressing force is controlled by a piezoelectric actuator, in feedforward or feedback control, typically by a nonlinear algorithm, e.g. a neural network [107].

In aircraft technology, active controllers have been developed for manoeuvre and gust load alleviation, as well as for wing flutter control [96], and for noise and vibration reduction in helicopters [54, 130]. Currently, intense research is devoted to the application of modern compound materials with embedded actuators and sensors (keywords are: bi-functional elements, adaptive or smart structures, adaptronics [101, 204]) where technical problems are encountered, among others, by the fact that sensor and actuator materials such as piezoceramics, piezopolymers, electro and magnetostrictive materials, shape memory alloys, electro and magnetorheological fluids (smart materials) are not constructional materials with a mechanical strength sufficient for load-bearing structures. Much information on research in the field of smart materials and structures is published in specialised journals such as *Smart Structures and Materials* or *Journal of Intelligent Material Systems and Structures*.

### 1.3.3  Active mounts

Possibilities for active noise control in road vehicles were discussed in Section 1.2.9. The predominant sources of interior noise are engine and wheel vibrations which propagate as structure-borne sound through the car body and finally radiate as airborne sound into the cabin. It is therefore reasonable to develop active engine mounts and active shock absorbers which are stiff enough to carry the static load, but dynamically resilient so that vibrations are not transmitted. Piezoceramic actuators are suited for excursions in the submillimetre range [73], for larger amplitudes and forces at frequencies of a few Hertz hydraulic and pneumatic actuators are available [278]. However, the commercial use of engine and wheel mounts has been impeded so far by technical problems. Compact and robust combinations of conventional rubber mounts with electrodynamically driven hydraulics have been constructed as active hydromounts for a wide frequency range [157, 330], but the stroke and power required for cars at low frequencies cannot yet be fulfilled by active hydromounts of reasonable size [254].

Active control technology has been applied for improved vibration isolation of tables for optical experiments, scanning microscopes, vibration sensitive semiconductor manufacturing stages etc. Commercial products are offered by companies such as Newport (USA), Halcyonics (Germany) and Integrated Dynamics Engineering (IDE, Germany), the latter company also offers active compensation systems for electromagnetic strayfields which is important, e. g., for high-resolution electron microscopes.

The performance of hydraulic shock absorbers can be improved by applying electrorheological fluids (ERF) [210]. ERF are fine suspensions of polarisable small dielectric particles in an unpolar basic fluid, e. g. polyurethane in low-viscosity silicone oil [15]. Their viscosity can be adjusted reversibly between watery and pasty by applying electrical fields of several kV/mm.

Also suitable are magnetorheological fluids (MRF), suspensions of small ferromagnetic particles in a fluid, requiring a magnetic field for the viscosity to be changed. The field is usually applied by electromagnets which demand high electric current instead of high voltage [289].

### 1.3.4   Civil engineering structures

Wind-induced swaying of modern tall, high-rise buildings can amount to amplitudes of several metres in the upper storeys. These can be damped by tuned mass dampers (TMD) acting as resonance absorbers: masses of about 1 per cent of the total mass of the building are placed in the uppermost storey and coupled to the building structure through springs and dampers. Their performance is raised by actively enhancing the relative motion. A prominent example where such an active TMD has been installed is the Citycorp Center in New York [236]. Less additional mass is required for aerodynamic appendages, protruding flaps that can be swivelled and utilise wind forces like sails to exert cancelling forces on the building [46].

Many research activities in the USA, Canada and particularly Japan aim at the development of active earthquake protection for buildings where, however, severe technical problems have still to be solved [146].

For slim structures such as antenna masts, bridges etc. tendon control systems have been constructed for the suppression of vibrations by controlled tensile forces acting in different diagonal directions [275].

### 1.3.5   Active and adaptive optics

The quality of pictures taken with optical or radio astronomical mirror telescopes depends essentially on the precision to which the optimal mirror shape is maintained. Modern swivelling large telescopes suffer from deformation under their own weight which is compensated for more efficiently by active shape control than by additional stiffeners which inevitably increase the mass of the structure. This technology is called active optics [260].

Although these motions are very slow (frequencies below 10 Hz) and therefore easy to control, adaptive optics have solved the more complicated problem of controlling picture blurring from atmospheric turbulence, the so-called seeing which fluctuates at frequencies of about 1000 Hz. The mirror surface rests on a matrix of piezoceramic actuators which are adjusted by an adaptive multichannel controller so that a reference star is optimally focused. If no reference star exists in the vicinity of the observed object an artificial guide star can be created by resonance scattering of an intense laser beam from sodium atoms in about 100 km height [68, 98]. Adaptive optics have improved the optical resolution of the best telescopes by a factor of 10 to 50. This technology, originally conceived in the 1950s, was developed in the USA during the 1970s for the military SDI project and was not declassified before 1991 when civil research had reached almost the same state [57]. Meanwhile, adaptive optical mirrors also find industrial applications such as laser cutting and welding [26].

### 1.3.6 Noise reduction by active structural control

Active control of structural vibrations and active control of sound fields have developed almost independently, including differing adaptive control concepts (mostly feedforward in acoustics, mostly feedback in vibration). It has only been in recent years that the two fields have become connected. Many noise problems result from radiation of structure-borne sound, e. g. in the interior of cars and aircraft, on ships, by vibrating cladding panels of machines etc. Here comes into action a concept known under the acronym ASAC (Active Structural Acoustic Control) [102] where noise reduction is not attained by superimposing airborne sound onto the disturbing noise field but by controlling the vibrating structure itself. This is possible by suitably placed and controlled actuators to suppress the structural vibration, although this is not necessarily the optimal solution. Acoustically relevant are mainly plate bending waves which due to their frequency dispersion ($c_B \propto \sqrt{\omega}$) are nonradiating for $\omega < \omega_g$ and radiating above the critical frequency $\omega_g$ at which the bending wave velocity equals the sound velocity in the surrounding medium. When $c_B < c_0$ the acoustical short circuit between adjacent wave crests and troughs yields a weak sound radiation into the far field, but for $c_B > c_0$ a very effective radiation results. The proportionality factor in $c_B \propto \sqrt{\omega}$ contains the flexural stiffness so that its modification shifts the critical frequency and can turn radiating modes into nonradiating ones (modal restructuring). Much work has been done to investigate how, e. g. by laminates from sheet metal and piezolayers as sensors and actuators, adaptive structures can be constructed which can suppress in propeller aircraft etc. the abovementioned fuselage excitation by eddy threads, so enabling a substitution or at least a supplement to the more involved (and heavier) direct noise control by mirophone/loudspeaker systems [53, 101].

## 1.4   Active Flow Control

Coherent active control technology is also applied to fields other than sound and structural vibrations, among which the physics of fluid flow is gaining more and more importance. One of the many interactions of sound and flow is the transition from laminar flow of a slim gas flame into turbulence by insonification. Conversely, the turbulence of a flame has been suppressed actively by feedback control of a microphone/loudspeaker system [67].

Laboratory experiments have shown since about 1982 that the transition from laminar to turbulent flow can be delayed by controlling the Tollmien-Schlichting waves in the boundary layer, thereby providing drag reduction which is of great technical relevance. This can be achieved with thermal inputs [176] and acoustical or vibrational excitation [74, 90, 187].

Also, the dangerous surge and stall in compressors, resulting from a compressor instability, can be suppressed acoustically [126].

It is also interesting that the active control of sound fields – which is generally restricted to linear superposition – controls a highly nonlinear process in these applications [94].

An ionised gas stream in a combustion chamber (e. g. in a rocket) tends to produce unstable resonance oscillations which can be suppressed by an appropriately controlled electric d.c. current through the ionised gas, employing a feedback controller with a photoelectric cell as oscillation sensor [20].

A microelectromechanical system (MEMS) to be mounted on fan blades is presented in Reference [169], comprising a turbulence sensor, an integrated circuit and an actuator by which turbulence noise can either be reduced, or – in the case of heat exchangers – amplified in order to improve heat transfer. Recent experiments on this interesting technique have been described in Reference [134] where flight control of a delta wing aircraft is reported, and in Reference [33] where the laminar/turbulent transition along a wing profile in a wind tunnel is influenced.

Blade-vortex interaction causing impulsive noise from helicopters can be reduced by controlling flaps at the trailing edges [50]. In a further development, tip vortices of helicopter blades, aircraft foils or marine propellers can be reduced by fluid injection to the high-pressure side of the lifting body [221].

Dynamic stabilisation of jet-edge flow with various adaptive linear feedback control strategies has been experimentally verified [243].

Disturbing resonances in a large wind tunnel with free-jet test section (the so-called Göttingen model) can be suppressed by feedback ANC, employing multiple loudspeakers [334].

Active flow control can also provide a low-frequency high-intensity sound source, utilising aeroacoustic instability [164].

## 1.5  Conclusions

Coherent active control systems are commercially applied in acoustics in certain problem areas, but only in acoustically "simple" situations: small volume, one-dimensional sound propagation, quasiperiodic noise, isolated modes. There are clearly more applications in vibration technology, but there are also fields where the nonapplication of a well-developed technology is at first sight surprising, among these flutter control of aircraft wings which has been successfully tried for more than 30 years. Here the flutter limit would be shifted to a higher flight speed, but this is not practised from safety considerations: if such an active controller failed, and that cannot be excluded with complex systems such as these, the danger of wing fracture and hence an air crash would be too high. This is a general problem; precautions have to be taken in safety-relevant applications, where failure of the active control system must not have catastrophic consequences.

The general interest in active control of noise and vibration is steadily increasing, a fact which can also be concluded from the growing number of textbooks, special conferences and journal papers per year. Based on the author's collection of more than 12 000 references on active control of sound and vibration, Figure 1.8 shows a histogram of the number of publications grouped in five-year periods. An exponential increase is observed from the 1950s through the early 1980s with doubling every five years, followed by further growth at a reduced pace (approximately doubling every ten years). There are nearly twice as many papers on active vibration control than on active sound field control, and there are about 7 per cent patent applications. This is relatively high for a research topic and proves the considerable commercial interest [116, 117].

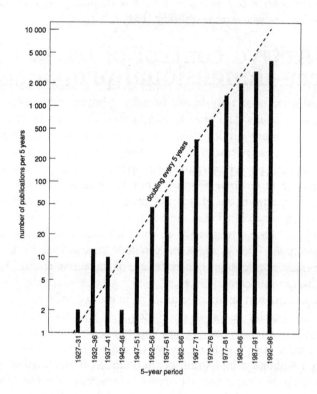

Figure 1.8    *Five-year cumulants of ANVC publications, based on the author's*
            *data files*

*Chapter 2*

# Active control of noise in three-dimensional propagation

# M. O. Tokhi[1] and K. Mamour[2]

[1] *Department of Automatic Control and Systems Engineering*
*The University of Sheffield, Sheffield, UK, Email: o.tokhi@sheffield.ac.uk*
[2] *Renishaw, UK*

*A coherent method for the analysis and design of active noise control systems, in a three-dimensional nondispersive propagation medium, is presented in this chapter. An analysis of single-input single-output and single-input multi-output control structures is provided. Conditions for the robust operation of such systems on the basis of optimum cancellation are determined. These include the physical extent of cancellation in the medium, relative stability of the inherent feedback loop and controller design. These conditions are interpreted as constraints on the geometric compositions of the system.*

## 2.1  Introduction

Active noise control (ANC) uses the intentional superposition of acoustic waves to create a destructive interference pattern such that a reduction of the unwanted sound occurs. This is realised by artificially generating cancelling (secondary) source(s) of sound through detecting the unwanted (primary) noise and processing it by an electronic controller, so that when the secondary wave is superimposed on the primary wave the two destructively interfere and cancellation occurs. In this manner, the interference of the component waves leads to a pattern of zones of cancellation and reinforcement in the medium; noise is cancelled in some regions and reinforced in others. The physical extent of zones of cancellation primarily depends upon the maximum frequency of the noise and separation between the sources. For a given maximum frequency of the noise, a reduction in the separation between the sources leads to an increase in the physical extent of cancellation. This is limited in practice by the physical dimensions of the sources. Moreover, the process of controller implementation, using analogue and/or digital techniques, can result in errors in realising the amplitude and phase characteristics of the controller [172]. Such errors will affect the level and physical extent of cancellation. However, a proper design methodology incorporating a suitable geometrical arrangement of cancelling sources can lead to a significant improvement in the level as well as physical extent of cancellation.

Acoustic feedback and reflected waves in an ANC system are found to significantly affect the performance of the system. This, incorporated with the geometrical arrangement of system components, has a major impact on the stability of the system and can lead to practical limitations in the design of the controller. Acoustic feedback, owing to reflected waves, is found to confuse the controller as to the exact level of the noise itself. This leads to system instability and/or poor system performance in some frequency bands. To solve this problem, in the case of the one-dimensional duct noise, for example, attempts have been made to use loudspeaker/microphone arrays so that the detector microphone detects the unwanted noise only, and that to provide a cancelling signal in the duct which propagates only in the downstream direction [2, 75, 174, 175, 286]. However, criteria based on relative stability measures of the system can lead to robust designs [302]. System geometry is also found to affect the practical realisation of the controller in an ANC system; certain geometrical arrangements of system components can lead to requiring the controller to have impractically large gains [299, 303].

The analysis and design presented in this chapter focus on the cancellation of noise of a compact source in a three-dimensional free-field propagation. These can be transformed and extended to cancellation of distributed sources and to other propagation media. The ANC system is realised within a single-input multi-output (SIMO) control structure for which the single-input single-output (SISO) structure is considered as a special case. The controller design relations are developed in the frequency-domain. These can, equivalently, be thought of either in the complex frequency $s$-domain or the $z$-domain. The corresponding practical realisation of the controller is carried out either in continuous-time or discrete-time using analogue or digital techniques accordingly.

The analysis focuses on ANC systems in stationary (steady-state) conditions. This corresponds to an ANC system with a fixed controller of the required characteristics under situations where substantial variations in the characteristics of the secondary source, transducers and other electronic equipment used do not occur. In an adaptive ANC system, this means that once a steady-state (stationary) condition has been reached the situation is equivalent to the case of the fixed controller. Therefore, in an adaptive ANC system the analysis applies to periods where a steady-state condition has been reached and substantial parameter variations do not occur. The stability and convergence of an adaptive controller in a time-varying (nonstationary/transient) nonlinear environment is difficult to analyse and guarantee. However, the problem of instability under such a situation can be avoided by designing a suitable supervisory level control within the adaptive mechanism [300, 301].

## 2.2   Active noise control structure

A schematic diagram of an SIMO feedforward ANC structure is shown in Figure 2.1$a$. The primary source emits unwanted acoustic signals (noise) into the medium. This is detected by the detector, processed by the controller and fed to a set of $k$ secondary sources. The secondary signals, thus generated, are superimposed upon the unwanted noise so that the level of noise is reduced at a set of $k$ observation points. The corresponding frequency-domain equivalent block diagram of Figure 2.1$a$ is shown in Figure 2.1$b$, where:

$\mathbf{E} = 1 \times 1$ matrix representing transfer characteristics of the acoustic path between the primary source and the detector

$\mathbf{F} = k \times 1$ matrix representing transfer characteristics of the acoustic paths between the secondary sources and the detector

$\mathbf{G} = 1 \times k$ matrix representing transfer characteristics of the acoustic paths between the primary source and the observers

$\mathbf{H} = k \times k$ matrix representing transfer characteristics of the acoustic paths between the secondary sources and the observers

$\mathbf{C} = 1 \times k$ matrix representing transfer characteristics of the controller

$\mathbf{L} = k \times k$ diagonal matrix representing transfer characteristics of the secondary sources

$\mathbf{M} = 1 \times 1$ matrix representing transfer characteristics of the detector

$\mathbf{M}_O = k \times k$ diagonal matrix representing transfer characteristics of the observers

$\mathbf{U}_D = 1 \times 1$ matrix representing the primary signal at the source point

$\mathbf{Y}_{DO} = 1 \times k$ matrix representing the primary signal at the observation points

$\mathbf{U}_C = 1 \times k$ matrix representing the secondary signals at the source points

$\mathbf{Y}_{CO} = 1 \times k$ matrix representing the secondary signal at the observation points

$\mathbf{U}_M = 1 \times 1$ matrix representing the detected signal

$\mathbf{Y}_O = 1 \times k$ matrix representing the combined primary and secondary signals at the observation points

Note in Figure 2.1 that if the observation points coincide with the detection point a feedback control structure (FBCS) is obtained. This type of structure, following Lueg's initial proposals [180], has been considered in numerous investigations [30, 48, 229, 230, 231, 262, 299, 303, 314, 315]. The feedforward control structure (FFCS), however, is more popular and has been employed and investigated in one-dimensional as well as three-dimensional enclosed and free fields [58, 59, 86, 132, 170, 172, 180, 215, 218, 249, 251, 299, 300, 301, 302, 303].

The objective in Figure 2.1 is to reduce the level of noise to zero at the observation points. This is the minimum variance design criterion in a stochastic environment, requiring the observed primary and secondary signals at each observation point to be equal in amplitudes and have a phase difference of 180°, i.e.:

$$\mathbf{Y}_{CO} = -\mathbf{Y}_{DO} \tag{2.1}$$

Using the block diagram in Figure 2.1b, $\mathbf{Y}_{DO}$ and $\mathbf{Y}_{CO}$ can be expressed as:

$$\mathbf{Y}_{DO} = \mathbf{U}_D \mathbf{G}$$

$$\mathbf{Y}_{CO} = \mathbf{U}_D \mathbf{EMCL}[\mathbf{I} - \mathbf{FMCL}]^{-1}\mathbf{H} \tag{2.2}$$

where $\mathbf{I}$ is the identity matrix. Substituting for $\mathbf{Y}_{DO}$ and $\mathbf{Y}_{CO}$ from eqn. (2.2) into eqn. (2.1) and simplifying yields the required controller transfer function as:

$$\mathbf{C} = \mathbf{M}^{-1}\mathbf{\Delta}^{-1}\mathbf{G}\mathbf{H}^{-1}\mathbf{L}^{-1} \tag{2.3}$$

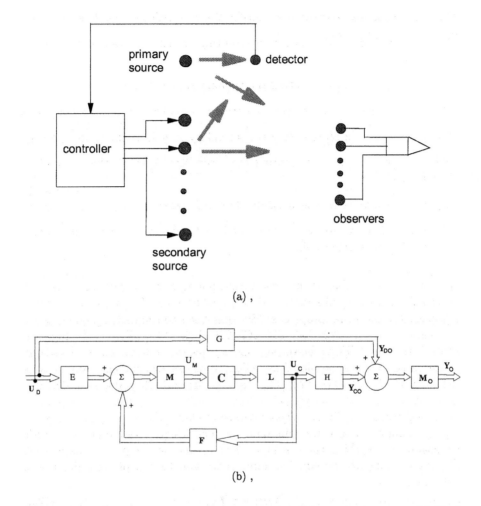

(a) ,

(b) ,

*Figure 2.1    Active noise control structure: (a) Schematic diagram; (b) block diagram*

where $\Delta$ is a $1 \times 1$ matrix given by:

$$\Delta = \mathbf{GH}^{-1}\mathbf{F} - \mathbf{E} \qquad (2.4)$$

This represents the required controller design relation for optimum cancellation of noise at the observation points.

## 2.3    Physical extent of cancellation

The interference of the component waves through a process of realisation of an ANC system, effectively, leads to a pattern of zones of cancellation and reinforcement in the medium; noise is cancelled in some regions and reinforced in others. The physical extent of zones of cancellation primarily depends upon the maximum frequency of the noise and separation between the sources. In practice, the separation is limited by the physical dimensions of the sources. In this section an investigation into the effects of system geometry on the physical extent of cancellation is presented.

A time-domain and a corresponding frequency-domain analysis of the process of phase cancellation leading to the basic conditions of cancellation have previously been considered [172, 174, 324]. A detailed parametric analysis and geometric description of the process of cancellation on the basis of an SISO structure has also been reported [305]. However, such an analysis and description to incorporate a multiple-source configuration is not given. Moreover, the effects of errors, resulting during an implementation process, on the physical extent of cancellation has not been investigated. In this section, the process of field cancellation from a frequency-domain viewpoint so that the analysis, for single tones of the component waves, provides a quantitative measure of the degree of cancellation in terms of amplitude and phase parameters of the waves, is given. In the case of broadband noise this provides relationships for the degree of cancellation as a function of frequency. The degree of cancellation is analysed in relation to source-related and geometry-related parameters of the system leading to a three-dimensional description of the pattern of cancellation and reinforcement in the medium. The description is initially provided on the basis of optimum cancellation at a finite set of observation points, and then extended to incorporate implementation errors in the controller.

### 2.3.1    The field cancellation factor

Consider the primary point source in Figure 2.1 emitting a wave $u_D(t)$, as function of time $t$, and the set of $k$ secondary sources, located at distances $r_i$ $(i = 1, 2, \ldots, k)$ relative to the primary source, emitting waves $u_{Ci}(t)$ into the medium, respectively. In propagating through a distance $r_{gq}$, the wave $u_D(t)$

results in a wave $y_{Dq}(t)$ at an arbitrary point $q$. Similarly, in propagating through a distance $r_{hiq}$, the secondary wave $u_{Ci}(t)$ results in a wave $y_{Ciq}(t)$ at the arbitrary point. Let the resultant field owing to the superposition of the primary and secondary waves at the point be denoted by $y_q(t)$. With $\omega$ representing the radian frequency, let:

$\mathbf{U}_D(j\omega) = 1 \times 1$ matrix representing $u_D(t)$ in the frequency domain,

$\mathbf{Y}_{Dq}(j\omega) = 1 \times 1$ matrix representing $y_{Dq}(t)$ in the frequency domain

$\mathbf{U}_C(j\omega) = 1 \times k$ matrix representing $u_C(t) = \{u_{Ci}(t)\}$ $(i = 1, 2, \ldots, k)$ in the frequency domain

$\mathbf{Y}_q(j\omega) =$ the signal $y_q(t)$ in the frequency domain

$S_{yyDq}(\omega) =$ the autopower spectral density of $y_{Dq}(t)$

$S_{yyq}(\omega) =$ the autopower spectral density of $y_q(t)$

$\mathbf{g}_q(j\omega) =$ the transfer characteristics of the acoustic path through $r_{gq}$

$\mathbf{h}_q(j\omega) = k \times 1$ matrix, $\{\mathbf{h}_{iq}(j\omega)\}$, representing the transfer characteristics of the acoustic paths through $\{r_{hiq}\}$ $(i = 1, 2, \ldots, k)$

The transfer functions $\mathbf{g}_q(j\omega)$ and $\mathbf{h}_{iq}(j\omega)$ for the propagation medium under consideration are given by:

$$\mathbf{g}_q(j\omega) = \frac{A}{r_{gq}} e^{-j\frac{\omega}{c}r_{gq}}$$

$$\mathbf{h}_{iq}(j\omega) = \frac{A}{r_{hiq}} e^{-j\frac{\omega}{c}r_{hiq}} \quad , \quad i = 1, 2, \ldots, k$$

(2.5)

where $A$ is a constant and $c$ is the speed of sound in the medium.

The autopower spectral densities of $y_{Dq}(t)$ and $y_q(t)$ can be written as [43, 276]:

$$S_{yyDq}(\omega) = |\mathbf{Y}_{Dq}(j\omega)|^2 = |\mathbf{U}_D(j\omega)\mathbf{g}_q(j\omega)|^2$$

$$S_{yyq}(\omega) = |\mathbf{Y}_q(j\omega)|^2 = |\mathbf{U}_D(j\omega)\mathbf{g}_q(j\omega) + \mathbf{U}_C(j\omega)\mathbf{h}_q(j\omega)|^2$$

(2.6)

Note that for cancellation of the primary wave to occur at the arbitrary point $q$ the condition

$$S_{yyq}(\omega) < S_{yyDq}(\omega) \tag{2.7}$$

must be satisfied.

For a quantitative description of cancellation, the field cancellation factor $K$ is defined as the ratio of the cancelled spectrum $S_{yyDq}(\omega) - S_{yyq}(\omega)$ to the primary spectrum $S_{yyDq}(\omega)$ that existed at the point prior to the superposition of the secondary waves with the primary wave [172, 305]:

$$K \triangleq 1 - \frac{S_{yyq}(\omega)}{S_{yyDq}(\omega)} \tag{2.8}$$

It follows from eqns (2.7) and (2.8) that, for the primary wave to be cancelled, $K$ must be between zero and unity:

$$0 < K \leq 1 \tag{2.9}$$

where $K = 1$ corresponds to complete (optimum) cancellation, $K = 0$ corresponds to no cancellation and if $K < 0$ then the situation corresponds to reinforcing the primary wave.

Substituting for $S_{yyDq}(\omega)$ and $S_{yyq}(\omega)$ from eqns (2.6) into eqn. (2.8), using eqn. (2.5) and simplifying yields:

$$K = -\sum_{i=1}^{k} (\alpha_i + 2\beta_i\sqrt{\alpha_i}) - 2\sum_{i=1}^{k}\sum_{j=1}^{k} \beta_{ij}\sqrt{\alpha_i\alpha_j} \quad , \quad i \neq j \tag{2.10}$$

where $\alpha_i$ is the power ratio of the secondary signal $y_{Ciq}(t)$ over the primary signal $y_{Dq}(t)$, $\beta_i$ is the cross-spectral density factor between $y_{Ciq}(t)$ and $y_{Dq}(t)$ and $\beta_{ij}$ is the cross-spectral density factor between the secondary signals $y_{Ciq}(t)$ and $y_{Cjq}(t)$:

$$\alpha_i = \frac{S_{yyCiq}(\omega)}{S_{yyDq}(\omega)} = \left(\frac{r_g}{r_{hi}}\right)^2 \alpha_{si} \quad , \quad \alpha_{si} = \frac{S_{uuCi}(\omega)}{S_{uuD}(\omega)}$$

$$\beta_i = \cos\left[\frac{\omega}{c}(r_g - r_{hi}) - \theta_i(\omega)\right]$$

$$\beta_{ij} = \cos\left[\frac{\omega}{c}(r_{hi} - r_{hj}) + \theta_i(\omega) - \theta_j(\omega)\right] \quad , \quad i \neq j$$

where $S_{yyCiq}(\omega)$, $S_{yyDq}(\omega)$, $S_{uuCi}(\omega)$ and $S_{uuD}(\omega)$ represent the autopower spectral densities of $y_{Ciq}(t)$, $y_{Dq}(t)$, $u_{Ci}(t)$ and $u_D(t)$, respectively, and $\theta_i(\omega)$ and $\theta_j(\omega)$ are the angles by which the secondary signals $y_{Ciq}(t)$ and $y_{Cjq}(t)$, respectively, lead the primary signal $y_{Dq}(t)$. Note that $\alpha_{si}$ represents the power ratio of the secondary signal $u_{Ci}(t)$ over the primary signal $u_D(t)$ at the source (before propagation) whereas $\alpha_i$ is the corresponding power ratio at the arbitrary point $q$.

Eqn. (2.10) gives a quantitative measure of the degree of cancellation in terms of the power ratios, interpreted as the relative amplitudes, and the cross-spectral density factors, interpreted as the relative phases, of the component waves. Note that the first summation on the right-hand side of this equation is due to the interference of the secondary fields with the primary field whereas the double summation term is due to the interference of the secondary fields with one another. Thus, for cancellation of the primary wave to be achieved at a point in the medium, the amplitudes and phases of the secondary waves should be adjusted relative to the primary wave such that the condition in eqn. (2.9) is satisfied.

### 2.3.2    Three-dimensional description of cancellation

Realisation of the required continuous frequency-dependent controller characteristics given in eqn. (2.3) within the ANC system of Figure 2.1 will ensure cancellation of the noise at and in the vicinity of all the observation points. In this manner, a region of cancellation is created around each observation point. Further away from these points, due, in principle, to a redistribution of the sound energy in the medium, cancellation will decrease and even reinforcement will occur. An investigation of the level and physical extent of cancellation around the observation points in relation to the number of cancelling sources, geometrical arrangement of system components and extent of errors in the realisation of the required controller is given in this Section.

Consider that the realisation of the controller transfer characteristics in Figure 2.1 accurately conforms to the design rule in eqn. (2.3). Using the block diagram in Figure 2.1$b$, the secondary signal $\mathbf{U}_C$ can be obtained as:

$$\mathbf{U}_C = \mathbf{U}_D \mathbf{EMCL} \left[ \mathbf{I} - \mathbf{FMCL} \right]^{-1} \qquad (2.11)$$

Substituting for $\mathbf{C}$ from eqn. (2.3) into eqn. (2.11), using eqn. (2.4) and simplifying, $s = j\omega$, yields:

$$\mathbf{U}_C(j\omega) = -\mathbf{U}_D(j\omega)\mathbf{G}(j\omega)\mathbf{H}^{-1}(j\omega) \qquad (2.12)$$

Substituting for $\mathbf{U}_C(j\omega)$ from eqn. (2.12) into eqn. (2.6), simplifying and using eqn. (2.8) yields the cancellation factor $K$ at the arbitrary point $q$ in the medium as:

$$K = 1 - \left| 1 + \mathbf{g}_q^{-1}(j\omega)\mathbf{G}(j\omega)\mathbf{H}^{-1}(j\omega)\mathbf{h}_q(j\omega) \right|^2 \qquad (2.13)$$

Eqn. (2.13) gives a quantitative measure of the degree of cancellation achieved with the ANC system at the arbitrary point $q$, under stationary (steady-state) conditions. It follows from this equation that, if no error is involved in implementing the controller, then the resultant cancellation at an arbitrary point in the medium is dependent only on the transfer characteristics of the acoustic paths from the primary and secondary sources to the observation points and to the arbitrary point in question. Therefore, for a given primary wave the amount and physical extent of cancellation is described by the geometrical arrangement of the sources and observers. A three-dimensional description of the pattern of zones of cancellation and reinforcement can thus be obtained by measuring/calculating the cancellation factor $K$ for given noise frequency and system geometry using eqn. (2.13).

A two-dimensional description of the interference pattern created by an SISO ANC structure with the primary source situated at point $P$, emitting a signal of wavelength $\lambda$, and the secondary source at point $S$ is shown in Figure 2.2. The detector is situated at point $E$, and the controller is designed for optimum cancellation to be achieved at the observation point $O$. The axes are graduated in the wavelength $\lambda$. The curves represent contours of constant $K$ and, thus, show the relative noise power after cancellation; e.g. the curve $K = 0.5$ shows the locus of points at which the noise is reduced by 50 per cent whereas the curve $K = -2$ shows the locus of points at which the noise is reinforced by 200 per cent. The curve $K = 0$ is the boundary between zones of cancellation ($0 < K \leq 1$) and reinforcement ($K < 0$). Note that at the observation point $K$ is unity, corresponding to optimum cancellation, whereas at the location of the secondary source $K$ is $-\infty$, corresponding to maximum reinforcement. Figures 2.2$a$ and 2.2$b$ correspond to the same geometric configuration of system components, with the separation between the sources reduced from 0.5$\lambda$ in Figure 2.2$a$ to 0.2$\lambda$ in Figure 2.2$b$. It is noted that the physical extent of cancellation is significantly larger in Figure 2.2$b$ than that in Figure 2.2$a$. This is due to the reduction in the spacing between the sources. The separation between the sources in Figure 2.2$c$ is the same as that in Figure 2.2$a$. The observation point in Figure 2.2$c$, however, is located at point $(-0.5\lambda, 0)$. It is noted that the physical extent of cancellation around the observation point is significantly smaller in Figure 2.2$c$ as compared to that in Figure 2.2$a$. Moreover, the overall level of noise in Figure 2.2$c$ is noticeably higher than that in Figure 2.2$a$. This has resulted from moving the observation point closer to the primary source and, relatively, away from the secondary source. The reverse situation will hold if the observation point is moved away from the primary source and, relatively, closer to the secondary source. It follows from the above that moving the observation point further away from the secondary source (relatively closer to the primary source) will lead to an overall increase in the level of noise in the medium and a reduction in the physical extent of cancellation around the observation point. A three-dimensional description of the patterns in Figure 2.2 can be obtained by revolving each diagram around a horizontal axis passing through the origin. In this manner, the relative levels of the noise within the regions of cancellation and reinforcement can be visualised.

To demonstrate the effect of employing additional secondary sources on the physical extent of cancellation, consider the ANC system with an SIMO control structure incorporating two secondary sources. A two-dimensional description of the interference pattern created with a primary source at $P(0,0)$, emitting a signal of wavelength $\lambda$, and two secondary sources at points $S$, is shown in Figure 2.3. The detector is situated at point $E$, and the controller is designed for optimum cancellation to be achieved at the two observation points indicated by $O$. Comparing the patterns in Figures 2.3a and 2.3b with those in Figures 2.2a and 2.2b, respectively, where the separations between the primary source and each secondary source are $0.5\lambda$ and $0.2\lambda$, reveals that the overall physical extent of cancellation around the observation points is significantly larger in Figures 2.3a and 2.3b than that in Figures 2.2a and 2.2b. This is achieved by incorporating an additional secondary source into the system. With the arrangement in Figure 2.3c, the resulting physical extent of cancellation around the observation points is still comparable to that in Figure 2.3a. However, by adopting the arrangement in Figure 2.3d, the physical extent of cancellation achieved around the observation points has increased significantly. A three-dimensional description of the patterns in Figure 2.3 can be obtained by revolving each about a straight line passing midway between the two observation points accordingly.

The results presented in Figures 2 and 3 are based on the assumption that the controller characteristics are free of errors. In practice, however, the process of controller implementation, using analogue and/or digital techniques, introduces errors in the amplitude and phase of the controller [170, 172]. Such errors will affect the amount and, hence, the physical extent of cancellation to a lessor or greater degree, depending on the extent of the error involved. This issue is addressed in the remainder of this Section.

Let the relative amplitude and phase errors introduced in the controller transfer function through the implementation process be represented by a $k \times k$

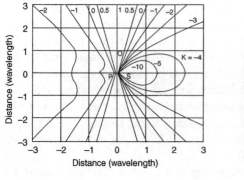

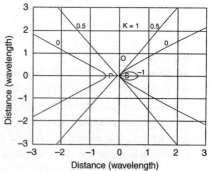

(a) $(E(-0.25\lambda, 0.0625\lambda), O(0, \lambda),$
$P(-0.25\lambda, 0), S(0.25\lambda, 0)$

(b) $(E(-0.1\lambda, 0.0625\lambda), O(0, \lambda),$
$P(-0.1\lambda, 0), S(0.1\lambda, 0)$

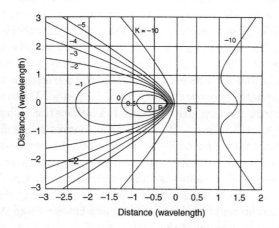

(c) $(E(-0.25\lambda, 0.0625\lambda), O(-0.5\lambda, 0), P(-0.25\lambda, 0), S(0.25\lambda, 0)$

**Figure 2.2**    *Interference pattern with an SISO ANC system, incorporating the optimal contoroller*

diagonal matrix $\mathbf{C_e}$. Thus, the transfer characteristics of the controller as implemented in Figure 2.1 will be equivalent to:

$$\mathbf{C} = \mathbf{M}^{-1} \left[ \mathbf{GH}^{-1}\mathbf{F} - \mathbf{E} \right]^{-1} \mathbf{GH}^{-1}\mathbf{L}^{-1}\mathbf{C_e} \qquad (2.14)$$

Substituting for **C** from eqn. (2.14) into eqn. (2.11) and simplifying yields

$$\mathbf{U}_C(j\omega) = -\mathbf{U}_D(j\omega)\mathbf{G}(j\omega)\mathbf{H}^{-1}(j\omega)\mathbf{T}(j\omega) \tag{2.15}$$

where **T** is a $1 \times k$ matrix given by:

$$\mathbf{T} = \left[\mathbf{I} - \mathbf{F}\left(\mathbf{GH}^{-1}\mathbf{F} - \mathbf{E}\right)^{-1}\mathbf{GH}^{-1}\right]\mathbf{C_e}\left[\mathbf{I} - \mathbf{F}\left(\mathbf{GH}^{-1}\mathbf{F} - \mathbf{E}\right)^{-1}\mathbf{GH}^{-1}\mathbf{C_e}\right]^{-1}$$

and **I** is the identity matrix.

Substituting for $\mathbf{U}_C(j\omega)$ from eqn. (2.15) into eqn. (2.6), simplifying and using eqn. (2.8) yields the cancellation factor $K$ at the arbitrary point $q$ in the medium as:

$$K = 1 - \left|1 - \mathbf{g}_q^{-1}\mathbf{GH}^{-1}\mathbf{Th}_q\right|^2 \tag{2.16}$$

Eqn. (2.16) gives a quantitative measure of the degree of cancellation achieved with the ANC system, under stationary (steady-state) conditions. The amount of cancellation is dependent on the system geometry and accuracy in implementation of the controller. A three-dimensional description of the pattern of zones of cancellation and reinforcement can thus be obtained by calculating the cancellation factor $K$ for given primary wave and system geometry.

Figure 2.4$a$ shows the interference pattern corresponding to the system in Figure 2.3$c$, for $\mathbf{C_e} = a_c \exp(j\phi_c)\mathbf{I}$ with $a_c = 0.8$ and $\phi_c = -30°$. It is noted that with this error the system is tuned to achieve optimum cancellation at points different from the observation points. Revolution of the diagram about a line passing midway between the observation points will result in the corresponding three-dimensional description. The cancellation factor $K$ corresponding to this system, at the observation points, as a function of the amplitude error $a_c$ and phase error $\phi_c$ is shown in Figure 2.4$b$. It is noted that the error tolerance in the controller phase increases with decreasing $a_c$. For the situation in Figure 2.4$a$, cancellation can still be achieved at the observation points for $-80° < \phi_c < 60°$.

Eqn. (2.16) can be used to determine the extent of the error in controller realisation that can be tolerated for cancellation to be achieved at the observation point(s). In this manner, with $c_{ei} = a_{ci} \exp(j\phi_{ci})$ representing the error in controller element $c_i$, the amplitude error $a_{ci}$ and phase error $\phi_{ci}$ can be tolerated as long as $K > 0$ at the observation point(s). Note in Figure 2.4$b$ that for $0 < a_c < 1$ the tolerance in $\phi_c$ increases by decreasing the value of $a_c$. This decrease in $a_c$ also results in a decrease in the maximum amount of cancellation achievable at the observation point.

Substituting for $K$ from eqn. (2.16) into eqn. (2.9) and simplifying, for an ANC system with a single secondary source, reveals that for cancellation to

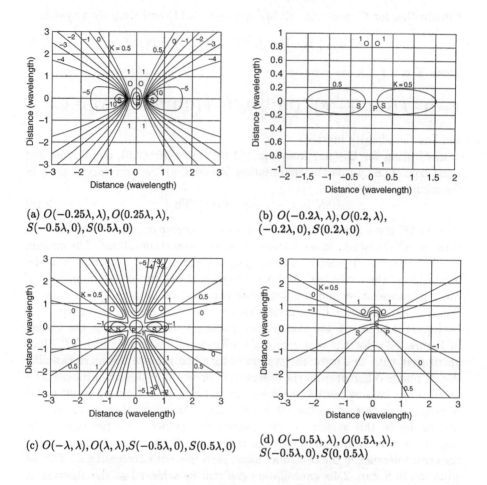

(a) $O(-0.25\lambda, \lambda), O(0.25\lambda, \lambda),$
$S(-0.5\lambda, 0), S(0.5\lambda, 0)$

(b) $O(-0.2\lambda, \lambda), O(0.2, \lambda),$
$(-0.2\lambda, 0), S(0.2\lambda, 0)$

(c) $O(-\lambda, \lambda), O(\lambda, \lambda), S(-0.5\lambda, 0), S(0.5\lambda, 0)$

(d) $O(-0.5\lambda, \lambda), O(0.5\lambda, \lambda),$
$S(-0.5\lambda, 0), S(0, 0.5\lambda)$

Figure 2.3    *Interference pattern with an SIMO ANC system incorporating the optimal controller, with the primary source at $P(0,0)$ and the detector at $E(0, 0.0625\lambda)$.*

be achieved at the observation point the amplitude and phase errors ($a_c$ and $\phi_c$) in controller realisation should satisfy the relation:

$$0 < a_c < 2\cos\phi_c \tag{2.17}$$

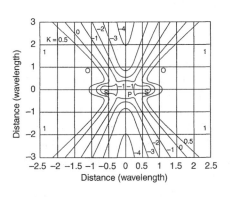

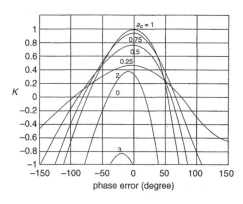

(a) Interference pattern with $a_c = 0.8$, $\phi_c = -30°$

(b) $K$ as function of error

Figure 2.4    *Effect of error in controler characteristics on degree of cancellation; with $E(0, 0.0625\lambda)$, $O(-\lambda, \lambda)$, $O(\lambda, \lambda)$, $P(0,0)$, $S(-0.5\lambda, 0)$, $S(0.5\lambda, 0)$.*

It further follows from eqn. (2.17) that for $a_c$ to be positive, $\cos \phi_c$ should be positive, or:

$$-\frac{\pi}{2} < \phi_c < \frac{\pi}{2}$$

Thus, for cancellation to be achieved at the observation point, the phase error $\phi_c$ should stay within $\pm\pi/2$ and the amplitude error $a_c$ should satisfy eqn. (2.17). This can be extended to multiple-source ANC systems in a similar manner as above.

It follows from the descriptions presented above that a decrease in either the frequency or separation between the sources leads to an increase in the physical extent of cancellation. Moreover, the extent of cancellation is considerably affected by the geometrical arrangement of sources and observers. Furthermore, errors in the controller transfer function result in a reduction in the level of cancellation (or even lead to noise reinforcement) at the observation point(s).

## 2.4   Limitations in the controller design

It follows from eqn. (2.3) that, for given detector and secondary sources with necessary electronic components, the controller characteristics required for optimum cancellation at the observation points are dependent on the characteristics of the acoustic paths from the primary and secondary sources to the detection and observation points. In particular, if the set of detection and observation points is such that $\Delta$ in eqn. (2.3) becomes zero then the critical situation of the infinite-gain controller (IGC) requirement arises. The locus of such points in the medium (as a practical limitation in the design of the controller) is, therefore, of crucial interest.

Under the situation of the IGC requirement eqn. (2.4), for $s = j\omega$, can be written as:

$$\Delta(j\omega) = \mathbf{G}(j\omega)\mathbf{H}^{-1}(j\omega)\mathbf{F}(j\omega) - \mathbf{E}(j\omega) = 0 \qquad (2.18)$$

where $\mathbf{E}(j\omega)$, $\mathbf{F}(j\omega)$, $\mathbf{G}(j\omega)$ and $\mathbf{H}(j\omega)$ represent the frequency responses of the corresponding acoustic paths in Figure 2.1. Note that eqn. (2.18) is given in terms of the characteristics of the acoustic paths in the system. This implies that the IGC requirement is a geometry-related problem in an ANC system. Therefore, an analysis of eqn. (2.18) will lead to the identification of loci of (detection and observation) points in the medium for which the IGC requirement holds. To obtain a solution for eqn. (2.18), an SISO system is considered first. The results obtained are then used and extended to the SIMO ANC system.

### 2.4.1   Single-input single-output structure

Let the ANC system in Figure 2.1 incorporate a single secondary source ($k = 1$) and the functions $\mathbf{E}(j\omega)$, $\mathbf{F}(j\omega)$, $\mathbf{G}(j\omega)$ and $\mathbf{H}(j\omega)$, in this case, be denoted by $\mathbf{e}(j\omega)$, $\mathbf{f}(j\omega)$, $\mathbf{g}(j\omega)$ and $\mathbf{h}(j\omega)$ with the associated distances as $r_e$, $r_f$, $r_g$ and $r_h$, respectively:

$$\mathbf{E}(j\omega) = \mathbf{e}(j\omega) = \frac{A}{r_e}e^{-j\frac{\omega}{c}r_e} \quad , \quad \mathbf{F}(j\omega) = \mathbf{f}(j\omega) = \frac{A}{r_f}e^{-j\frac{\omega}{c}r_f}$$

$$\tag{2.19}$$

$$\mathbf{G}(j\omega) = \mathbf{g}(j\omega) = \frac{A}{r_g}e^{-j\frac{\omega}{c}r_g} \quad , \quad \mathbf{H}(j\omega) = \mathbf{h}(j\omega) = \frac{A}{r_h}e^{-j\frac{\omega}{c}r_h}$$

where $A$ is a constant. Substituting for $\mathbf{E}(j\omega)$, $\mathbf{F}(j\omega)$, $\mathbf{G}(j\omega)$ and $\mathbf{H}(j\omega)$ from eqn. (2.19) into eqn. (2.18) and simplifying yields:

$$\left(\frac{r_e}{r_f}\right) e^{-j(r_f - r_e)\frac{\omega}{c}} = \left(\frac{r_g}{r_h}\right) e^{-j(r_h - r_g)\frac{\omega}{c}}$$

This equation is true if and only if the amplitudes as well as the exponents (phases) on either side of the equation are equal. Equating the amplitudes and the phases, accordingly, yields:

$$\frac{r_e}{r_f} = \frac{r_g}{r_h} = a \quad , \quad r_f - r_e = r_h - r_g \tag{2.20}$$

where $a$ is a positive real number representing the distance ratio. Eqn. (2.20) defines the locus of points for which $\Delta(j\omega) = 0$ and the controller is required to have an infinitely large gain. Note that this equation is in terms of the distances $r_e$, $r_f$, $r_g$ and $r_h$ only. Therefore, the critical situation of the IGC requirement is determined only by the locations of the detector and observer relative to the primary and secondary sources.

Eliminating $r_f$ and $r_h$ in eqn. (2.20) and simplifying yields:

$$r_e(a - 1) = r_g(a - 1) \tag{2.21}$$

Two possible situations, namely $a = 1$ and $a \neq 1$, can be considered separately.

For a unity distance ratio eqn. (2.21) yields the identity $0 = 0$. Therefore, substituting for $a = 1$ into eqn. (2.20) yields the locus of points for IGC requirement as:

$$\frac{r_e}{r_f} = 1 \quad \text{and} \quad \frac{r_g}{r_h} = 1 \tag{2.22}$$

If the locations of the primary and secondary sources are fixed then each relation in eqn. (2.22) defines a plane surface perpendicularly bisecting the line joining the locations of the primary and secondary sources [305]. This plane for the primary and secondary sources located at points $P(0, 0, 0)$ and $S(u_s, v_s, w_s)$, respectively, with a distance $r$ apart in a three-dimensional $UVW$ space is given by:

$$\frac{u}{\left(\frac{r^2}{2u_s}\right)} + \frac{v}{\left(\frac{r^2}{2v_s}\right)} + \frac{w}{\left(\frac{r^2}{2w_s}\right)} = 1$$

which intersects the $U$, $V$ and $W$ axes at points $\left(\frac{r^2}{2u_s}, 0, 0\right)$, $\left(0, \frac{r^2}{2v_s}, 0\right)$ and $\left(0, 0, \frac{r^2}{2w_s}\right)$, respectively. If both the detector and the observer are on this plane (the IGC plane), then the critical situation of eqn. (2.18) arises and the controller is required to have an infinitely large gain for optimum cancellation to be achieved at the observation point.

For a nonunity distance ratio, eqns (2.20) and (2.21) yield:

$$\frac{r_e}{r_f} = a \quad , \quad \frac{r_g}{r_h} = a \quad , \quad \text{and} \quad \frac{r_e}{r_g} = 1 \tag{2.23}$$

Each of the first two relations in eqn. (2.23) describes a spherical surface [305]. These surfaces for the primary and secondary sources located, respectively, at $P(0,0,0)$ and $S(u_s, v_s, w_s)$ are defined by:

$$\left[u + \frac{a^2 u_s}{1-a^2}\right]^2 + \left[v + \frac{a^2 v_s}{1-a^2}\right]^2 + \left[w + \frac{a^2 w_s}{1-a^2}\right]^2 = \left[\frac{ar}{1-a^2}\right]^2 \quad , \quad a \neq 1 \tag{2.24}$$

which has a radius $R = \frac{ar}{|1-a^2|}$ and centre along the line $PS$ at point:

$$Q\left(-\frac{a^2 u_s}{1-a^2}, -\frac{a^2 v_s}{1-a^2}, -\frac{a^2 w_s}{1-a^2}\right)$$

The third relation in eqn. (2.24) requires the equality of the distances $r_e$ and $r_g$. The locus of such points in the three-dimensional $UVW$ space (for, say, constant $r_e$) is a sphere with radius equal to $r_e$ and centre at the location of the primary source:

$$u^2 + v^2 + w^2 = r_e^2 \tag{2.25}$$

Therefore, the locus of points defined by eqn. (2.23) is given by the intersection of the two spheres in eqns (2.24) and (2.25). Such an intersection results in a circle (the IGC circle) located in a plane, which is at right angles with the line joining the centres of the spheres. The centre of the circle is the point of intersection of the plane and the line.

Manipulating eqns (2.24) and (2.25) yields the plane of the IGC circle as

$$\frac{u}{\left(\frac{B}{u_s}\right)} + \frac{v}{\left(\frac{B}{v_s}\right)} + \frac{w}{\left(\frac{B}{w_s}\right)} = 1 \tag{2.26}$$

where

$$B = \frac{1}{2}\left[r^2 - \left(\frac{1}{a^2} - 1\right) r_e^2\right] = \frac{1}{2}\left[r^2 - \left(r_f^2 - r_e^2\right)\right] \tag{2.27}$$

Eqn. (2.26) defines a plane surface on which the IGC circle is residing. It can be shown that the line $PS$ is at right angles with the plane of the IGC circle

[305].

The quantity $B$ in eqn. (2.27) gives a measure of the intersection of the plane in eqn. (2.26) with the coordinate axes and, thereby, with the line $PS$. It is evident from eqn. (2.27) that $B$ is dependent on $r$, $r_e$ and $r_f$ or, for constant $r$, is dependent on the location of the detector only. If $\theta$ denotes the angle between the lines $PE$ and $PS$ in a plane formed by these lines, where $E$ denotes the location of the detector, then the following holds:

$$r_f^2 = r^2 + r_e^2 - 2r_e r \cos \theta \qquad (2.28)$$

Substituting for $r_f^2$ from eqn. (2.28) into eqn. (2.27) yields:

$$B = r r_e \cos \theta$$

Therefore, as the detection point varies the limits for $B$ are found to be:

$$|B| < r_e r$$

The radius $r_e$ of the IGC circle, accordingly, is given by:

$$r_c = r_e \sin \theta \quad , \quad 0 \le \theta \le \pi$$

Thus, the maximum value $r_{c\,\max}$ of the radius is $r_e$ and occurs when the plane of the IGC circle intersects the line $PS$ at point $P$. Movement of the plane to either side of point $P$ will lead to a decrease in the radius. At the extreme cases where the line $PE$ is aligned with the line $PS$ ($\theta$ is either $0°$ or $180°$) the radius $r_c$ is zero. In general, for constant values of the angle $\theta$ the radius $r_c$ is directly proportional to the distance $r_e$ between the primary source and the detector. This implies that for $r_c$ to be minimised the detector is required to be placed as close to the primary source as possible.

It follows from the above that the requirement of an infinitely large gain controller is directly linked with the locations of the detector and the observer relative to the primary and secondary sources. This derives from the dependence of the controller characteristics on the transfer characteristics of the acoustic paths from the detector and observer to the primary and secondary sources which demand a particular controller transfer function for a particular set of detection and observation points in the medium. The above analysis reveals that certain sets of detection and observation points in the medium exist, which for optimum cancellation require the controller to have an infinitely large gain. These form the loci of the IGC requirement as follows

(i) If the detector and observer are equidistant from the sources the locus is a plane surface, which perpendicularly bisects the line joining the locations of the primary and secondary sources.

(*ii*) If the detector and observer are not equidistant from the sources the locus is a circle, with centre along the line $PS$ joining the locations of the primary and secondary sources, and on a plane that is parallel with that in (*i*). The radius of the circle is equal to the distance between the detector and the line $PS$.

Note in Figure 2.1$a$ that if the observer and the detector coincide with one another then the FBCS is obtained. In such a process the distances $r_g$ and $r_h$ effectively become equal to the distances $r_e$ and $r_f$, respectively. This in terms of the transfer functions $\mathbf{E}$, $\mathbf{F}$, $\mathbf{G}$ and $\mathbf{H}$ and the distances $r_e$, $r_f$, $r_g$ and $r_h$ corresponds to:

$$r_g = r_e \quad \text{or} \quad \mathbf{G} = \mathbf{E}$$

$$(2.29)$$

$$r_h = r_f \quad \text{or} \quad \mathbf{H} = \mathbf{F}$$

Projecting the above into the controller design relation in eqn. (2.3), the corresponding controller design relation for the FBCS is obtained.

Substituting for $\mathbf{G}$ and $\mathbf{H}$ from eqn. (2.29) into eqn. (2.4) and simplifying yields $\Delta = 0$ corresponding to the critical situation of the IGC requirement discussed above. Therefore, for optimum cancellation of noise, the FBCS will always require a controller with an infinitely large gain. With a practically acceptable compromise between system performance and controller gain, and careful consideration of system stability, reasonable amounts of cancellation of the noise can be achieved with this structure.

## 2.4.2    Single-input multi-output structure

Let the ANC system in Figure 2.1 incorporate $k$ secondary sources. Thus, the controller transfer characteristics, $\mathbf{C}$, in eqn. (2.3) will represent a $1 \times k$ matrix:

$$\mathbf{C} = \begin{bmatrix} c_1 & c_2 & \dots & c_k \end{bmatrix}$$

where $c_i$ ($i = 1, 2, \dots, k$) represents the required controller transfer function along the secondary path from the detector to secondary source $i$. In this manner, the controller is realised in an SIMO form. Let the functions $\mathbf{E}(j\omega)$, $\mathbf{F}(j\omega)$, $\mathbf{G}(j\omega)$ and $\mathbf{H}(j\omega)$ be represented as:

$$\mathbf{E}(j\omega) = \mathbf{e}(j\omega)$$

$$\mathbf{F}(j\omega) = \begin{bmatrix} \mathbf{f}_1(j\omega) & \mathbf{f}_2(j\omega) & \ldots & \mathbf{f}_k(j\omega) \end{bmatrix}^T \tag{2.30}$$

$$\mathbf{G}(j\omega) = \begin{bmatrix} \mathbf{g}_1(j\omega) & \mathbf{g}_2(j\omega) & \ldots & \mathbf{g}_k(j\omega) \end{bmatrix}$$

$$\mathbf{H}(j\omega) = \begin{bmatrix} \mathbf{h}_{11}(j\omega) & \mathbf{h}_{12}(j\omega) & \ldots & \mathbf{h}_{1k}(j\omega) \\ \mathbf{h}_{21}(j\omega) & \mathbf{h}_{22}(j\omega) & \ldots & \mathbf{h}_{2k}(j\omega) \\ \ldots & \ldots & \ldots & \ldots \\ \mathbf{h}_{k1}(j\omega) & \mathbf{h}_{k2}(j\omega) & \ldots & \mathbf{h}_{kk}(j\omega) \end{bmatrix}$$

where

$$\mathbf{e}(j\omega) = \frac{A}{r_e}e^{-j\frac{\omega}{c}r_e} \quad , \quad \mathbf{f}_i(j\omega) = \frac{A}{r_{fi}}e^{-j\frac{\omega}{c}r_{fi}}$$

$$\tag{2.31}$$

$$\mathbf{g}_i(j\omega) = \frac{A}{r_{gi}}e^{-j\frac{\omega}{c}r_{gi}} \quad , \quad \mathbf{h}_{im}(j\omega) = \frac{A}{r_{him}}e^{-j\frac{\omega}{c}r_{him}}$$

$i = 1, 2, \ldots, k$, $m = 1, 2, \ldots, k$, $A$ is a constant and $r_e$, $r_{fi}$, $r_{gi}$ and $r_{him}$ are the distances of the acoustic paths with transfer characteristics $\mathbf{e}(j\omega)$, $\mathbf{f}_i(j\omega)$, $\mathbf{g}_i(j\omega)$ and $\mathbf{h}_{im}(j\omega)$, respectively. Substituting for $\mathbf{E}(j\omega)$, $\mathbf{F}(j\omega)$, $\mathbf{G}(j\omega)$ and $\mathbf{H}(j\omega)$ from eqn. (2.30) into eqn. (2.18) and manipulating yields the set of solutions:

$$\frac{\mathbf{f}_i}{\mathbf{e}} = \frac{\mathbf{h}_{i1}}{\mathbf{g}_1} = \frac{\mathbf{h}_{i2}}{\mathbf{g}_2} = \ldots = \frac{\mathbf{h}_{ik}}{\mathbf{g}_k} \quad , \quad i = 1, 2, \ldots, k,$$

$$m = 1, 2, \ldots, k \tag{2.32}$$

$$\frac{\mathbf{f}_m}{\mathbf{f}_i} = \frac{\mathbf{h}_{m1}}{\mathbf{h}_{i1}} = \frac{\mathbf{h}_{m2}}{\mathbf{h}_{i2}} = \ldots = \frac{\mathbf{h}_{mk}}{\mathbf{h}_{ik}} \quad , \quad i \neq m$$

Substituting for the functions in eqn. (2.32) from eqn. (2.31) accordingly and simplifying yields:

$$\frac{r_e}{r_{fi}} = \frac{r_{g1}}{r_{hi1}} = \frac{r_{g2}}{r_{hi2}} = \ldots = \frac{r_{gk}}{r_{hik}} = a_i$$

$$, \quad i = 1, 2, \ldots, k \tag{2.33}$$

$$r_{fi} - r_e = r_{hi1} - r_{g1} = r_{hi2} - r_{g2} = \ldots = r_{hik} - r_{gk}$$

and

$$\frac{r_{fi}}{r_{fm}} = \frac{r_{hi1}}{r_{hm1}} = \frac{r_{hi2}}{r_{hm2}} = \ldots = \frac{r_{hik}}{r_{hmk}} = a_{im} \quad , \quad i = 1, 2, \ldots, k,$$

$$m = 1, 2, \ldots, k$$

$$r_{fm} - r_{fi} = r_{hm1} - r_{hi1} = r_{hm2} - r_{hi2} = \ldots = r_{hmk} - r_{hik} \quad , \quad i \neq m \tag{2.34}$$

where $a_i$ and $a_{im}$ are positive real numbers representing distance ratios. Eqns. (2.33) and (2.34) define the loci of detection and observation points for which $\Delta(j\omega) = 0$ and the controller, for optimum cancellation of noise at the observation points, is required to have an infinitely large gain in each secondary path.

Eqn. (2.33) describes the locus of detection and observation points relative to the location of the primary source and secondary source $i$ $(i = 1, 2, \ldots, k)$. It follows from the analysis presented above that, if the primary source and secondary source $i$ are located at points $P$ and $S_i$, respectively, in the medium, then the locus for unity distance ratio $(a_i = 1)$ is a plane perpendicularly bisecting the line $PS_i$. However, if $a_i \neq 1$ the locus is a circle, with centre along a straight line joining points $P$ and $S_i$ in a plane, which is at right angles with this line. The radius of the circle $r_{ci}$ is given by:

$$r_{ci} = r_e \sin \theta_i \quad , \quad 0 \le \theta_i \le \pi$$

where, assuming that the detector is located at point $E$ in the medium, $\theta_i$ is the angle between the lines $PE$ and $PS_i$ in a plane formed by these lines.

Eqn. (2.34) describes the locus of detection and observation points relative to secondary sources $i$ $(i = 1, 2, \ldots, k)$ and $m$ $(m = 1, 2, \ldots, k)$, located at points $S_i$ and $S_m$ respectively. In this case, as follows from the previous analysis, the locus, for unity distance ratio $(a_{im} = 1)$, is a plane perpendicularly bisecting the line $S_iS_m$. For a nonunity distance ratio $(a_{im} \neq 1)$, however, the locus is a circle, with centre along a straight line joining the points $S_i$ and $S_m$, in a plane that is at right angles with this line. The radius $r_{cim}$ of this circle is given by:

$$r_{cim} = r_{fi} \sin \theta_{im} \quad , \quad 0 \le \theta_{im} \le \pi$$

where, assuming the detector is located at point $E$, $\theta_{im}$ is the angle between the lines $S_iE$ and $S_iS_m$.

It follows from the above that in an SIMO ANC system the locus of detection and observation points leading to the IGC requirement is defined in relation to the locations of the primary source considered with each secondary source as well as each secondary source considered with any other secondary source. In this manner, for a system with $k$ secondary sources a total of $\sum_{i=1}^{k} i$ pairs of sources can be identified. Among these, the primary source considered with each secondary source leads to $k$ pairs, whereas the remaining $\sum_{i=1}^{k} i$ pairs are formed by considering the secondary sources with one another. In each case, assuming the two sources in question are located at points $X$ and $Y$, the following two situations lead to the IGC requirement:

(*i*) When the detector and all observers are equidistant from points $X$ and $Y$. This defines the locus of detection and observation points as a plane (the IGC plane), which perpendicularly bisects the line $XY$.

(*ii*) When the distance ratios from point $X$ to the detector and observer $m$ ($m = 1, 2, \ldots, k$) and to each pair of observation points as well as the distance ratios from point $Y$ to the detector and observer $m$ ($m = 1, 2, \ldots, k$) and to each pair of observation points are each equal to unity. This defines the locus of detection and observation points as a circle (the IGC circle), with centre along a straight line passing through points $X$ and $Y$, on a plane, which is at right angles with this line.

Note that in an FBCS, where both the detection and observation points co-incide with one another, the situation described in (*i*) above corresponds to the detection point being on the IGC plane. In an FFCS, however, this corresponds to the situation when the detection point and all the observation points are on the IGC plane. With the situation described in (*ii*), on the other hand, an FBCS always satisfies the requirement. In an FFCS, however, it is possible to minimise the region of space occupied by the IGC circle by a proper geometrical arrangement of system components.

## 2.5 System stability

As noted in Figure 2.1, the secondary signals reaching the detection point form (positive) feedback loops, which can cause the system to become unstable. Therefore, an analysis of the system from a stability point of view is important at a design stage. For practical systems a measure of absolute stability is not useful; a system that has an extremely long and oscillatory transient response is unlikely to be accepted. In this respect, a measure of relative stability can provide a more acceptable design criterion. This can be achieved, through the utilisation of the Nyquist's stability criterion, in terms of gain and phase margins.

Substituting for $\mathbf{C}$ from eqn. (2.3) into eqn. (2.11) and simplifying yields:

$$\mathbf{U}_C = \mathbf{U}_D \mathbf{E} \Delta^{-1} \mathbf{G} \mathbf{H}^{-1} \left[ \mathbf{I} - \mathbf{F} \Delta^{-1} \mathbf{G} \mathbf{H}^{-1} \right]^{-1}$$

This is the closed-loop system equation between $\mathbf{U}_D$ and $\mathbf{U}_C$ with the characteristic equation, using eqn. (2.4), given as:

$$Ch.Eq = \left| \mathbf{I} - \mathbf{F} \left[ \mathbf{G} \mathbf{H}^{-1} \mathbf{F} - \mathbf{E} \right]^{-1} \mathbf{G} \mathbf{H}^{-1} \right| \qquad (2.35)$$

Eqn. (2.35) is the required relation for analysing the system from a stability viewpoint. Note that this is expressed in terms of the transfer characteristics of the acoustic paths in Figure 2.1. Therefore, the stability of the system is affected by the geometrical arrangement of system components. To explore this further, and for simplicity purposes, the system in Figure 2.1 is considered within an SISO structure. Thus, substituting for the functions $\mathbf{E}$, $\mathbf{F}$, $\mathbf{G}$ and $\mathbf{H}$ in eqn. (2.35) as $\mathbf{e}$, $\mathbf{f}$, $\mathbf{g}$ and $\mathbf{h}$ respectively and simplifying yields:

$$Ch.Eq = 1 + \mathbf{X}$$

where

$$\mathbf{X} = \frac{1}{\dfrac{\mathbf{eh}}{\mathbf{fg}} - 1} \qquad (2.36)$$

Thus, according to the Nyquist stability criterion, for $s = j\omega$, the stability of the system is expressed in terms of the polar plot of the frequency response function $X(j\omega)$. If this is represented in terms of a magnitude $B(\omega)$ and a phase $\theta(\omega)$ as:

$$\mathbf{X}(j\omega) = B(\omega)e^{j\theta(\omega)} \qquad (2.37)$$

Then, for the system to be stable the magnitude of $X(j\omega)$, at some frequency $\omega$ for which $\theta(\omega) = -(2n+1)\pi$ $(n = 0, 1, \ldots)$, should be less than unity [63]:

$$B(\omega) < 1 \quad \text{when} \quad \theta(\omega) = -(2n+1)\pi \quad , \quad n = 0, 1, \ldots \qquad (2.38)$$

where the negative angle (clockwise) indicates the direction of approach towards the $\pi$ axis on a polar plot of $\mathbf{X}(j\omega)$. This can be expressed graphically by following the polar plot of $\mathbf{X}(j\omega)$ from $\omega = 0$ to $\omega = \infty$ and observing each crossing of the $\pi$ axis. If the point 1 lies on the left-hand side then the system is considered to be stable, whereas if the point 1 lies on the right-hand side then the system is unstable.

The gain margin of the closed-loop system is defined, at some frequency $\omega$ for which the phase $\theta(\omega)$ is 180°, as the additional gain $k_g$ for the system to become unstable. In terms of the amplitude transfer function $B(\omega)$, $k_g$ is given as:

$$k_g = \frac{1}{B(\omega)} \quad \text{when} \quad \theta(\omega) = -\pi \qquad (2.39)$$

Thus, it follows from eqns (2.38) and (2.39) that for a system to be stable the gain margin $k_g$ must be greater than unity. A gain margin less than unity will mean that the system is unstable.

The phase margin of the closed-loop system is defined, at some frequency

$\omega$ for which $B(\omega) = 1$, as the additional phase $k_\theta$ that is required to make the system unstable. This in terms of the phase $\theta(\omega)$ is given by:

$$k_\theta = \theta(\omega) + \pi \quad \text{when} \quad B(\omega) = 1 \tag{2.40}$$

Thus, it follows from eqns. (2.38) and (2.40) that the phase margin $k_\theta$ at a frequency $\omega$ for which $B(\omega) = 1$ is the amount of phase shift that would just produce instability. For minimum-phase systems to be stable the phase margin must be positive. A negative phase margin will mean that the system is unstable.

Substituting for **e**, **f**, **g** and **h**, for $s = j\omega$, from eqn. (2.19) into eqn. (2.36), simplifying and using eqn. (2.37) yields $B(\omega)$ and $\theta(\omega)$ as:

$$B(\omega) = \left[Q^2(\omega) + 1 - 2Q(\omega)\cos\phi(\omega)\right]^{-0.5}$$

$$\tag{2.41}$$

$$\theta(\omega) = \tan^{-1}\left[\frac{Q(\omega)\sin\phi(\omega)}{1 - Q(\omega)\cos\phi(\omega)}\right] + 2m\pi \quad , \quad m = 0, \pm 1, \ldots$$

where

$$\frac{\mathbf{e}(j\omega)\mathbf{h}(j\omega)}{\mathbf{f}(j\omega)\mathbf{g}(j\omega)} = Q(\omega)e^{j\phi(\omega)}$$

$$Q(\omega) = \frac{r_f r_g}{r_e r_h} \tag{2.42}$$

$$\phi(\omega) = \frac{2\pi}{\lambda}(r_{gh} - r_{ef})$$

with $r_{ef} = r_f - r_e$, $r_{gh} = r_h - r_g$ and $\lambda$ represents the signal wavelength.

Eqn. (2.42) gives the magnitude $Q(\omega)$ and the phase $\phi(\omega)$ in terms of the distances from the detection and observation points to the primary and secondary sources and the signal wavelength only. Using these relations the stability of the ANC system can be determined in terms of the locations of the detector and observer with respect to the primary and secondary sources in the three-dimensional propagation medium.

### 2.5.1    Gain margin

Substituting for $\theta(\omega)$ from eqn. (2.41) into eqn. (2.39) and simplifying yields:

$$\phi(\omega) = \begin{cases} (2n+1)\pi & \text{for} \quad Q(\omega) \geq 1 \\ n\pi & \text{for} \quad 0 < Q(\omega) < 1 \end{cases} \quad , \quad n = 0, \pm 1, \ldots \tag{2.43}$$

Substituting for $\phi(\omega)$ from eqn. (2.42) into eqn. (2.43) and simplifying yields:

$$r_{gh} - r_{ef} = \begin{cases} (2n+1)\dfrac{\lambda}{2} & \text{for} \quad Q(\omega) \geq 1 \\[2mm] n\left(\dfrac{\lambda}{2}\right) & \text{for} \quad 0 < Q(\omega) < 1 \end{cases}, \quad n = 0, \pm 1, \ldots \quad (2.44)$$

Eqns (2.43) and (2.44) give the necessary conditions under which $\theta(\omega) = 180°$ and, thus, the gain margin $k_g$ is given by eqn. (2.39). Substituting for $\phi(\omega)$ according to eqn. (2.43) into eqn. (2.41), using eqn. (2.39) and simplifying yields the gain margin as:

$$k_g = \begin{cases} 1 - Q(\omega) & \text{for } \phi(\omega) = 2n\pi \quad \text{and } 0 < Q(\omega) < 1 \\ 1 + Q(\omega) & \text{for } \phi(\omega) = (2n+1)\pi \quad \text{and } Q(\omega) > 0 \end{cases}, \quad n = 0, \pm 1, \ldots$$
$$(2.45)$$

To find the specified regions in a three-dimensional $UVW$ space with the primary source, secondary source, detector and observer located at points $P(0,0,0)$, $S(u_s, v_s, w_s)$, $E(u_e, v_e, w_e)$ and $O(u_o, v_o, w_o)$, respectively, corresponding to eqn. (2.45) and analyse system stability in these regions consider the two cases of $Q(\omega) \geq 1$ and $Q(\omega) < 1$.

**The case of $Q(\omega) \geq 1$**

Substituting for $Q(\omega) \geq 1$ into eqn. (2.42) and simplifying yields:

$$\frac{r_e}{r_f} \leq a$$

where $a$ is a positive real number denoting the distance ratio:

$$a = \frac{r_g}{r_h} \quad (2.46)$$

It can be shown that eqn. (2.46) defines a family of spheres in the three-dimensional $UVW$ space as [305]:

$$\left[u_o + \frac{a^2 u_s}{1 - a^2}\right]^2 + \left[v_o + \frac{a^2 v_s}{1 - a^2}\right]^2 + \left[w_o + \frac{a^2 w_s}{1 - a^2}\right]^2 = \left[\frac{ar}{1 - a^2}\right]^2 \quad (2.47)$$

where $r$ is the distance between the primary and secondary sources. Eqn. (2.47), for $a = 0$ corresponds to point $P$ (location of the primary source) and for $a = \infty$ to point $S$ (location of the secondary source). If $a = 1$, then eqn. (2.47) defines a plane surface, which perpendicularly bisects the line $PS$ joining the locations of the primary and secondary sources:

$$\frac{u_o}{\left(\frac{r^2}{2u_s}\right)} + \frac{v_o}{\left(\frac{r^2}{2v_s}\right)} + \frac{w_o}{\left(\frac{r^2}{2w_s}\right)} = 1$$

It follows from the above that for $Q(\omega) \geq 1$ the detection point should remain inside the sphere defined by eqn. (2.47). This, using eqn. (2.36) yields the locus of detection points $E$ as:

$$\left[u_e + \frac{a^2 u_s}{1 - a^2}\right]^2 + \left[v_e + \frac{a^2 v_s}{1 - a^2}\right]^2 + \left[w_e + \frac{a^2 w_s}{1 - a^2}\right]^2 \leq \left[\frac{ar}{1 - a^2}\right]^2 \quad , \quad a < 1$$

$$\frac{u_e}{\left(\frac{r^2}{2u_s}\right)} + \frac{v_e}{\left(\frac{r^2}{2v_s}\right)} + \frac{w_e}{\left(\frac{r^2}{2w_s}\right)} \leq 1 \quad , \quad a = \mathbf{(2.48)}$$

$$\left[u_e - \frac{a^2 u_s}{a^2 - 1}\right]^2 + \left[v_e - \frac{a^2 v_s}{a^2 - 1}\right]^2 + \left[w_e - \frac{a^2 w_s}{a^2 - 1}\right]^2 \geq \left[\frac{ar}{a^2 - 1}\right]^2 \quad , \quad a > 1$$

Therefore, if the observer and the detector are restricted to the regions of the $UVW$ space defined by eqns (2.47) and (2.48), respectively, then $Q(\omega) \geq 1$. In terms of eqns (2.44) and (2.45) this means that for such observation and detection points where the distance difference $r_{gh} - r_{ef}$ is an odd multiple of half the signal wavelength (implying that $\phi(\omega)$ is an odd multiple of $\pi$ and $\theta(\omega)$ is $-180°$) the gain margin $k_g$ assumes values that are either equal to or greater than 2. This implies that, under such a situation, at locations of the observation and detection points for which $\phi(\omega)$ is an odd multiple of $\pi$ the system will be stable.

**The case of $Q(\omega) < 1$**

Substituting for $Q(\omega) < 1$ into eqn. (2.42) and using eqn. (2.46) yields:

$$\frac{r_e}{r_f} > a \qquad (2.49)$$

It follows from the results obtained above, for $Q(\omega) \leq 1$, that eqn. (2.49), for the observation and detection points, defines the region of the three-dimensional $UVW$ space which is the complement of that given in eqns (2.47) and (2.48); i.e. for the observer restricted to the locus defined by eqn. (2.47) the detector is to be restricted to the following:

$$\left[u_e + \frac{a^2 u_s}{1-a^2}\right]^2 + \left[v_e + \frac{a^2 v_s}{1-a^2}\right]^2 + \left[w_e + \frac{a^2 w_s}{1-a^2}\right]^2 > \left[\frac{ar}{1-a^2}\right]^2 \quad , \quad a < 1$$

$$\frac{u_e}{\left(\frac{r^2}{2u_s}\right)} + \frac{v_e}{\left(\frac{r^2}{2v_s}\right)} + \frac{w_e}{\left(\frac{r^2}{2w_s}\right)} > 1, \; a = \quad (2.50)$$

$$\left[u_e - \frac{a^2 u_s}{a^2-1}\right]^2 + \left[v_e - \frac{a^2 v_s}{a^2-1}\right]^2 + \left[w_e - \frac{a^2 w_s}{a^2-1}\right]^2 < \left[\frac{ar}{a^2-1}\right]^2 \quad , \quad a > 1$$

Thus, if the observer and detector are restricted to their respective regions defined by eqns (2.47) and (2.50), then at locations for which the distance difference $r_{gh} - r_{ef}$ in eqn. (2.44) is an integral multiple of the signal wavelength (implying that $\phi(\omega)$ is an integral multiple of $\pi$ and $\theta(\omega)$ will be $-180°$) the gain margin of the closed-loop system is given by eqn. (2.45) as:

$$k_g = \begin{cases} 1 - Q(\omega) & \text{for} \quad \phi(\omega) = 2n\pi \\ 1 + Q(\omega) & \text{for} \quad \phi(\omega) = (2n+1)\pi \end{cases} \quad , \quad n = 0, \pm 1, \ldots$$

From which the ranges of the gain margin are obtained as:

$$\begin{aligned} k_g < 1 \quad &\text{for} \quad \phi(\omega) = 2n\pi \\ k_g > 1 \quad &\text{for} \quad \phi(\omega) = (2n+1)\pi \end{aligned} \quad , \quad n = 0, \pm 1, \ldots \quad (2.51)$$

Eqn. (2.51) implies that for such detection and observation points in the region of the three-dimensional $UVW$ space defined by eqn. (2.49) for which $\phi(\omega)$ is an even multiple of $\pi$ the gain margin is less than unity and hence the system will be unstable. However, if at such points $\phi(\omega)$ is an odd multiple of $\pi$, then the gain margin is greater than unity and hence the system will be stable.

### 2.5.2   Phase margin

Substituting for $B(\omega)$ from eqn. (2.41) into eqn. (2.40) and simplifying Yields:

$$0 < Q(\omega) \leq 2 \quad (2.52)$$

and the range of $\phi(\omega)$ as:

$$\frac{(4n-1)\pi}{2} < \phi(\omega) < \frac{(4n+1)\pi}{2} \quad \text{for} \quad 0 < Q(\omega) \leq 2 \quad , \quad n = 0, \pm 1, \ldots$$
$$(2.53)$$

Substituting for $\phi(\omega)$ from eqn. (2.42) into eqn. (2.53) and simplifying yields the allowable distance difference $r_{gh} - r_{ef}$ as:

$$(4n-1)\frac{\lambda}{4} < r_{gh} - r_{ef} < (4n+1)\frac{\lambda}{4} \quad \text{for} \quad 0 < Q(\omega) \leq 2 \quad , \quad n = 0, \pm 1, \ldots$$
$$(2.54)$$

Eqn. (2.54), or equivalently eqn. (2.53), is the necessary condition for $B(\omega) = 1$ in eqn. (2.40). Under this condition the phase angle $\theta(\omega)$ follows from eqn. (2.41). Using eqns (2.40), (2.41) and (2.53) yields the phase margin as:

$$k_\theta = \begin{cases} \tan^{-1}\frac{Q(\omega)\left[4-Q^2(\omega)\right]^{0.5}}{2-Q^2(\omega)} + (2m+1)\pi & \text{for} \quad 0 \leq \phi(\omega) < \frac{(4n+1)\pi}{2} \\[2mm] -\tan^{-1}\frac{Q(\omega)\left[4-Q^2(\omega)\right]^{0.5}}{2-Q^2(\omega)} + (2m+1)\pi & \text{for} \quad \frac{(4n-1)\pi}{2} < \phi(\omega) < 0 \end{cases}$$

where $0 < Q(\omega) \leq 2$ and $m$ and $n$ are integer numbers. To find the corresponding region of the three-dimensional $UVW$ space in terms of the locations of the observation and detection points, eqn. (2.52) is considered.

Substituting for $Q(\omega)$ from eqn. (2.42) into eqn. (2.52), using eqn. (2.46) and simplifying yields:

$$\frac{r_e}{r_f} \geq 0.5a$$

This yields the locus of detection points $E$ as:

$$\left[u_e + \frac{a^2 u_s}{4-a^2}\right]^2 + \left[v_e + \frac{a^2 v_s}{4-a^2}\right]^2 + \left[w_e + \frac{a^2 w_s}{4-a^2}\right]^2 \geq \left[\frac{2ar}{4-a^2}\right]^2 \quad , \quad a < 2$$

$$\frac{u_e}{\left(\frac{r^2}{2u_s}\right)} + \frac{v_e}{\left(\frac{r^2}{2v_s}\right)} + \frac{w_e}{\left(\frac{r^2}{2w_s}\right)} \geq 1, \ a = \quad (2.55)$$

$$\left[u_e - \frac{a^2 u_s}{a^2-4}\right]^2 + \left[v_e - \frac{a^2 v_s}{a^2-4}\right]^2 + \left[w_e - \frac{a^2 w_s}{a^2-4}\right]^2 \leq \left[\frac{2ar}{a^2-4}\right]^2 \quad , \quad a > 2$$

It follows from the above that, for a minimum-phase situation, the system will be stable for locations of detector and observer in the regions defined by eqn. (2.55), where the phase margin assumes positive values; however, for negative values of $k_\theta$ the system will be unstable.

The analysis presented above demonstrates that system stability is affected by the geometric arrangement of system components. This has led to identifying regions of the propagation medium as loci of detection and observation points for which the system will be unstable. For simplicity reasons, an SISO structure has been considered in the above analysis. The results thus obtained can be extended, in a similar manner, to an SIMO structure through consideration of the feedback loops in the system in a unified manner to affect the location of the detection point and all the observation points relative to the primary and secondary sources.

## 2.6  Conclusions

A coherent analysis and design of ANC systems in a three-dimensional nondispersive propagation medium have been presented. The relation between the transfer characteristics of the required controller and the geometrical arrangement of system components has been studied, and conditions interpreted as geometrical constraints in the design of ANC systems have been derived and analysed.

The interference of the component waves in an ANC system effectively leads to a pattern of zones of cancellation and reinforcement. An increase in the physical extent of cancellation, for a given maximum frequency of the noise, can be achieved by decreasing the separation between the primary and secondary sources. In practice, this is limited due to the physical dimensions of the sources. However, it has been demonstrated that, in practice, the physical extent of cancellation can be significantly increased through a proper design incorporating a suitable geometrical arrangement of system components.

In practice, the process of controller implementation, using analogue and/or digital techniques, results in errors in the amplitude and phase of the controller. Such errors affect the amount and, hence, the physical extent of cancellation to a lessor or greater degree, depending on the extent of the error involved. Although, the general trend of the pattern of zones of cancellation and reinforcement appear similar to the situation when there are no errors in the controller characteristics, the physical extent of cancellation is affected by such errors and the level of cancellation is reduced at the observation point(s).

The dependence of the required controller characteristics on the characteristics of the acoustic paths in the system, arising from geometrical arrangement of system components, can sometimes lead to practical difficulties in the design of the controller and to instability in the system. In particular, specific arrangements of system components have been identified as loci of detection and observation points relative to the sources, which lead to the critical situation of the IGC requirement.

In an ANC system incorporating a single primary source and $k$ secondary sources, realised within an SIMO structure, the locus of detection and observation points leading to the IGC requirement is defined in relation to the locations of the primary source considered with each secondary source as well as each secondary source considered with any other secondary source; for $k = 1$, only the primary and secondary sources are considered. In this manner, with an FBCS, where both the detection and observation points coincide with one another, the conditions for the IGC requirement are always met. With an FFCS, however, it is possible to avoid the locus of detection and observation points, which lead to the IGC requirement by a proper geometrical arrangement of system components.

The dependence of the controller characteristics, for optimal cancellation of the noise, on the characteristics of system components and geometry can sometimes lead to practical difficulties in system stability. For given system components and controller transfer function the stability and relative stability of the system have been found to be dependent on the observer and detector locations relative to the primary and secondary sources in the medium. Loci of observation and detection points have been identified in the medium that can cause the system to become unstable. Thus, it has been demonstrated that stable operation of the system can be assured by a proper ANC system design incorporating a suitable geometric arrangement of system components.

*Chapter 3*

# Adaptive methods in active control

## S. J. Elliott

*Institute of Sound and Vibration Research, University of Southampton,*
*Southampton, UK*
*Email: sje@isvr.soton.ac.uk*

*In this chapter adaptive feedforward and fixed feedback controllers are reviewed and their robust stability conditions contrasted for active control applications.*

## 3.1 Introduction

Active control systems, and particularly active sound control systems, have a number of distinguishing features compared with other control problems. The disturbance to be rejected often has broadband and narrowband components, with the latter often being nonstationary. The plant is generally high order, well damped and nonminimum phase, and its response can change rapidly, due for example to the movement of people within an enclosure. The narrowband disturbance components are often attenuated with a feedforward controller which is made adaptive to track the nonstationarities. The filtered-reference LMS algorithm, which is generally used to adapt such controllers, is particularly robust to changes in plant response. If the plant is subject to unstructured multiplicative uncertainty, the algorithm is stable provided the upper bound of the uncertainty is less than unity. The algorithm can be made even more robust by introducing a leakage term into the update equation. Feedback controllers can be used to control the broadband components of the disturbance, but their performance bandwidth is limited by propagation delay in the plant. Such feedback controllers must be robustly stable to plant changes, and the internal model control (IMC) architecture allows the condition for robust stability to be related back to that for the adaptive feedforward system. The feedback controller can be made adaptive to nonstationary disturbances by adapting the control filter in the IMC arrangement. The robust stability condition can be maintained during adaptation by incorporating a constraint into the adaptation algorithm, which can be implemented with the least loss of performance by using an adaptation algorithm operating in the discrete frequency domain.

Adaptive methods are used in active control systems to automatically adjust the response of the controller to compensate for changes in the response of the plant or in the form of the excitation. Active systems for the control of sound and vibration are generally designed to reject the disturbances generated by the original, primary, course, rather than to track a command signal, and thus the design of the controller which gives the best performance will depend on the statistical properties of the disturbance. If the disturbance is nonstationary, as it is in many applications of active control, the controller must be adapted to track the changes in the properties of the disturbance if good control is to be maintained. In many applications of active control, particularly active sound control, some of the changes which occur in the response of the plant are too rapid to identify effectively without introducing an unacceptably high level of identification noise. Externally generated identification noise is generally necessary to measure the plant response in active control since the disturbance is not sufficiently rich. For rapid changes in plant response the level of identification noise required to accurately track the change can contribute more to the perceived acoustic output than the original disturbance, which rather defeats the object of the controller. If the plant response can only be identified under nominal conditions, but varies about this nominal response in a way which cannot be measured, then it is necessary for the control loop and the adaptation algorithm which compensates the control loop, for changes in the disturbance, to be robust to these changes. In this chapter, adaptation methods for active control systems will be described which are robust to changes in the plant response. Because the controller can be made adaptive, feedforward control is no longer open loop and is widely used in active control when an independent reference signal is available, which is well correlated with the disturbance. If the controller is implemented as a digital FIR filter, or an array of filters for the multichannel case, it can be adapted to minimise a quadratic cost function using the filtered-reference LMS algorithm. The convergence properties and robustness of this algorithm

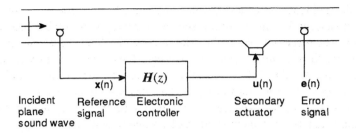

*Figure 3.1    Example of the application of a single-channel feedforward control system to the active control of plane waves of sound in a duct.*

are briefly described in both the single channel and multichannel cases. In applications where no independent reference signals are available active control systems can be implemented using feedback control. In this case there are two dangers from changes in the plant response: first that the feedback loop itself may become unstable, and second that the adaptation algorithm may become unstable. Both of these possibilities will be considered and it will be demonstrated how the adaptation algorithm can be modified to minimise a quadratic cost function with the constraint that the feedback loop remains robustly stable.

## 3.2    Feedforward control

In this section we consider the performance and stability of adaptive control algorithms for feedforward control systems. The case of a single-channel system is considered first, with a single reference signal, secondary actuator and error sensor, but with a general broadband excitation. Such a control system is typically used in the active control of sound in ducts, for example. Second, the behaviour is considered of a system with multiple actuators and error sensors but only when it is excited by relatively narrowband excitation about a known frequency. Such systems are used for the active control of sound in propeller aircraft and helicopters.

### 3.2.1    Single-channel feedforward control

The application of a single-channel feedforward control system to the active control of plant sound waves in a duct is illustrated in Figure 3.1. The waveform of the incident disturbance, which is assumed to be stochastic and broadband, is measured by the detection sensor, a microphone in this case, and fed forward to the electronic controller the output of which drives the secondary actuator, which is a loudspeaker in this case. This arrangement was first suggested by Lueg in 1936 [180]. In order to monitor the performance and provide a feedback signal for the adaptation of the controller an error sensor is located further downstream to measure the residual soundfield. Typically, the electronic controller is implemented digitally, since it is easier to adapt a digital filter, and as well as the usual data converters, effective antialiasing filters and reconstruction filters are included in the controller to prevent aliasing which would otherwise be audible. Also, the possible acoustic feedback path from the secondary actuator back to the detection sensor can be compensated for inside the controller, by using a digital filter to model this response and using an arrangement similar to the echo canceller used in telecommunications systems, as described for example in Reference [235]. The block diagram of the single channel feedforward control system can now be drawn as in Figure 3.2, from which the sampled error signal can be written as:

$$e(n) = d(n) + \mathbf{g}^T \mathbf{u}(n) \tag{3.1}$$

where $d(n)$ is the disturbance, $\mathbf{g}$ is the vector of impulse response coefficients of the plant and $u(n)$ is the vector of past values of the input signal to the plant. If the control filter is linear and time invariant, the ordering of the control filter and plant in Figure 3.2,($a$) can be notionally transposed and the block diagram redrawn as in Figure 3.2,($b$), in which case the sampled error signal can be written in the alternative form:

$$e(n) = d(n) + \mathbf{w}^T \mathbf{r}(n) \tag{3.2}$$

where $\mathbf{w}$ is the vector of coefficients of the FIR control filter and $\mathbf{r}(n)$ is the vector of past values of the reference signal filter by the plant response. Eqn. (3.2) allows a cost function equal to the mean square error signal:

$$J = E[e^2(n)] \tag{3.3}$$

where $E$ denotes the expectation operator, to be written as a quadratic function of the coefficients of the control filter. This may be differentiated with respect to the filter coefficients and the instantaneous value used to update the control filters at each sample in a stochastic gradient algorithm called the filtered-reference LMS algorithm

$$\mathbf{w}(n+1) = \mathbf{w}(n) - \alpha \mathbf{r}(n)e(n) \tag{3.4}$$

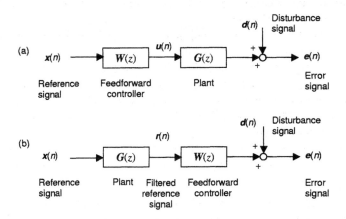

*Figure 3.2* (a) Block diagram of a single-channel feedforward control system. (b) An equivalent block diagram for a linear time-invariant controller.

as described for example in Reference [337], where $\alpha$ is a convergence coefficient. In a practical implementation the filtered reference signal must be generated by passing the reference signal through an estimate of the plant response, as shown in Figure 3.3, to generate the estimated filtered reference signal $\hat{r}(n)$, and the practical form of the filtered-reference LMS algorithm becomes:

$$w(n+1) = w(n) - \alpha \hat{r}(n)e(n) \tag{3.5}$$

The convergence behaviour of this algorithm can be analysed by using eqn. (3.2) for $e(n)$ in eqn. (3.5), assuming $\mathbf{w}(n)$ and $\mathbf{r}(n)$ are independent and taking expectations to give:

$$E(\mathbf{w}(n+1) - \mathbf{w}_\infty) = [I - \alpha E(\hat{r}(n)\hat{r}^T(n))]E[\mathbf{w}(n) - \mathbf{w}_\infty] \tag{3.6}$$

where $\mathbf{w}_\infty$ is the steady-state value of the filter coefficients if the algorithm is stable, which is given by:

$$\mathbf{w}_\infty = \left[E(\mathbf{r}(n)\mathbf{r}(n)^T)\right]^{-1} E(\mathbf{r}(n)d(n)) \tag{3.7}$$

The stability of the algorithm depends on the eigenvalues of the matrix which in general are complex. The condition for stability is that the real part of each

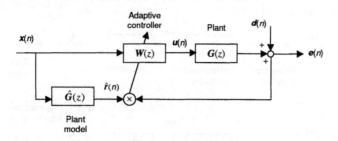

Figure 3.3    *Block diagram of the practical implementation of the filtered-reference LMS algorithm, in which the plant model is used to generate the filtered-reference signal.*

of these eigenvalues must be positive [269], i.e.:

$$\mathrm{Re}[\mathrm{Eig}[E(\mathbf{r}(n)\mathbf{r}(n)^T)]] > 0 \tag{3.8}$$

In general, it is not straightforward to relate the eigenvalues of this matrix to the difference between the responses of the physical plant and the plant model. It has, however, been shown [325] that a sufficient condition for convergence is that the ratio of transfer functions of the plant model and plant, $\frac{\hat{G}(z)}{G(z)}$, is strictly positive real (SPR) which implies that stability of the algorithm is assured providing:

$$\mathrm{Re}[\hat{G}(e^{j\omega T})G(e^{j\omega T})] > 0, \quad \text{for all } \omega T \tag{3.9}$$

If the plant model is only in error by a phase shift $\phi(\omega T)$, so that $\hat{G}(e^{j\omega T}) = G(e^{j\omega T})\phi(\omega T)$, then eqn. (3.9) reduces to:

$$\cos(\phi(\omega T)) > 0, \quad \text{for all } \omega T \tag{3.10}$$

Stability is thus assured provided the phase error is less than 90° at each frequency, which is a generalisation of a previous result for a single frequency reference signal [70]. Alternatively if the plant model can be assumed to be equal to the plant response under nominal conditions, but the plant itself is assumed to be subject to multiplicative uncertainty, the magnitude of which is bounded, then:

$$G(e^{j\omega T}) = G_0(e^{j\omega T}) \qquad (3.11)$$

and

$$\hat{G}(e^{j\omega T}) = G_0(e^{j\omega T})(1 + \Delta(e^{j\omega T})) \qquad (3.12)$$

where

$$|\Delta(e^{j\omega T})| \le B(e^{j\omega T}) \qquad (3.13)$$

In this case the stability condition in eqn. (3.9) becomes:

$$1 + \text{Re}[\Delta(e^{j\omega T})] > 0 \qquad (3.14)$$

If the multiplicative uncertainty is unstructured, so that its phase is unknown, then the worst-case condition for stability will be where the phase of $\Delta(e^{j\omega T})$ is 180° and its magnitude is equal to its upper bound. In this case the condition for the stability of the filtered-reference LMS algorithm reduces to:

$$B(e^{j\omega T}) < 1 \qquad (3.15)$$

which can be directly compared with the condition for stability of a conventional feedback system, as described below. The filtered-reference LMS algorithm can be made more robust to differences between the plant and plant model by introducing some leakage into the adaptation eqn, which then becomes:

$$\mathbf{w}(n+1) = (1 - \alpha\beta)\mathbf{w}(n) - \alpha\hat{r}(n)e(n) \qquad (3.16)$$

which minimises a modified quadratic cost function given by:

$$J = E[e^2(n)] + \beta E[\mathbf{w}^T(n)\mathbf{w}(n)] \qquad (3.17)$$

The convergence condition given by eqn. (3.9) is modified by the leakage term to be of the form:

$$\text{Re}[\hat{G}'(e^{j\omega T})G(e^{j\omega T})] + \beta > 0 \quad \text{for all} \ \ \omega T \qquad (3.18)$$

If the plant model is again only in error by a phase shift $\phi(\omega T)$, then eqn. 3.18 can be used to show that a sufficient condition for stability is:

$$cos(\phi(\omega T)) > \frac{-\beta}{|G(e^{j\omega T})|^2} \quad \text{for all} \ \ \omega T \qquad (3.19)$$

So that if $|G(e^{j\omega T})|^2$ is particularly small at some frequencies, in particular much less than $\beta$, then the right-hand side of eqn. (3.19) will be less than 1 and the algorithm is stable for any phase error in the plant model at this frequency. Alternatively if we assume an unstructured multiplicative uncertainty as in eqns (3.11), (3.12) and (3.13), then the sufficient condition for stability, assuming again the worst-case uncertainty so that $\Delta(e^{j\omega T}) = -B(e^{j\omega T})$, is that:

$$B(e^{j\omega T}) < 1 + \frac{\beta}{|G_0(e^{j\omega T})|^2} \quad \text{for all} \quad \omega T \tag{3.20}$$

If we again consider the situation in which $|G_0(e^{j\omega T})|^2$ is much less than $\beta$ at some frequencies, then the upper bound on the multiplicative uncertainty could be much greater than unity at these frequencies and yet the algorithm would remain stable. The value of $\beta$ thus allows a trade-off to be made between performance, low $\beta$, and robust stability, high $\beta$. A more precise control over this trade-off could be achieved if the filtered-reference LMS algorithm was implemented in the frequency domain and $\beta$ was allowed to vary as a function of frequency. A more complete analysis of the filtered-reference LMS algorithm, which includes the dynamics of the plant, can be performed for sinusoidal reference signals, which shows that the behaviour of the adaptive feedforward algorithm can be exactly represented by an equivalent feedback control system in which the response of the equivalent feedback controller has a peak at the frequency of the reference signal [269]. The equivalent feedback control system thus has a notch at this frequency in its response from the disturbance input to the error output. The maximum bandwidth of this notch, which gives the maximum bandwidth over which the disturbance can be effectively controlled, is equal to the reciprocal of the delay in the plant [269], which is a condition also encountered below for conventional feedback control systems. Active feedforward control systems are widely used to control narrowband disturbances caused by rotating or reciprocating sources, because a reference signal at the fundamental frequency can typically be obtained from a tachometer signal. Active sound control systems with multiple secondary loudspeakers and multiple error microphones have been used to give good control of the low-frequency tonal noise in the passenger cabin of propeller aircraft for some time [83] and are now widely used commercially [45]. Such systems need to be adaptive to track nonstationarities in the primary field during different flight conditions and typically use a fixed model $\hat{G}$, of the matrix of plant responses at the operating frequency $G$, in their adaptation equation. The stability condition for such systems, with effort weighting parameter $\beta$, now takes the form [9]:

$$\text{Re}\left[\text{Eig}[G^H G + \beta I]\right] > 0 \tag{3.21}$$

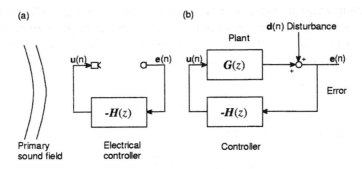

Figure 3.4     (a) Schematic of a feedback controller for the active suppression of an acoustic disturbance. (b) Its equivalent block diagram.

which only has to be satisfied at the excitation frequency. Very small values of $\beta$ ensure that the control system is stable for the changes in plant response typically encountered in practice, due to people moving around in the passenger cabin for example, without significantly degrading the performance.

## 3.3     Feedback control

### 3.3.1     Fixed feedback controllers

When no external reference signal is available that is well correlated with the disturbance, then the output of the error sensor must be used to drive the secondary actuator via a negative feedback controller, as illustrated in Figure 3.4. The ratio of the output of the error sensor after control to that before control is equal to the sensitivity function of the feedback controller, which is given by:

$$S(j\omega) = \frac{E(j\omega)}{D(j\omega)} = \frac{1}{1 + G(j\omega)H(j\omega)} \qquad (3.22)$$

We shall see in the following section how such an arrangement can be viewed as being equivalent to a feedforward system with an internal reference signal, but for now we will briefly review the robust stability conditions for the feedback loop. With the plant under nominal conditions the stability of the closed loop is determined by the well known Nyquist criterion applied to the polar plot of the open-loop frequency response. A simple geometric construction in the Nyquist plane can also be used to derive a condition for the robust stability of a fixed feedback controller when the plant is subject to unstructured multiplicative uncertainty of the form given in eqns (3.12) and (3.13), which can be written as [95]:

$$|T(j\omega)B(j\omega)| < 1 \quad \text{for all } \omega \qquad (3.23)$$

where $T(j\omega)$ is the complementary sensitivity function, which for the feedback controller shown in Figure 3.4 is equal to:

$$T(j\omega) = \frac{G(j\omega)H(j\omega)}{1 + G(j\omega)H(j\omega)} \qquad (3.24)$$

## 3.4  Internal model control

An interesting way of looking at the action of the feedback controller is to assume that it is implemented using internal model control [212], which is illustrated in Figure 3.5 for a sampled time implementation, to be consistent with the discussion of feedforward controllers above. If we assume that the plant is under nominal conditions and that the plant model is equal to the plant response under these conditions, so that $G(z) = \hat{G}(z) = G_0(z)$, it can be seen from Figure 3.5 that the block diagram reduces to an equivalent feedforward system for which the sensitivity function is equal to:

$$S(z) = \frac{E(z)}{D(z)} = 1 + W(z)G_0(z) \qquad (3.25)$$

Under these conditions the output of the plant model exactly cancels the output of the physical plant, and so the signal driving the controller is equal to the disturbance, $\hat{d}(n) = d(n)$, and this acts as an internal reference signal for this arrangement. The complementary sensitivity function $T(z)$ is equal to $1S(z)$ which from eqn. (3.25) is given by:

$$T(z) = -W(z)G_0(z) \qquad (3.26)$$

The robust stability condition, eqn. (3.23), can thus be cast in a particularly simple form for an internal model control system as:

$$|W(e^{j\omega T})G_0(e^{j\omega T})B(e^{j\omega T})| < 1 \quad \text{for all } \omega T \qquad (3.27)$$

Eqn. (3.27) illustrates the fact that if the uncertainty $B(e^{j\omega T})$ is particularly large at some frequencies then the gain of the control filter, $|W(e^{j\omega T})|$, must

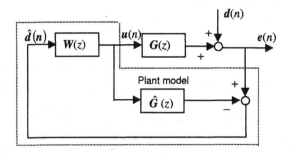

*Figure 3.5*    *Block diagram of a negative feedback controller implemented using internal model control.*

be reduced correspondingly in order for the robust stability condition in eqn. (3.27) to be satisfied. If the control filter $W(z)$ was an FIR device the optimum $H_2$ controller, which minimises the mean square error, can be derived using conventional Wiener theory to calculate the coefficients of $W(z)$ under nominal conditions using eqn. (3.25). We have already seen how introducing a term similar to control effort into the quadratic cost function being minimised, eqn. (3.17), can reduce the magnitude of the control filter's frequency response at the expense of a slightly increased mean square error. One way of minimising the mean square error while ensuring that the robust stability condition given by eqn. (3.27) is satisfied would thus be to derive the Wiener filter which minimised eqn. (3.17) for a sequence of increasingly large values of $\beta$ until eqn. (3.27) is satisfied at all frequencies. Another approach, which would give better performance than the weighted $H_2$ approach above, is to take a discrete grid of frequency points, i.e. $z = e^{j2\pi k/N}$, $k = 0, 1, \ldots, N-1$, and solve the convex optimisation problem of minimising an $H_2$ performance criterion equal to the mean square error, which is given by the sum of the values of the error signal's power spectral density, $s_{ee}(k)$, in each frequency bin:

$$\min_{W(k)} \sum_{k=1}^{N-1} s_{ee}(k) = \sum_{k=1}^{N-1} |1 + W(k)G_0(k)|^2 S_{dd}(k) \tag{3.28}$$

while maintaining the robust stability constraint that:

$$|W(k)G_0(k)B(k)| < 1 \quad \text{for all} \quad \omega T \tag{3.29}$$

as suggested by Boyd *et al.* [35].

This method has been implemented using sequential quadratic programming by Rafaely and Elliott [19] to derive the frequency response of a feedback controller which minimises the $H_2/H_\infty$ control problem defined by eqns (3.28) and (3.29) for an active acoustic headrest. A low-order controller was then fitted to this frequency response and implemented in practice. An advantage of this approach is that all the parameters required for the controller design, such as the nominal plant response, $G_0(k)$, the disturbance power spectrum, $S_{dd}(k)$, and the multiplicative plant uncertainty, $B(k)$, can be directly measured from simple experiments on the physical system under control. Also, if the control loop is made robust to plant uncertainty, it will also tend to be robust to errors introduced into the controller response when reducing its order. The bandwidth over which significant reductions can be achieved with a feedback control system is inherently limited to be less than the reciprocal of the delays in open loop. In an active sound control system the performance bandwidth is thus fundamentally limited by the propagation delay between the secondary actuator and error sensor, but any additional delay in the controller will further reduce this bandwidth. Feedback systems designed to control relatively broadband disturbances are thus typically implemented using analogue components to minimise this delay. In some active noise control applications, such as inside helicopters, there are many narrowband components in the acoustic disturbance spectrum as well as the broadband component. The frequencies of many of these narrowband components can also change with time and it is inconvenient to obtain external reference signals for each of them. Whereas the frequency response of the optimal controller for the broadband noise is a relatively smooth function of frequency, which can readily be implemented with analogue components, that for the narrowband components will only have a high gain at the frequencies of the main tonal peaks, and in order to make this adaptive, to track the nonstationarity, it is most convenient to implement this part of the controller digitally. The extra delay introduced in the antialiasing filters, data converters and digital processing will not affect the performance for these disturbance components since their bandwidth is small. This reasoning leads to the design of a combined analogue/digital controller for such applications, with a fixed analogue controller applied directly round the plant to attenuate the broadband component of the disturbance, and an outer adaptive digital controller to attenuate the narrowband components. The design of such an adaptive digital controller, which maintains robust stability, is discussed in Section 3.4.1.

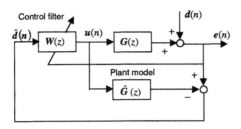

*Figure 3.6    A simple form of adaptive feedback controller in which the control filter in an IMC arrangement is adjusted to maintain performance in the face of nonstationary disturbances.*

### 3.4.1  Adaptive feedback control

Adaptive feedback controllers are generally designed to compensate for changes in the plant response, and generally require the injection of a probe signal, or identification noise, into the system [1]. In active control applications the plant response often changes relatively quickly and the levels of identification noise need to be relatively high if these changes are to be successfully tracked. Such levels of identification noise can increase the output of the plant to the extent that the attenuation in the disturbance can be negated. It may still be necessary to make the feedback controller adaptive, however, to maintain good performance for nonstationary disturbances, but if the plant response cannot be accurately identified at every instant the adaptation algorithm must be designed to ensure that the feedback controller is always robustly stable to these plant changes. The internal model control arrangement provides a convenient framework within which such an adaptive controller can be designed. Consider the adaptive feedback controller shown in Figure 3.6, in which an FIR control filter in an IMC arrangement is adapted to minimise the mean square error.

There are now two levels of feedback round the system, that due to the feedback controller and that due to the adaptation algorithm, and in general it is very difficult to analyse the interaction between these two loops. If the plant model is perfect, the arrangement reduces to an adaptive feedforward system, for which the stability conditions have been derived above. The sensitivity function of the controller is then given by eqn. (3.25), which shows that the mean square error is a quadratic function of the coefficients of the FIR filter $W(z)$, and thus the adaptive filtering algorithms described above can be used directly. It is, however, unrealistic to expect that the plant model will always be perfect since otherwise it could be perfectly compensated for and a feedback controller would not be necessary. If the plant model is not assumed to be perfect, the sensitivity function of the internal model control system shown in Figure 3.5 is:

$$S(z) = \frac{1 + G(z)W(z)}{1 - (G(z) - \hat{G}(z))W(z)} \tag{3.30}$$

The stability of the feedback loop is assured if $|(G(e^{j\omega T}) - \hat{G}(e^{j\omega T}))W(e^{j\omega T})|$ is less than unity, which is equivalent to the condition for robust stability in eqn. (3.27) above. If we assume that $|G(e^{j\omega T}) - \hat{G}(e^{j\omega T})W(e^{j\omega T})|$ is considerably less than unity, however, then the system reduces to the feedforward system, for which the mean square error is a quadratic function of the filter coefficients. Between these two conditions it has been found that the mean square error appears to be a convex function of the filter coefficients even if it is not a quadratic function [18], and so in principle the gradient descent algorithms such as the filtered reference LMS can still be used to adapt the control filter $W(z)$. In practice, however, the system may drift very close to the stability boundary of the feedback loop and stochastic variations in the filter coefficients may push the system into instability. A sensible strategy for adapting the control filter to minimise the mean square error, while avoiding the danger of instability in the feedback loop, would be to use the leaky filtered reference LMS algorithm, eqn. (3.16), and use the leakage parameter $\beta$ to prevent the frequency response of the control filter from becoming large enough to approach the robust stability condition. A far more precise control of this constraint can be obtained by implementing the adaptive algorithm in the frequency domain, in which case individual values of $\beta$ can be used in each frequency bin [81]. By monitoring the parameter $|W(k)G_0(k)B(k)|$ in each frequency bin during convergence, it can be detected when the robust stability condition, eqn. (3.28), is approached and the relevant value of $\beta$ then increased to prevent the controller from converging too close to this constraint [81].

## 3.5  Conclusions

There are various characteristic features of the active control problem, particularly the problem of active sound control, which can be listed as:

(i) Disturbance rejection rather than set-point following 2) disturbance may have both narrowband and broadband components and is generally non-stationary

(ii) Plant response is high order, well damped and non-minimum phase, including delays.

(iii) Plant response may change over short time periods.

(iv) Objective is minimising a mean square ($H_2$) output.

Because the main object is to reduce the mean-square output of the system, e.g. the mean-square output of a pressure microphone which determines the sound pressure level, high levels of identification noise cannot be used to track the rapid changes in the plant response. The control methods that are used must thus be robust to these changes in the plant.

If an external reference signal is available adaptive feedforward controllers are widely used to control the narrowband disturbances. The adaptation algorithm is often of the filtered-reference LMS type in which the update is generated by multiplying the instantaneous error signal by the reference signal filtered by an internal model of the plant. For slow adaptation rates the stability of this algorithm is assured provided the ratio of the transfer function of the internal plant model to that of the plant is strictly positive real. This implies that the phase error between the plant model and the physical plant must always be less than 90° or can alternatively be interpreted as requiring that the unstructured multiplicative uncertainty has an upper bound of less than unity. If a leakage term is introduced into the adaptation algorithm both of these conditions are relaxed and the system becomes even more robust to differences between the response of the internal plant model and that of the physical plant. Adaptive feedforward systems are used to control the tonal components of the sound in the passenger cabins of propeller aircraft [45, 83].

The broadband components of the disturbance can sometimes be controlled with a feedback controller, although the bandwidth over which control can be achieved is fundamentally limited by the propagation time between the secondary actuator and error sensor. The internal model control (IMC) architecture of the feedback controller allows a direct comparison of the performance of the feedback and feedforward controllers. It also allows the robust stability condition to be simply expressed as an upper bound on the frequency response of the control filter within the IMC controller. A straightforward design strategy can then be developed in the discrete frequency domain for the calculation of the frequency response of the controller which minimises the mean-square error while maintaining the robust stability constraint. This $H_2/H_\infty$ design method only uses parameters which can be directly measured from the plant and disturbance. Such a method has been used in the design of a feedback controller for an active headrest [19]. The IMC architecture also inspires a strategy for an adaptive feedback controller, which maintains good performance in the face of nonstationary disturbances but is robust to plant variations. If such an adaptive controller is implemented digitally, the delay in the processor will further increase the delay in the loop and thus reduce the bandwidth over which control can be achieved. It may thus be advisable to use a combination of an inner analogue control loop for the stationary broadband disturbances and an adaptive digital outer control loop or nonstationary narrowband disturbances. These features have been tested on an active headset, in which the secondary loudspeaker and error sensor were located inside the earshell [17].

# Part II

# Recent algorithmic developments

*Chapter 4*

# Multichannel active noise control: stable adaptive algorithms

## T. Shimizu, T. Kohno, H. Ohmori and A. Sano

*Department of System Design Engineering, Keio University*
*Keio, Japan, Email:sano@sano.elec.keio.ac.jp*

*This chapter presents stable adaptive schemes in two cases: in the first case reference microphones are available to detect unwanted primary noise, and in the second case no reference microphones are available. The latter case requires no causality condition but its availability is limited to periodic primary noises. In both cases, we propose stability-assured adaptive algorithms to update the adaptive feedforward controllers.*

## 4.1 Introduction

Active noise control, which suppresses unwanted noises generated from primary sound sources at the objective points by producing artificial sounds from secondary sound sources, has recently found more and more applications in improving industrial and living environments, since it can complement traditional passive technologies and attain better performance on attenuation of low-frequency noises [80, 103, 161, 171, 280]. Adaptive feedforward control schemes are essential to achieve the control performance even in the presence of model uncertainties and variations in sound path dynamics and microphones.

We will investigate stable adaptive schemes in two cases: in the first case reference microphones are available to detect unwanted primary noise, and in the second case no reference microphones are available. The latter case requires no causality condition but its availability is limited to periodic primary noises. In both cases, we propose stability-assured adaptive algorithms to update the adaptive feedforward controllers. Conventionally, a variety of filtered-$x$ adaptive algorithms have been employed to adjust the weights of adaptive feedforward controllers implemented by FIR-type adaptive filters [29, 77, 78, 85, 92, 209]. However, FIR filters sometimes display a serious problem of instability, so that the step size of the filtered-$x$ LMS algorithm should be chosen carefully, especially in multichannel systems. In the single channel case, we have investigated the stability of adaptive control algorithms based on strictly positive realness properties in the time-domain [151, 226], frequency-domain [150, 226] and wavelet transform domain [253].

In multichannel active noise control with a multiple number of primary noise sources, secondary sources, reference microphones and error microphones, an ordinary filtered-$x$ LMS adaptive algorithm can hardly attain both stability and quick adaptation by choosing step sizes adequately, since the path matrices and controllers have nondiagonal elements. In this chapter, on the assumption that the secondary channel dynamics are known *a priori* and not changeable, an LMS-type robust adaptive algorithm for updating the feedforward controllers is proposed which can assure the stability of the adaptation and keep each cancelling errors within a tight bound when each upper bound of the uncertainty terms due to disturbances and unmodelled dynamics of sound propagation is known *a priori*.

Next, two identification-based approaches are investigated to deal with a case in which both the primary and secondary channel matrices are unknown and uncertainly changeable. In the approaches, we first identify equivalent primary and secondary channel matrices in an online manner, then give two schemes to calculate the adaptive feedforward controllers corresponding to the identified path matrices. Finally, the effectiveness of the proposed adaptive algorithms is investigated and compared with each other and with the conventional filtered-$x$ algorithm in active noise control experiments.

## 4.2  Multichannel active noise control problems

Figure 4.2 illustrates a multichannel active noise control system with reference microphones (Case 1) and without reference microphones (Case 2).

The primary noises are generated from the $N_s$ sources, which are denoted by $s(k) \in \mathcal{R}^{N_s}$, and are detected by $N_r$ reference microphones in case 1. The detected signals $r(k) \in \mathcal{R}^{N_r}$ are the inputs to the $N_c \times N_r$ adaptive feedforward controller matrix $\hat{C}(z, k)$, where $N_c$ is the number of secondary loudspeakers,

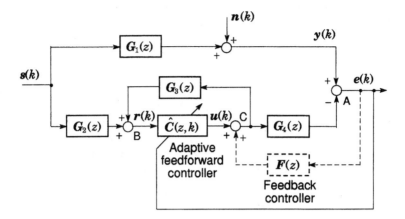

(a) Case 1: control system with reference microphones

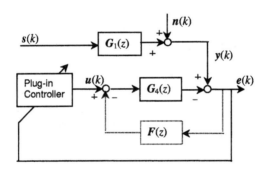

(b) Case 2: control system without reference microphones

Figure 4.1    *Two types of multichannel adaptive active noise control system*

which produce artificial sounds to cancel the unwanted noises $s(k)$ at the $N_c$ objective points. The cancelling errors are detected as $e(k) \in \mathcal{R}^{N_c}$ by the $N_c$ error microphones. All of the primary and secondary sounds propagating through different path dynamics are sensed by the reference and error microphones. Here $G_1(z) \in \mathcal{Z}^{N_c \times N_s}$ and $G_2(z) \in \mathcal{Z}^{N_r \times N_s}$ represent the primary channel dynamics from the primary noises $s(k)$ to the reference microphones and error microphones, respectively. $G_3(z) \in \mathcal{Z}^{N_r \times N_c}$ and $G_4(z) \in \mathcal{Z}^{N_c \times N_c}$ are the secondary channel dynamics from the secondary control sounds $u(k)$ to the reference and error microphone sets, respectively, where $G_3(z)$ is referred to as the feedback channel dynamics.

In case 2, since the reference microphones are not used, $G_2(z)$ and $G_3(z)$ are not needed in the implementation of the plug-in type of adaptive controller $\hat{C}(z,k)$, and the causality assumption is not also required. However, it is assumed that the primary noises consist of multiple sinusoids with known or unknown frequencies. The parameters in $\hat{C}(z,k)$ are updated by using only available signals $e(k)$ and $u(k)$ as well as prior knowledge.

## 4.3    Structure and Algorithms

### *4.3.1    Error system description in Case 1*

In Figure 4.2, the structure of the active noise control system is described by:

$$e(k) = G_1(z)s(k) - G_4(z)u(k) + n(k) \qquad (4.1a)$$
$$r(k) = G_2(z)s(k) + G_3(z)u(k) \qquad (4.1b)$$
$$u(k) = C(z)r(k) \qquad (4.1c)$$

where $n(k) \in \mathcal{R}^{N_c}$ denote the disturbance noises which enter the system through the error microphones. In order to realise active noise cancelling by using only accessible signals $r(k)$, $u(k)$ and $e(k)$ in case 1, the unknown source noise $s(k)$ should be eliminated from eqn. (4.1). By assuming $N_s \leq N_r$, the error is described as:

$$\begin{aligned}
e(k) &= \bar{G}_1(z)r(k) - \bar{G}_4(z)u(k) + n(k) \qquad (4.2a) \\
&= \bar{G}_4(z)[\bar{G}_4^{-1}(z)\bar{G}_1(z)r(k) - u(k)] + n(k) \\
&= \bar{G}_4(z)[\bar{G}_4^{-1}(z)\bar{G}_1(z) - C(z)]r(k) + n(k) \\
&= \bar{G}_4(z)[\bar{C}^*(z) - C(z)]r(k) + n(k) \qquad (4.2b)
\end{aligned}$$

where

$$\bar{G}_1(z) \equiv G_1(z)[G_2^T(z)G_2(z)]^{-1}G_2^T(z) \qquad (4.3a)$$
$$\bar{G}_4(z) \equiv G_4(z) + G_1(z)[G_2^T(z)G_2(z)]^{-1}G_2^T(z)G_3(z) \qquad (4.3b)$$

where $\bar{C}^*(z)$ in (4.2) is defined by

$$\bar{C}^*(z) = \bar{G}_4^{-1}(z)\bar{G}_1(z) \qquad (4.4)$$

If $\bar{C}^*(z)$ is stable and the feedforward controller $C(z)$ is chosen as eqn. (4.4), and the causality holds, then the error can be cancelled perfectly. However, the path dynamics can hardly be modelled precisely and are uncertainly changeable, so adaptive approaches become essential.

Eqn. (4.2) gives an error system which expresses the relation between the cancelling errors $e(k)$ and parameter errors of the adaptive controller. Multiple FIR type adaptive filters are employed to realise stable controllers $\hat{C}(z,k)$ instead of $C(z)$, which is given by an $N_c \times N_r$ polynomial matrix the elements of which are expressed as:

$$\hat{C}_{ij}(z,k) = \hat{c}_{ij}^{(1)}(k)z^{-1} + \hat{c}_{ij}^{(2)}(k)z^{-1} + \cdots + \hat{c}_{ij}^{(m_{ij})}(k)z^{-m_{ij}} \qquad (4.5)$$

Thus the adaptive feedforward control input $u(k)$ is given by

$$u(k) = \hat{C}(z,k)r(k) \qquad (4.6)$$

where the $i$th element $u_i(k)$ of the control input eqn. (4.6) is expressed by

$$u_i(k) = \sum_{j=1}^{N_r} \hat{C}_{ij}(z,k)r_j(k) = \sum_{j=1}^{N_r} \phi_{ij}^T(k)\hat{c}_{ij}(k) = \phi_i^T(k)\hat{\theta}_i(k) \qquad (4.7)$$

where $\hat{c}_{ij}(k) = [\hat{c}_{ij}^{(1)}(k), \hat{c}_{ij}^{(2)}(k), \cdots, \hat{c}_{ij}^{(m_{ij})}(k)]^T$, $\phi_{ij}(k) = [r_j(k-1), r_j(k-2),$ $\cdots, r_j(k-m_{ij})]^T$, $\hat{\theta}_i(k) = [\hat{c}_{i1}^T(k), \hat{c}_{i2}^T(k), \cdots, \hat{c}_{iN_r}^T(k)]^T$ and $\phi_i(k) = [\phi_{i1}^T(k), \phi_{i2}^T(k),$ $\cdots, \phi_{iN_r}^T(k)]^T$.

Further, the control input $u(k)$ can be rewritten into a vector-matrix form as

$$\begin{aligned} u(k) &= [\phi_1^T(k)\hat{\theta}_1(k), \phi_2^T(k)\hat{\theta}_2(k), \cdots, \phi_{N_c}^T(k)\hat{\theta}_{N_c}(k)]^T \\ &= \text{Diag}[\phi_1^T(k), \phi_2^T(k), \cdots, \phi_{N_c}^T(k)] [\hat{\theta}_1^T(k), \hat{\theta}_2^T(k), \cdots, \hat{\theta}_{N_c}^T(k)]^T \\ &= \Phi^T(k)\hat{\theta}(k) \end{aligned} \qquad (4.8)$$

where $\hat{\theta}(k) = [\hat{\theta}_1^T(k), \hat{\theta}_2^T(k), \cdots, \hat{\theta}_{N_c}^T(k)]^T$ and $\Phi(k) = \text{Diag}[\phi_1(k), \phi_2(k), \cdots, \phi_{N_c}(k)]$.

Similarly, let the parameter vector $\theta$ corresponding to the controller eqn. (4.4) be denoted by $\theta^*$, then we have:

$$\bar{C}^*(z)r(k) = \Phi^T(k)\theta^* + \Delta C(z)r(k) \qquad (4.9)$$

where $\Delta C(z)$ denotes a modelling error matrix.

Thus the error system can be rewritten into a parameterised as:

$$e(k) = \bar{G}_4(z)[\bar{C}^*(z) - \hat{C}(z,k)]r(k) + n(k)$$
$$= \bar{G}_4(z)[\Phi^T(k)(\theta^* - \hat{\theta}(k))] + \xi(k) \qquad (4.10)$$

where $\xi(k)$ is an uncertain vector due to the disturbances and modelling error.

The equivalent secondary channel dynamics $\bar{G}_4(z)$ does not satisfy the strictly positive real (SPR) property. However, if $\bar{G}_4(z)$ is known, the error system eqn. (4.10) can be transformed to an SPR error system, from which we can derive a stability-assured adaptive algorithm. It can be noticed from eqn(4.3) that $\bar{G}_4(z)$ includes all channel dynamics $G_1(z)$‘$G_4(z)$; then we make the following assumptions:

*Assumption 4.1:* the secondary channel dynamics $G_3(z)$ and $G_4(z)$ are known a priori.

*Assumption 2:* each component of the uncertain term $\xi(k)$ is bounded as

$$|\xi_i(k)| \le \bar{\beta}_i, \quad i = 1, 2, \cdots, N_c \qquad (4.11)$$

where each upper bound $\bar{\beta}_i$ is known a priori.

Assumption 4.1 that $G_3(z)$ is known is equivalent to $G_3(z) = 0$, since the feedback effect from $G_3(z)$ can be cancelled by subtracting $G_3(z)u(k)$ from the detected signal $r(k)$ in an algorithm. Then the error system eqn. (4.10) can be reduced to

$$e(k) = G_4(z)[\Phi^T(k)(\theta^* - \hat{\theta}(k))] + \xi(k) \qquad (4.12)$$

Assumption 4.2 is effective in evaluation of upper bounds of the cancelling error $e_i(k)$ in a strictly tight way, thus conservativeness can be avoided.

### 4.3.2  Robust adaptive algorithm

A robust adaptive algorithm with assured stability is given for the two error systems eqns (4.12) and (4.22). By introducing an auxiliary signal vector $e_a(k)$ and an extended error vector $\eta(k)$ as:

$$e_a(k) = [G_4(z)\Phi^T(k)]\hat{\theta}(k) - G_4(z)[\Phi^T(k)\hat{\theta}(k)] \qquad (4.13)$$
$$\eta(k) = e(k) - e_a(k) \qquad (4.14)$$

both error systems can be rewritten as:

$$\eta(k) = \Psi^T(k)(\theta^* - \hat{\theta}(k)) + \xi(k) \qquad (4.15)$$
$$\Psi^T(k) = G_4(z)\Phi^T(k) \qquad (4.16)$$

which satisfies the SPR property. Then we can give the robust adaptive algorithm with dead zone decision by which the controller parameter vector can be updated in a stable manner.

*Robust adaptive algorithm:*

$$\hat{\boldsymbol{\theta}}(k+1) = \hat{\boldsymbol{\theta}}(k) + \gamma(k)\boldsymbol{\Psi}(k)\bar{\boldsymbol{\eta}}(k) \tag{4.17a}$$

$$\gamma(k) \leq \frac{2\alpha\bar{\boldsymbol{\eta}}^T(k)\bar{\boldsymbol{\eta}}(k)}{d + \bar{\boldsymbol{\eta}}^T(k)\boldsymbol{\Psi}^T(k)\boldsymbol{\Psi}(k)\bar{\boldsymbol{\eta}}(k)} \tag{4.17b}$$

$$\boldsymbol{\eta}(k) = \boldsymbol{e}(k) + \boldsymbol{G}_4(z)\boldsymbol{u}(k) - \boldsymbol{\Psi}^T(k)\hat{\boldsymbol{\theta}}(k) \tag{4.17c}$$

$$\bar{\boldsymbol{\eta}}(k) = [\bar{\eta}_1, \ \bar{\eta}_2, \ \cdots, \ \bar{\eta}_{N_c}]^T \tag{4.17d}$$

$$\bar{\eta}_i(k) = \begin{cases} \eta_i(k) - \bar{\beta}_i & : \quad \eta_i(k) > \bar{\beta}_i \\ 0 & \quad |\,\eta_i(k)\,| \leq \bar{\beta}_i \\ \eta_i(k) + \bar{\beta}_i & : \quad \eta_i(k) < -\bar{\beta}_i \end{cases} \tag{4.17e}$$

where $0 < \alpha < 1$, and $d > 0$ is a positive small constant.

The convergency of the updated parameter vector and the boundedness of the cancelling error vector in the above algorithm are given in the following.

*Property of adaptive algorithm in eqn. (4.17):*

$$\lim_{k \to \infty} \hat{\boldsymbol{\theta}}(k) = \text{constant} \tag{4.18a}$$

$$\lim_{k \to \infty} \sup |e_i(k)| = \bar{\beta}_i \tag{4.18b}$$

*Comparison with ordinary filtered-x algorithms:*
Ordinary LMS-type filtered-x algorithms are described by:

$$\hat{\boldsymbol{\theta}}(k+1) = \hat{\boldsymbol{\theta}}(k) + \gamma\boldsymbol{\Psi}(k)\boldsymbol{e}(k) \tag{4.19a}$$

$$\boldsymbol{\Psi}^T(k) = \boldsymbol{G}_4(z)\boldsymbol{\Phi}^T(k) \tag{4.19b}$$

However, the algorithm cannot assure the stability for arbitrary step size even if there are no modelling errors and disturbances, so the step size $\gamma$ should be carefully selected by compromising the stability and convergence rate. The proposed algorithm eqn. (4.17) has the following distinctive features, unlike ordinary methods:

(i) The parameter vector $\hat{\boldsymbol{\theta}}(k)$ is adaptively adjusted according to $\bar{\boldsymbol{\eta}}(k)$, unlike the conventional filtered-x algorithm which uses the actual cancelling error $\boldsymbol{e}(k)$.

(ii) The step size $\gamma(k)$ is variable in accordance with the errors and signals, and does not include matrix inversion (as described later) but only scalar normalisation, which can reduce computational complexity.

(iii) Each worst cancelling error can be estimated independently in a tighter bound by using the upper bound of each component of the vector $\boldsymbol{\xi}(k)$ as given in eqn. (4.11), rather than its Euclidean norm such as $\|\boldsymbol{\xi}(k)\| \leq \beta$ which tends to overestimate the cancelling errors.

### 4.3.3  Error system description in Case 2

In case 2, it is assumed that the primary source noises consist of a multiple number of sinusoids with known frequencies. Then the causality condition is not needed in its implementation without using the reference microphones. It follows from Figure 1 (b) that the cancelling error $\boldsymbol{e}(k)$ is expressed as:

$$\boldsymbol{e}(k) = \boldsymbol{G}_1(z)\boldsymbol{s}(k) - \boldsymbol{G}_4(z)\boldsymbol{u}(k) + \boldsymbol{n}(k) \qquad (4.20)$$

and the $i$th primary source noise is expressed by:

$$s_i(k) = \sum_{j=1}^{M}[\alpha_{ij}\sin\omega_{ij}k + \beta_{ij}\cos\omega_{ij}k] = \boldsymbol{\phi}_i^T(k)\boldsymbol{g}_i \qquad (4.21)$$

where $\boldsymbol{g}_i(k) = [\alpha_{i1}, \beta_{i1}, \ldots, \alpha_{iM}, \beta_{iM}]^T$ and $\boldsymbol{\phi}_i(k) = [\sin\omega_{i1}k, \cos\omega_{i1}k, \ldots, \sin\omega_{iM}k, \cos\omega_{iM}k]$. The amplitudes of the sinusoids are unknown but their frequencies are known *a priori*. $M$ is set to an upper bound of the number of sinusoids. In the case of unknown frequencies, a frequency estimation algorithm can be used at the same time. $\boldsymbol{G}_1(z)$ is unknown but $\boldsymbol{G}_4(z)$ is assumed to be known.

Then, the cancelling error eqn. (4.20) can be rewritten into:

$$\begin{aligned} \boldsymbol{e}(k) &= \boldsymbol{G}_4(z)[\boldsymbol{\Phi}^T(k)\boldsymbol{\theta}^* - \boldsymbol{u}(k)] + \boldsymbol{\xi}(k) \\ &= \boldsymbol{G}_4(z)[\boldsymbol{\Phi}^T(k)(\boldsymbol{\theta}^* - \hat{\boldsymbol{\theta}}(k))] + \boldsymbol{\xi}(k) \end{aligned} \qquad (4.22)$$

where $\boldsymbol{\Phi}(k) = \mathrm{Diag}[\boldsymbol{\phi}_1(k), \cdots, \boldsymbol{\phi}_{N_c}]$, $\hat{\boldsymbol{\theta}}(k) = [\hat{\boldsymbol{\theta}}_1(k), \cdots, \hat{\boldsymbol{\theta}}_{N_c}(k)]$. Thus the adaptive control is given by:

$$\boldsymbol{u}(k) = \boldsymbol{\Phi}^T(k)\hat{\boldsymbol{\theta}}(k) \qquad (4.23)$$

It can be noticed that the error system of eqn. (4.22) in Case 2 has the same structure as does eqn. (4.12) in Case 1. Therefore, the robust adaptive algorithm given in eqn. (4.17) can also be applied to Case 2.

Other adaptive approaches are also applicable to Case 2 with periodic primary noise sources, based on the plug-in adaptive control approach [207, 225]. Adaptive feedback control schemes are also applicable in the periodic noise case, by estimating their frequencies in an online manner.

## 4.4 Identification-based adaptive control in case 1

*4.4.1 Identification of equivalent primary and secondary channel matrices*

When $G_4(z)$ is uncertain, an alternative approach adopting identification-based algorithms will be required. Since the cancelling error is given by eqn. (4.2), the equivalent primary and secondary channel matrices $\bar{G}_1(z)$ and $\bar{G}_4(z)$ can be identified as FIR models by using the accessible data $r(k)$, $u(k)$ and $e(k)$ as:

$$\bar{G}_1(z) = H_1(z) + \Delta_1(z) \tag{4.24a}$$

$$\bar{G}_4(z) = H_4(z) + \Delta_4(z) \tag{4.24b}$$

where $\Delta_n(z)$ are the truncated errors and $H_n(z)$ are given for $n = 1$ and 4 as:

$$H_n(z) = \begin{bmatrix} H_{n11}(z) & \cdots & H_{n1N_r}(z) \\ \vdots & \ddots & \vdots \\ H_{nN_c1}(z) & \cdots & H_{nN_cN_r}(z) \end{bmatrix}$$

where

$$H_{nij}(z) = h_{nij}^{(1)} z^{-1} + \cdots + h_{nij}^{(\bar{m}_{nij})} z^{-\bar{m}_{nij}}$$

By using the above notations, the error system of eqn. (4.2) can be described by:

$$e(k) = H_1(z)r(k) - H_4(z)u(k) + \Delta_1(z)r(k) - \Delta_4(z)u(k) + n(k)$$
$$= \Omega^T(k)h + \zeta(k) \tag{4.25}$$

where

$$h = [h_1^T, \cdots, h_{N_c}^T]^T$$
$$h_i^T = [h_{1i1}^T, \cdots, h_{1iN_r}^T, h_{4i1}^T, \cdots, h_{4iN_c}^T]$$
$$h_{1ij} = [h_{1ij}^{(1)}, \cdots, h_{1ij}^{(\bar{m}_{ij})}]^T$$
$$h_{4ij} = [h_{4ij}^{(1)}, \cdots, h_{4ij}^{(\bar{n}_{ij})}]^T$$
$$\Omega^T(k-1) = Diag[\omega_1^T(k-1), \cdots, \omega_{N_c}^T(k-1)]$$
$$\omega_i(k) = \left[\omega_{i1}^{rT}(k), \cdots, \omega_{iN_r}^{rT}(k), \omega_{i1}^{uT}(k), \cdots, \omega_{iN_c}^{uT}(k)\right]^T$$
$$\omega_{ij}^r(k) = [r_j(k-1), \cdots, r_j(k-\bar{m}_{ij})]^T$$
$$\omega_{ij}^u(k) = [u_j(k-1), \cdots, u_j(k-\bar{n}_{ij})]^T$$
$$\zeta(k) = \Delta_1(z)r(k) - \Delta_4(z)u(k) + n(k)$$

The estimate of the FIR model $\boldsymbol{h}$ is updated by the robust LMS algorithm with the proposed scalar normalisation as:

$$\hat{\boldsymbol{h}}(k+1) = \hat{\boldsymbol{h}}(k) + \mu(k)\boldsymbol{\Omega}(k)\boldsymbol{\varepsilon}(k) \tag{4.26a}$$

$$\mu(k) \leq \frac{2\alpha'\bar{\boldsymbol{\varepsilon}}^T(k)\bar{\boldsymbol{\varepsilon}}(k)}{d' + \bar{\boldsymbol{\varepsilon}}^T(k)\boldsymbol{\Omega}^T(k)\boldsymbol{\Omega}(k)\bar{\boldsymbol{\varepsilon}}(k)} \tag{4.26b}$$

$$\boldsymbol{\varepsilon}(k) = \boldsymbol{e}(k) - \boldsymbol{\Omega}(k)\hat{\boldsymbol{h}}(k) \tag{4.26c}$$

$$\bar{\boldsymbol{\varepsilon}}(k) = [\bar{\varepsilon}_1(k), \bar{\varepsilon}_2(k), \cdots, \bar{\varepsilon}_{N_c}(k)]^T \tag{4.26d}$$

$$\bar{\varepsilon}_i(k) = \begin{cases} \varepsilon_i(k) - \bar{\beta}_i : & \varepsilon_i(k) > \bar{\beta}_i \\ 0 & |\varepsilon_i(k)| \leq \bar{\beta}_i \\ \varepsilon_i(k) + \bar{\beta}_i : & \varepsilon_i(k) < -\bar{\beta}_i \end{cases} \tag{4.26e}$$

where let $\boldsymbol{\varepsilon}(k)$ be referred to as the identification error, and $\bar{\beta}_i$ denotes an upper bound satisfying $|\zeta_i(k) \leq \bar{\beta}_i$, and let $0 < \alpha' < 1$ and $d' > 0$.

The persistently exciting (PE) property of the control sound from the secondary loudspeaker is needed in the above identification. When the PE condition is not satisfied, a dither signal vector $\boldsymbol{d}(k)$ can be introduced and added to the control signal vector $\boldsymbol{u}(k)$ in the following way:

$$u_i(k) = u_i(k) + d_i(k) \tag{4.27a}$$

$$d_i(k) = f\boldsymbol{e}^T(k-1)\boldsymbol{v}_i(k) \tag{4.27b}$$

where $\boldsymbol{v}_i(k)$ is a zero mean white noise vector satisfying the following property:

$$E[\boldsymbol{v}_i(k)\boldsymbol{v}_j^T(m)] = I\delta_{ij}\delta_{km}$$

In the scheme, the white noise $\boldsymbol{v}_i(k)$ is amplitude-modulated by the cancelling error $\boldsymbol{e}^T(k-1)$. Therefore, when the magnitude of the cancelling error is large, the effort of the dither signal will be large, resulting in high identification accuracy, and when the magnitude of the cancelling error is small, the influence of the dither signal will become small correspondingly to preserve the cancelling performance.

### 4.4.2 Identification-based adaptive controller

Two approaches are discussed for designing the adaptive feedforward controller based on the identified channel matrices.

#### 4.4.2.1 Indirect design of adaptive control

By using the identified model of the channel matrices, the adaptive feedforward controller can be calculated by eqn. (4.4) in an online manner as:

$$\hat{\boldsymbol{C}}(z,k) = \hat{\boldsymbol{H}}_4^{-1}(z)\hat{\boldsymbol{H}}_1(z) \tag{4.28}$$

The calculation needs the inversion of the matrix. However, if the inversion formula for $\hat{\boldsymbol{H}}_4^{-1}(z)$ can be implemented, the controller is obtained *via* only complex manipulations with the FFT and inverse FFT. The details will be given in the experimental study in a case with $N_s = N_r = N_c = 2$.

*4.4.2.2 Error system updating scheme*
If $\bar{\boldsymbol{G}}_1(z)$ and $\bar{\boldsymbol{G}}_4(z)$ are identified as $\hat{\boldsymbol{H}}_1(z)$ and $\hat{\boldsymbol{H}}_4(z)$, the error system of eqn. (4.10) can also be replaced by:

$$e(k) = \hat{\boldsymbol{H}}_4(z)[\boldsymbol{\Phi}^T(k)(\boldsymbol{\theta}^* - \hat{\boldsymbol{\theta}}(k))] + \boldsymbol{\xi}(k) \qquad (4.29)$$

As already stated, the above error system can also be rewritten into

$$\boldsymbol{\eta}(k) = \boldsymbol{\Psi}^T(k)(\boldsymbol{\theta}^* - \hat{\boldsymbol{\theta}}(k)) + \boldsymbol{\xi}(k) \qquad (4.30a)$$
$$\boldsymbol{\Psi}^T(k) = \hat{\boldsymbol{H}}_4(z)\boldsymbol{\Phi}^T(k) \qquad (4.30b)$$

Then the robust adaptive algorithm is given by replacing eqn. $\boldsymbol{G}_4(z)$ in (4.17) with $\hat{\boldsymbol{H}}_4(z)$.

Updating $\hat{\boldsymbol{H}}_4(z)$ in the error system and the robust adaptive algorithm eqn. (4.17)at every instant by the identified channel dynamics cannot always assure the stability of the adaptation. We consider two kinds of error: one is the cancelling error norm $\|e(k)\|$ or its increasing rate $\|e(k) - e(k-1)\|$, and the other is the identification error norm $\|\varepsilon(k)\|$. Then we give the decision rule for updating the error system as follows:

(i) Minimum interval for updating the error system eqn. (4.30) should be chosen to be more than the order of the model $\hat{\boldsymbol{H}}_4(z)$.

(ii) If the moving average of the identification error $\|\varepsilon(k)\|$ is smaller than a threshold value $C_I$ and the moving average of the increasing rate of cancelling error $\|e(k) - e(k-1)\|$ is larger than a threshold $C_E$, then the error system should be updated.

## 4.5 Experimental results using the proposed adaptive algorithms

Figure 4.2 illustrates the experimental setup for the adaptive active noise control in a room. Sound reflections on the room walls are suppressed passively as much as possible. Two primary loudspeakers A and B, two reference microphones A and B, two secondary loudspeakers C and D, and two error microphones C and D are placed initially at indicated locations, configuring a multichannel active noise control system with the dimensions of $N_s = N_r = N_c = 2$.

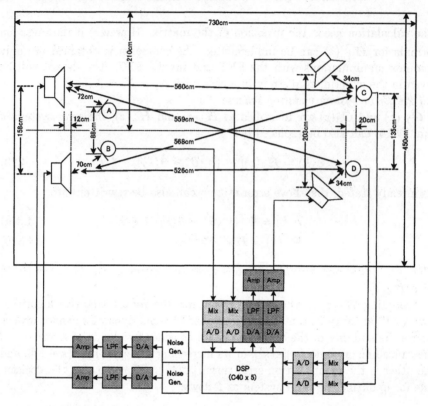

*Figure 4.2    Experimental setup for adaptive active noise control in a room*

To realise the multichannel adaptive algorithm, five DSP (TMS320-C40) chips
are used for performing parallel computation, where the sampling period is
chosen as 1 ms. The power spectra of the primary sources are limited in the
low-frequency range from 50 to 450 Hz. The order of the FIR adaptive controllers is set at 64, which is selected according to off-line identification results
of each channel. To investigate the proposed robust direct adaptive algorithm
and the identification-based indirect adaptive algorithm, as well as the conventional filtered-$x$ algorithm, the experiments are performed with the following
two channel dynamics change cases:
*Case 1:* two reference microphones are moved continuously from the initial
positions (about 70 cm from the primary loudspeakers) to the end positions
(about 140 cm from each corresponding loudspeaker) in between 10 and 30
seconds. These movements of the reference microphones will result in large

changes in $G_2(z)$ and moderate changes in $G_3(z)$, and no changes in $G_1(z)$ and $G_4(z)$.

*Case 2:* two error microphones are moved continuously from the initial positions (34 cm from the secondary loudspeakers) to the end positions (70 cm from the loudspeakers) in between 10 and 30 seconds. These movements of the error microphones will result in large change in $G_4(z)$ and moderate changes in $G_1(z)$, and no changes in $G_2(z)$ and $G_3(z)$.

The main difference between Case 1 and Case 2 lies in whether the channel dynamic $G_4(z)$ is changed or not. In Case 1 the secondary channel dynamic $G_4(z)$ is not changeable, and therefore we can apply all the proposed algorithms, the stability-assured robust adaptive algorithm given by eqn. (4.17), the indirect adaptive algorithm given by eqn. (4.26) and eqn. (4.28) and the error system updating algorithm given by eqn. (4.26) and eqn. (4.30). In Case 2 $G_4(z)$ is changeable and the robust adaptive algorithm in eqn. (4.17) cannot be applicable.

Figure 4.5 shows the cancelling error at one objective point obtained in Case 1, and Figure 4.5 gives the cancelling error obtained in Case 2. The adaptive control algorithms adopted corresponding to each subfigure (*a*) to (*e*) in Figs.4.5 and 4.5 are described as follows:

(*a*)*No control:* no active noise control is done.

(*b*)*Normalized filtered-x algorithm:* although normalisation is employed for improving stability, stability is not assured in this algorithm, which is described by:

$$\hat{\boldsymbol{\theta}}(k+1) = \hat{\boldsymbol{\theta}}(k) + \gamma[\boldsymbol{I} + \gamma\boldsymbol{\Psi}^T(k)\boldsymbol{\Psi}(k)]^{-1}\boldsymbol{\Psi}(k)e(k) \tag{4.31}$$

In the experiment, the step size is selected to be as large as $\gamma = 10^3$ for attaining fast convergence. A model identified prior to the experiment is used as $G_4(z)$ in the error system. However, since $G_4(z)$ is changed during the interval from 10 to 30 seconds, the algorithm may not be stability assured.

(*c*)*Robust adaptive algorithm (proposed):* the algorithm is given by eqn. (4.17). In the experiment, the parameters in eqn (4.17) are selected as $\alpha = 0.8$, $d = 10^{-6}$, $\bar{\beta}_1 = \bar{\beta}_2 = 0.002$. To get satisfactory control performance for compromising both cancellation and stability, it is better to select $\bar{\beta}_i$ as $1/5 \sim 1/10$ of the average absolute value of the cancelling error $e_i(k)$ in the steady-state when $\bar{\beta}_i = 0$ .

(*d*) *Indirect adaptive control algorithm (proposed):* path dynamics $H_1(z)$ and $H_2(z)$ are identified in an online manner by use of eqn. (4.26), where the dither signals are also added to the control sound $\boldsymbol{u}(k)$ according to eqn. (4.27). The magnitude of the dither signal is selected as 10 per cent of that of the cancelling error in steady state. By using the identified results, the adaptive feedforward controller can be obtained by eqn. (4.28) as follows:

$$\hat{H}_1(z,k) = \begin{bmatrix} \hat{H}_{111}(z,k) & \hat{H}_{112}(z,k) \\ \hat{H}_{121}(z,k) & \hat{H}_{122}(z,k) \end{bmatrix}$$

$$\hat{H}_4(z,k) = \begin{bmatrix} \hat{H}_{411}(z,k) & \hat{H}_{412}(z,k) \\ \hat{H}_{421}(z,k) & \hat{H}_{422}(z,k) \end{bmatrix}$$

$$\hat{C}(z,k) = \begin{bmatrix} \hat{C}_{11}(z,k) & \hat{C}_{12}(z,k) \\ \hat{C}_{21}(z,k) & \hat{C}_{22}(z,k) \end{bmatrix}$$

$$= \hat{H}_4^{-1}(z,k)\,\hat{H}_1(z,k) \tag{4.32}$$

Thus it gives that:

$$\begin{bmatrix} \hat{C}_{11}(z,k) & \hat{C}_{12}(z,k) \\ \hat{C}_{21}(z,k) & \hat{C}_{22}(z,k) \end{bmatrix} = \frac{1}{\mathrm{Det}\hat{H}_4(z,k)}$$
$$\cdot \begin{bmatrix} \hat{H}_{422}\hat{H}_{111} - \hat{H}_{412}\hat{H}_{121} & \hat{H}_{422}\hat{H}_{112} - \hat{H}_{412}\hat{H}_{122} \\ -\hat{H}_{421}\hat{H}_{111} + \hat{H}_{411}\hat{H}_{121} & -\hat{H}_{421}\hat{H}_{112} + \hat{H}_{411}\hat{H}_{122} \end{bmatrix} \tag{4.33}$$

where

$$\mathrm{Det}\hat{H}_4(z,k) = \hat{H}_{411}(z,k)\hat{H}_{422}(z,k) - \hat{H}_{412}(z,k)\hat{H}_{421}(z,k)$$

In order to calculate the controller parameters, the FFT and the inverse FFT techniques should be applied. First, take the FFT of the parameters of the FIR models $\hat{H}_{1ij}(z,k)$ and $\hat{H}_{2il}(z,k)$, respectively. Then execute arithmetic computation in the frequency domain according to eqn (4.33), and give the frequency domain controller as $\hat{C}_{nm}(e^{j\omega_i}) = N_{nm}(e^{j\omega_i})/D_{nm}(e^{j\omega_i})$. To avoid the ill condition in the inverse calculation, an appropriate regularised constant $\alpha > 0$ is introduced and the above computation is modified by:

$$\hat{C}_{nm}(e^{j\omega_i}) = \frac{N_{nm}(e^{j\omega_i})D_{nm}^*(e^{j\omega_i})}{|D_{nm}(e^{j\omega_i})|^2 + \alpha}$$

Finally, apply the inverse FFT into $\hat{C}_{nm}(e^{j\omega_i})$ calculated in the frequency domain to obtain the controllers $\hat{C}_{nm}(z,k)$ in the time domain, perform complex algebraic manipulations in eqn. (4.33), and take the inverse FFT of the results to obtain the controllers in the time domain. It should be noted that the regularisation constant $\alpha$ does not need to be introduced for all of the frequencies, but only for the cases in which $|D_{nm}^*(e^{j\omega_i})|^2$ are very small. Moreover, to update a complete controller for each control period $k$, a computational amount of the above arithmetic calculation and $2^2$ FFT and inverse FFT are needed.

(e)*Error system updating scheme (proposed):* the adaptive algorithm is given by eqns (4.26) and (4.30). Similar to (d), we first identify $H_1(z)$ and $H_2(z)$ in an online manner, then we update the error system in an appropriate timing. The timing is decided as follows: if the moving average of the identification error $\|\varepsilon(k)\|$ is smaller than a threshold value $C_I$ and the moving average of the increasing rate of cancelling error $\|e(k) - e(k-1)\|$ is larger than a threshold $C_E$, the error system may be updated by replacing the error dynamics with the recently identified $\hat{G}_4(z)$.

In Case 1, where $G_4(z)$ is known *a priori* and unchangeable, all of the adaptive algorithms can give fast convergence after the start of control, but when the reference microphones are moved, the filtered-$x$ algorithm (b) diverged about 10 seconds after the movement, although $G_4(z)$ is available and the algorithm is normalised. On the other hand, since the proposed adaptive algorithm (c) can assure the update stability when $G_4(z)$ is known and unchangeable, it can maintain its high control performance. For the two identification-based adaptive algorithms (d) and (e), since the generalized primary and secondary channel dynamics in the on-line manner, the stable control results can be achieved, but comparing with (c) where a relatively accurate secondary channel dynamics $G_4(z)$ is known, their control performances become worse.

In Case 2, where the secondary channel dynamics $G_4(z)$ is uncertainly changeable, as the algorithms (b) and (c) use prior information on $G_4(z)$, both algorithms diverged at 14 seconds after the error microphones moved. On the other hand, since the adaptive algorithms (d) and (e) make use of the on-line identification results of the generalised primary and secondary channel dynamics, their stable control performance can be maintained. Owing to real-time identification, algorithm (d) has high adaptability, but it needs a large amount of FFT and inverse FFT computation. Comparing with (d), the cancelling performance of (e) is worse, but the amount of computation is much less. Figure 4.5 illustrates the moving average of the squares cancelling error and identification error obtained by algorithm (e). It shows that the cancelling error increased greatly after the error microphones were moved, while the identification error changed a little because the identification excited by the dither signals is well performed. Thus, the error system dynamics are updated and the adaptation divergence is controlled.

## 4.6   Conclusions

Adaptive controller construction and a robust adaptive algorithm with assured stability for realising multichannel adaptive ANC, which involves multiple of primary noises, secondary control sources, reference and error microphones have successfully been achieved. The proposed algorithm can keep each cancelling error independently within a tight bound and does not need to calculate the inverse matrix when the secondary channel dynamics are known *a priori*. In the case when the secondary channel dynamics are uncertain, identification-based adaptive algorithms have also been proposed. The effectiveness of the proposed algorithms has been validated in the experiments.

## 4.7 Appendix: proof of the theorem

Consider a candidate for the Lyapunov function as:

$$V(k) \equiv \tilde{\boldsymbol{\theta}}^T(k)\tilde{\boldsymbol{\theta}}(k) = \|\tilde{\boldsymbol{\theta}}(k)\|^2 \tag{4.34}$$

where $\tilde{\boldsymbol{\theta}}(k) \equiv \boldsymbol{\theta}^* - \hat{\boldsymbol{\theta}}(k)$. Then the time difference of $V(k)$ can be written:

$$\Delta V(k) = \|\tilde{\boldsymbol{\theta}}(k+1)\|^2 - \|\tilde{\boldsymbol{\theta}}(k)\|^2 \tag{4.35}$$

Rewriting eqn. (4.17) gives:

$$\tilde{\boldsymbol{\theta}}(k+1) = \tilde{\boldsymbol{\theta}}(k) - \gamma(k)\boldsymbol{\Psi}(k)\bar{\boldsymbol{\eta}}(k) \tag{4.36}$$

Substituting eqn. (4.36) into eqn. (4.35) and using eqns (4.15) and (4.17) leads to:

$$
\begin{aligned}
\Delta V &= -2\gamma\tilde{\boldsymbol{\theta}}^T\boldsymbol{\Psi}\bar{\boldsymbol{\eta}} + \gamma^2(\boldsymbol{\Psi}\bar{\boldsymbol{\eta}})^T(\boldsymbol{\Psi}\bar{\boldsymbol{\eta}}) \\
&= -2\gamma(\boldsymbol{\eta} - \boldsymbol{\xi})^T\bar{\boldsymbol{\eta}} + \gamma^2\bar{\boldsymbol{\eta}}^T\boldsymbol{\Psi}^T\boldsymbol{\Psi}\bar{\boldsymbol{\eta}} \\
&= -2\gamma(\boldsymbol{\eta} - \boldsymbol{\xi})^T\bar{\boldsymbol{\eta}} + 2\alpha\gamma\bar{\boldsymbol{\eta}}^T\bar{\boldsymbol{\eta}}\frac{\bar{\boldsymbol{\eta}}^T\boldsymbol{\Psi}^T\boldsymbol{\Psi}\bar{\boldsymbol{\eta}}}{d + \bar{\boldsymbol{\eta}}^T\boldsymbol{\Psi}^T\boldsymbol{\Psi}\bar{\boldsymbol{\eta}}}
\end{aligned}
\tag{4.37a}
$$

$$
\begin{aligned}
&\leq -2\gamma(\boldsymbol{\eta} - \boldsymbol{\xi})^T\bar{\boldsymbol{\eta}} + 2\gamma\bar{\boldsymbol{\eta}}^T\bar{\boldsymbol{\eta}} \\
&= -2\gamma\sum_{i=1}^{N_c}[(\eta_i - \xi_i)\bar{\eta}_i - \bar{\eta}_i^2]
\end{aligned}
\tag{4.37b}
$$

where the time index $k$ is omitted for the purpose of simplicity. By making use of the following inequality [234]:

$$(\eta_i - \xi_i)\bar{\eta}_i \geq \bar{\eta}_i^2 \tag{4.38}$$

it follows from eqn. (4.37) that $\Delta V \leq 0$ can be confirmed. Therefore it can be concluded that the algorithm in eqn. (4.17) can ensure that:

$$\|\tilde{\boldsymbol{\theta}}(k)\|^2 \leq \|\tilde{\boldsymbol{\theta}}(k-1)\|^2 \leq \cdots \leq \|\tilde{\boldsymbol{\theta}}(0)\|^2 \tag{4.39}$$

or the sequence $\{\| \tilde{\boldsymbol{\theta}}(k) \|^2\}_0^\infty$ is a nonincreasing sequence bounded below by zero and thus converges. Hence it can be obtained that:

$$\lim_{k \to \infty} \left( \| \tilde{\boldsymbol{\theta}}(k+1) \|^2 - \| \tilde{\boldsymbol{\theta}}(k) \|^2 \right) = 0 \tag{4.40}$$

Furthermore, combining (4.35) and (4.37) gives:

$$
\begin{aligned}
\| \tilde{\boldsymbol{\theta}}(k+1) \|^2 &= \| \tilde{\boldsymbol{\theta}}(k) \|^2 - 2\gamma(k)[\boldsymbol{\eta}(k) - \boldsymbol{\xi}(k)]^T \bar{\boldsymbol{\eta}}(k) \\
&+ 2\alpha\gamma(k)\bar{\boldsymbol{\eta}}^T(k)\bar{\boldsymbol{\eta}}(k) - \frac{2d\alpha\gamma(k)\bar{\boldsymbol{\eta}}^T(k)\bar{\boldsymbol{\eta}}(k)}{d + \bar{\boldsymbol{\eta}}^T(k)\boldsymbol{\Psi}^T(k)\boldsymbol{\Psi}(k)\bar{\boldsymbol{\eta}}(k)} \\
&= \| \tilde{\boldsymbol{\theta}}(k) \|^2 - 2\gamma(k) \left\{ [\boldsymbol{\eta}(k) - \boldsymbol{\xi}(k)]^T \bar{\boldsymbol{\eta}}(k) \right. \\
&\left. - \alpha\bar{\boldsymbol{\eta}}^T(k)\bar{\boldsymbol{\eta}}(k) + d\gamma(k)/2 \right\}
\end{aligned}
\tag{4.41}
$$

Then according to eqn. (4.40), from eqn. (4.41) it can be obtained that:

$$\lim_{k \to \infty} \gamma(k) \left\{ [\boldsymbol{\eta}(k) - \boldsymbol{\xi}(k)]^T \bar{\boldsymbol{\eta}}(k) - \alpha\bar{\boldsymbol{\eta}}^T(k)\bar{\boldsymbol{\eta}}(k) + d\gamma(k)/2 \right\} = 0 \tag{4.42}$$

Moreover, it follows from eqn. (4.17) that:

$$[\eta_i(k) - \xi_i(k)]\bar{\eta}_i(k) = \bar{\eta}_i^2(k) + c_i(k) \, | \, \bar{\eta}_i(k) \, | \tag{4.43}$$

where

$$c_i(k) \geq \bar{\beta}_i - | \, \xi_i(k) \, | \geq 0 \tag{4.44}$$

Therefore, eqn. (4.42) can be expressed as:

$$
\begin{aligned}
0 &= \lim_{k \to \infty} \gamma(k) \left\{ \sum_{i=1}^{N_c} \left[ [\eta_i(k) - \xi_i(k)]\bar{\eta}_i(k) - \alpha\bar{\eta}_i^2(k) \right] + d\gamma(k)/2 \right\} \\
&= \lim_{k \to \infty} \gamma(k) \left\{ \sum_{i=1}^{N_c} \left[ \bar{\eta}_i^2(k) + c_i(k) \, | \, \bar{\eta}_i(k) \, | - \alpha\bar{\eta}_i^2(k) \right] + d\gamma(k)/2 \right\} \\
&= \lim_{k \to \infty} \gamma(k) \left\{ \sum_{i=1}^{N_c} \left[ (1 - \alpha)\bar{\eta}_i^2(k) + c_i(k) \, | \, \bar{\eta}_i(k) \, | \right] + d\gamma(k)/2 \right\}
\end{aligned}
\tag{4.45}
$$

Since it follows that $(1 - \alpha) > 0$, $c_i(k) \geq 0$, $d > 0$ and $\gamma(k) \geq 0$, it is clear that eqn. (4.45) is equivalent to:

$$
\begin{cases}
\lim_{k \to \infty} \gamma(k) = 0 \\
\lim_{k \to \infty} \bar{\eta}_i(k) = 0
\end{cases}
\tag{4.46}
$$

Since $\{r_i(k) : i = 1, 2, \cdots, N_c\}$ are bounded and $\gamma(k)$ is a zero-order variable of $\bar{\boldsymbol{\eta}}(k)$, it can be seen from eqn. (4.17) that the second equation of eqn. (4.46) is just the solution of the first. Therefore, the solution of eqn. (4.46) can be summarised by:

$$\lim_{k\to\infty} \bar{\eta}_i(k) = 0 \tag{4.47}$$

Based on the result from eqn. (4.17), it can be shown that $\hat{\boldsymbol{\theta}}(k)$ is convergent, which gives eqn. (4.18), and from eqn. (4.17) it can be obtained that $\eta_i(k)$ is bounded by $\bar{\beta}_i(k)$, as:

$$\lim_{k\to\infty} \sup |\eta_i(k)| = \bar{\beta}_i \tag{4.48}$$

As a result of $\hat{\boldsymbol{\theta}}(k)$ converging to a constant vector, the two terms of the right-hand side in eqn. (4.13) will be cancelled in the steady-state due to the fact that the calculation sequence of the converged parameter vector $\hat{\boldsymbol{\theta}}(k)$, or a constant vector, relating to the operator $\boldsymbol{G}_4(z)$ can be changed, which indicates that $\boldsymbol{e}_a(k)$ approaches zero and in return $\boldsymbol{\eta}(k)$ approaches $\boldsymbol{e}(k)$ according to eqn (4.14):

$$\lim_{k\to\infty} \eta(k) = \lim_{k\to\infty} [e(k) - e_a(k)]$$
$$= \lim_{k\to\infty} e(k) \tag{4.49}$$

Then eqn. (4.49) can be obtained from eqn (4.49) and (4.48).

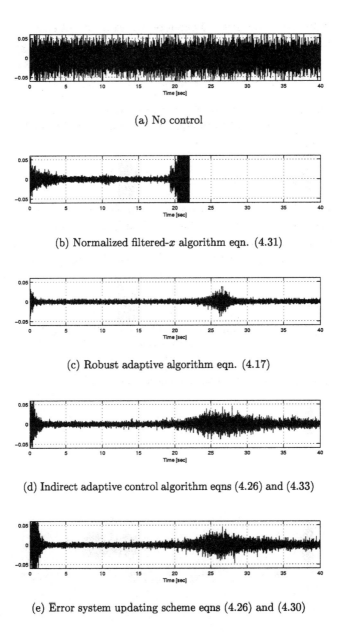

(a) No control

(b) Normalized filtered-$x$ algorithm eqn. (4.31)

(c) Robust adaptive algorithm eqn. (4.17)

(d) Indirect adaptive control algorithm eqns (4.26) and (4.33)

(e) Error system updating scheme eqns (4.26) and (4.30)

Figure 4.3    *Changes in channel dynamics by movement of reference micro-*
*phones: Comparison of canceling error $e_1(k)$ obtained by various*
*algorithms*

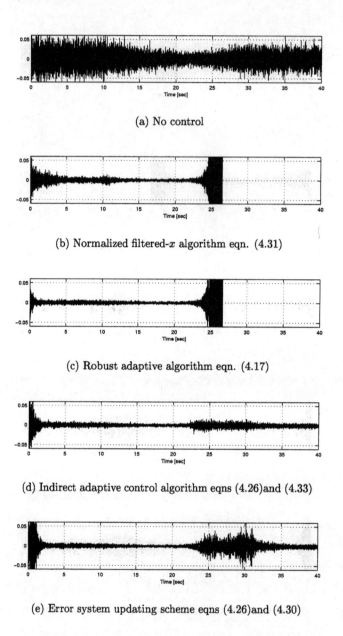

(a) No control

(b) Normalized filtered-$x$ algorithm eqn. (4.31)

(c) Robust adaptive algorithm eqn. (4.17)

(d) Indirect adaptive control algorithm eqns (4.26)and (4.33)

(e) Error system updating scheme eqns (4.26)and (4.30)

Figure 4.4    *Changes in channel dynamics by movement of error microphones:*
*Comparison of cancelling error $e_1(k)$ obtained by various algo-*
*rithms*

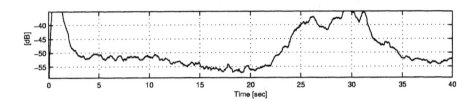

(a) Cancelling error norm by eqns (4.26) and (4.30)

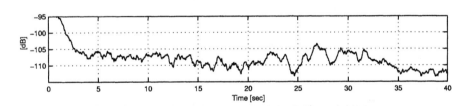

(b) Identification error norm by eqns (4.26) and (4.30)

Figure 4.5    *Cancelling error and identification error obtained by error system updating scheme (e)*

*Chapter 5*

# Adaptive harmonic control: tuning in the frequency domain

## S. M. Veres and T. Meurers

*School of Engineering Sciences, University of Southampton*
*Southampton, UK, Email:s.m.veres@soton.ac.uk*

*This chapter discusses adaptive control based on control at individual harmonics for cancellation of periodic disturbances. It is shown that the original idea, which has been around for a long time, can be developed much further to produce robust and highly adaptable controllers. First, a review is given of the most accessible literature, after which frequency-selective RLS and LMS methods are presented. Simulations illustrate the effectiveness of the method.*

## 5.1 Introduction

As in some applications the plant can be considered linear, a fundamental idea is to separate the control problem at each relevant harmonic of the disturbance if that is dominated by a discrete set of frequencies. Then, by linearity there is no interaction between the control solutions found at different harmonics. This idea can be described as follows for a single tonal active vibration or sound control problem. At frequency $\omega$ denote the detection signal by $x(\omega)$, the control signal by $u(\omega)$, the additive disturbance by $d(\omega)$ and the error signal by $e(\omega)$. Figure 5.1 describes the problem of producing $u(\omega)$ so that $e(\omega)$ is as near to zero as possible.

Let $\omega_1$, $\omega_2$, ..., $\omega_n$ be a finite set of frequencies at which $x$ or $d$ have relevant harmonic components. The objective is to design a $W$ such that

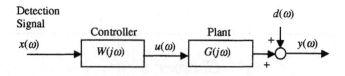

*Figure 5.1    Block diagram for the harmonic control problem*

$e = d + GWx = 0$ at these frequencies. At each frequency the harmonic components of the signals, and also the response of the controller and the plant dynamics, can be characterised by single complex numbers, which will be denoted by $x(\omega_k)$, $u(\omega_k)$, $d(\omega_k)$, $e(\omega_k)$ and $W(j\omega_k)$, $G(j\omega_k)$, respectively. Assuming that there is no feedback from the actuator to the detection, the component of the error signal can be expressed as:

$$e(\omega_k) = d(\omega_k) + W(j\omega_k)G(j\omega_k)x(\omega_k), \quad k = 1, 2, \ldots, n$$

Figure 5.1 shows the block diagram of the harmonic control system. For cancellation the controller has to satisfy:

$$W(j\omega_k) = -\frac{d(\omega_k)}{x(\omega_k)G(j\omega_k)}, \quad k = 1, 2, \ldots n. \tag{5.1}$$

Control of periodic sound at individual harmonics has been around for some time and Reference [219] gives a summary of the subject. William Conover built an electronic system to achieve this cancellation for the harmonics of electrical transformers [58] in 1956. To satisfy these conditions in Conover's work the transfer function $W$ is physically realised by bandpass filters, phase shifters and amplifiers. Figure 5.2 shows a block diagram of Conover's system. The analogue circuits offered little flexibility in the selected frequencies, the amplifiers (amplitude control) and phase shifters were tuned manually. Later digital control solutions were sought when computer technology made that a possibility. Various adaptive digital feedforward control methods have been proposed, among them the most widely used today is the filtered-$x$ LMS method.

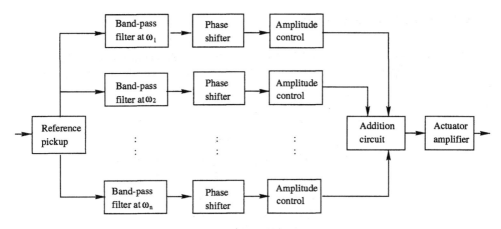

*Figure 5.2*   *Block diagram of Conover's harmonic controller*

This chapter presents a study of adaptive frequency selective solutions to periodic disturbance cancellation. The control schemes suggested are pure feedback control schemes, no detection signal is needed, as these are frequently unavailable in practical problems. Achievable performance is excellent compared to pure linear feedback solutions and the advantages can be summarised as follows:

(i) For periodic disturbances with a nearly discrete spectrum the performance is far better than that which is achievable by linear feedback systems.

(ii) A high degree of adaptability for dynamical changes of the plant is achieved.

(iii) Stability is ensured despite a the high level of adaptation.

Compared to standard LMS-type feedforward control an obvious advantage is that there is no need for a detection signal which is strongly correlated with the disturbance. The only disadvantage of the scheme presented is that the disturbance has to be dominated by a discrete spectrum (possibly slowly time varying). This is a requirement that is satisfied by a large number of practical vibration and sound control problems, hence the usefulness of the approach.

## 5.2   Problem formulation

The plant to be controlled is a linear time-varying two-input/two-output system as presented in Figure 5.3:    It is described by stable $2 \times 2$ transfer

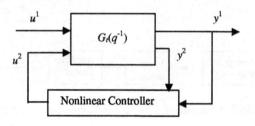

*Figure 5.3    Block diagram of the basic control scheme*

functions:
$$\mathbf{G}_t(q^{-1}) = w_0^k + w_1^k q^{-1} + w_2^k q^{-2} + \ldots \, t > 0 \tag{5.2}$$
so that at each time instant $t$:

$$\mathbf{y}(t) = \sum_{i=0}^{\infty} w_i^k \mathbf{u}(t-i), \, t > 0; \quad \mathbf{y} = [y^1 \, y^2]^T, \, \mathbf{u} = [u^1 \, u^2]^T \tag{5.3}$$

and in matrix form we can also write:

$$\mathbf{y}(t) = \mathbf{G}_k(q^{-1})\mathbf{u}(t), \, k > 0 \tag{5.4}$$

where
$$\mathbf{G}_k(q^{-1}) = \begin{bmatrix} G_k^{11}(q^{-1}) & G_k^{12}(q^{-1}) \\ G_k^{21}(q^{-1}) & G_k^{22}(q^{-1}) \end{bmatrix} \tag{5.5}$$

The following situation is practically important:

(i) $u^1$ is an unmeasured excitation of the vibration.

(ii) $u^2$ is a control input.

(iii) $y^1$ is a measurable output which is to be regulated to zero.

(iv) $y^2$ is a measurable output.

The regulation problem of $y^1$ to zero, described under conditions (i)–(iv), is difficult or in general impossible to solve using mainstream adaptive control theory because of the unmeasured $u^1$. In the following an RLS and an LMS approach will be looked at to remedy this situation. To solve these problems some simplifying assumptions will be made and the ideas of harmonic control will be used.

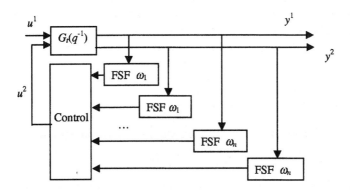

*Figure 5.4    Frequency selective filtering (FSF) at a set of frequencies $\omega_1 \omega_2, \ldots, \omega_n$*

(i) Regulation of $y^1$ to zero is to be achieved at a given set of frequencies

$$\omega_1, \omega_2, \ldots, \omega_n \in [0, 2\pi]$$

(ii) The plant is slowly time varying so that $\|\mathbf{G}_t - \mathbf{G}_{t+1}\|_1 \leq \iota$, $t > 0$ with $\iota > 0$ known *a priori*.

(iii) The outputs are measurable with a given accuracy $\rho > 0$.

A possible scheme to achieve this is as follows. Each output is led through a frequency selective filter (FSF) for each frequency $\omega_1, \omega_2, \ldots, \omega_n$ as shown in the block diagram in Figure 5.4. The signals coming from the filters are processed by the controller and a suitable input is synthesised. The main result of this section is dealing with the problem of how to compute a suitable control input in the frequency domain.

First, some of the surrounding details will be clarified. The frequency selective filters are of the form:

$$H_i(q) = H_i^1(q)H_i^1(q)H_i^1(q)$$

$$H_i^1(q) = \frac{(1-r)(q+1)(q-1)}{q^2 - 2r\cos(\omega_i T)q + r^2}$$

where $r < 1$, $r \sim 1$. The advantage of frequency selective filters is that the settling time of the filters can typically be smaller than the length of an FFT batch which could be used instead. Typically, the settling time of a satisfactory FSF can be around $40 \sim 60$ sampling periods and FFTs should be based on 256 samples to have similar accuracy.

Notations:

$$\mathbf{y}^1(e^{j\omega_i}), \ \mathbf{y}^2(e^{j\omega_i}), \ \mathbf{u}^1(e^{j\omega_i}), \ \mathbf{u}^2(e^{j\omega_i}), \ i = 1, 2, \ldots, n$$

will be used for the steady-state complex amplitudes of the sine waves:

$$H_i(q)y^1, \ H_i(q)y^2, \ H_i(q)u^1, \ H_i(q)u^2, \ i = 1, 2, \ldots, n$$

respectively. Then (A) means that the control objective is to achieve:

$$y^1(e^{j\omega_i}) = 0, \ \ i = 1, 2, \ldots, n.$$

The control signal will be computed as a linear combination of phase-shifted sine-waves (an alternative method could be based on a robust $H_\infty$ controller but there is more danger of instability in that case because of possibly false uncertainty assessment of the frequency selective model used). Let the infinite time horizon be split into periods of $N$ sampling instants, so that the $k$th period is denoted by:

$$T(k) = [kN + 1, kN + N]$$

In many practical systems it might be realistic to assume that there is an average transfer function $\mathbf{G}^k(q^{-1})$ valid over $T(k)$ such that:

$$\|\mathbf{G}^k(e^{-j\omega_i}) - \mathbf{G}_t(e^{-j\omega_i})\| \le \nu, \ \ t \in T(k)$$

with *a priori* known $\nu > 0$, where $\|G\|$ denotes $\max |G_{lm}|$. Similarly, for the filtered signals:

$$H_i(q)y^1, \ H_i(q)y^2, \ H_i(q)u^1, \ H_i(q)u^2, \ \ i = 1, 2, \ldots, n,$$

the respective average complex amplitudes over time period $T(k)$ will be denoted by:

$$y_k^1(e^{j\omega_i}),\ y_k^2(e^{j\omega_i}),\ u_k^1(e^{j\omega_i}),\ u_k^2(e^{j\omega_i}),\quad i = 1, 2, \ldots, n$$

$u_k^2(e^{j\omega_i})$, $i = 1, 2, \ldots, n$, actually represent the components of the control input so that $u_k^2$ is defined as the sum of sine waves with complex amplitudes $u_k^2(e^{j\omega_i})$, $i = 1, 2, \ldots, n$ during period $T(k)$. This operation is clearly not linear filtering.

For further use let the components of $\mathbf{G}^k(e^{j\omega_i})[u^1(e^{j\omega_i})\ 1]^T$ be denoted by:

$$\begin{bmatrix} g_k^0(i) & g_k^1(i) \\ h_k^0(i) & h_k^1(i) \end{bmatrix} = \mathbf{G}_k(e^{j\omega_i}) \begin{bmatrix} u^1(e^{j\omega_i}) \\ 1 \end{bmatrix},\quad i = 1, 2, \ldots, n$$

## 5.3  A frequency selective RLS solution

Let $\mathbf{x}_k$, $\mathbf{d}_k$ and $\mathbf{e}_k$ denote complex numbers representing the harmonic content of the signals $y^2$, $G^{11}u_1$ and $y_1$, during time period $T(k)$, as described above at given frequency $\omega$ and indicated in Figure 5.6 . Assume that $G^{12}$ changes slowly and good estimates $\hat{\mathbf{g}}_k$ can be calculated for its complex gain at frequency $\omega$. (Let's postpone discussion of the estimability of $G^{12}$ for a little while.) First a simple filtered-$x$ algorithm can be introduced for each frequency $\omega$ of interest. Introduce $\mathbf{r}_k = \mathbf{x}_k \hat{\mathbf{g}}_k$. For any complex number $\mathbf{c}$ the notation $\mathbf{c} = c^1 + c^2 j$ will be used. Then the equation:

$$\mathbf{w}_k \mathbf{r}_k + \mathbf{d}_k + \mathbf{n}_k = \mathbf{e}_k \tag{5.6}$$

can be rewritten in the form:

$$e_k^1 = r^1 w_k^1 - r^2 w_k^2 + d_k^1 + n_k^1,\ e_k^2 = r^2 w_k^1 + r^1 w_k^2 + d_k^2 + n_k^2 \tag{5.7}$$

$$T(k) = \{\ t \mid kN < \ t \ < kN{+}N{+}1\ \}$$

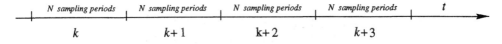

| $N$ sampling periods | $N$ sampling periods | $N$ sampling periods | $N$ sampling periods | $t$ |
|---|---|---|---|---|
| $k$ | $k+1$ | $k+2$ | $k+3$ | |

Figure 5.5    *Periodic sectioning of the time scale for compensator design and frequency response estimation*

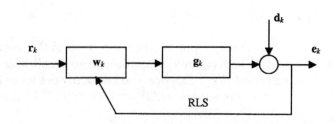

Figure 5.6    *Block diagram of the RLS solution at a single frequency*

for the output errors $\mathbf{e}_k$ which are measured with error/noise $n_k^1$ and $n_k^2$. To find a suitable controller gain $\mathbf{w}_k$ represented by the coefficients $w_k^1$ and $w_k^2$, the following exponentially weighted criterion can be optimised:

$$\sum_{i=1}^{k} \lambda^{k-i} [(e_i^1)^2 + (e_k^2)^2]$$

by the usual RLS algorithm with a forgetting factor $\lambda < 1$. This can be carried out in two stages for each $k \geq 1$:

$$\tilde{w}_k = w_{k-1} + \tilde{K}_k e_k^1, \quad \tilde{K} = P_{k-1}\tilde{\phi}(\lambda + \tilde{\phi}_k^T P_{k-1}\tilde{\phi}_k)^{-1}, \quad \tilde{P}_k = (I - \tilde{K}_k \tilde{\phi}_k^T)P_{k-1}/\lambda$$

$$w_k = \tilde{w}_k + K_k e_k^2, \quad K_k = \tilde{P}_k \phi (1 + \phi_k^T \tilde{P}_k \phi_k)^{-1}, \quad P_k = (I - K_k \phi_k^T)\tilde{P}_k$$

(5.8)

where the notations:

$$w_k = \begin{bmatrix} w_k^1 \\ w_k^2 \end{bmatrix}, \quad \tilde{\phi}_k = \begin{bmatrix} -r_k^1 \\ r_k^2 \end{bmatrix}, \quad \phi_k = \begin{bmatrix} -r_k^2 \\ -r_k^1 \end{bmatrix}$$

are used and $\tilde{K}_k$, $\tilde{w}_k$, $\tilde{P}_k$ are only intermediate variables in an updating step. Hence $w_k$, $P_k$ are computed recursively with initial conditions dependent on *a priori* knowledge of the plant. The following simulation illustrates how this algorithm works.

Let:

$$G = G(e^{j\omega T}) = \begin{bmatrix} 1.2 - 0.2j & 0.6 - 0.05j + g_k \\ 0.8 + 0.4j & 0.9 - 0.1j \end{bmatrix}$$

where $g_k$ is time-varying gain, $g_k = c(\cos 0.02k + j \sin 0.02k)$ and $c = 0.15$, the dynamics are time varying.

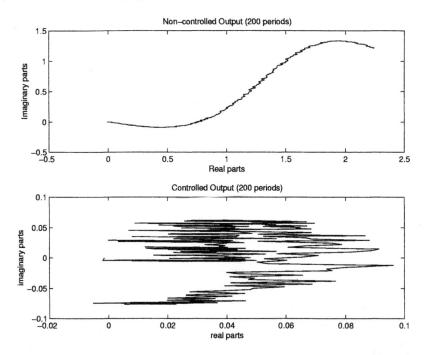

*Figure 5.7*    *Complex amplitudes of the controlled and uncontrolled outputs for 200 periods: (a) Non-controlled output. (b) Controlled output*

The forgetting factor is set to $\lambda = 0.7$ and the detection signal is a fixed constant $\mathbf{r}_k = 1$, therefore no measured detection is used. As the forgetting factor is low, reasonable performance is achieved as shown in Figures 5.7 and 5.8. For a plant model the fixed and incorrect:

$$\hat{G} = \hat{G}(e^{j\omega T}) = \begin{bmatrix} 1.6 - 0.2j & 0.5 - 0.05j \\ 0.9 + 0.4j & 0.94 - 0.1j \end{bmatrix}$$

is used.

The simulation illustrates that the method can work as the complex amplitudes of the controlled output are small in Figures 5.7 and 5.8. Conditions of stability with regard to $\hat{G}$ and $\mathbf{x}_k$ are difficult to establish. The frequency selective LMS solution presented in the next Section easily lends itself to stability conditions.

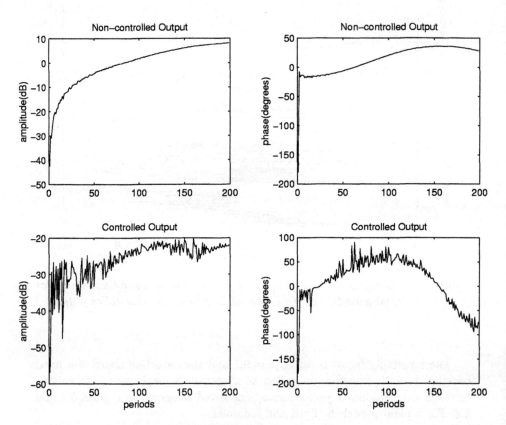

Figure 5.8    *Amplitudes (dB) and phase (°) of uncontrolled outputs for 200 periods: (a) Non-controlled output. (b) Controlled output*

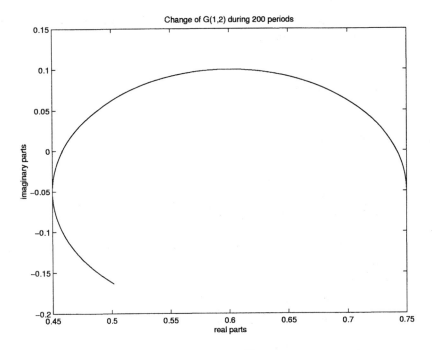

**Figure 5.9**  *Change of the complex gain $G^{12}$ during 200 periods*

## 5.4  A frequency selective LMS solution

An adaptive law will be derived for the input complex amplitude at a single frequency and then its stability will be analysed in view of the uncertainty of the transfer function of the plant. It will be proved that the adaptation law is convergent under large relative uncertainty of the plant transfer function.

Let the input complex amplitude during time period $T(k)$ be denoted by $u_k \overset{def}{=} u_r^k + u_i^k j$. Similarly, the disturbance and output amplitudes will be denoted by $d_k \overset{def}{=} d_r^k + d_i^k j$ and $y_k \overset{def}{=} y_r^k + y_i^k j$, respectively. Then, at a single frequency the model of the plant equation will be:

$$y_k = g u_k + d_k$$

where $g$ is a complex number representing the plant transfer at the given frequency.

The objective will be to bring $y_k$ to zero and therefore an instantaneous criterion function for control will be:

$$c_k \stackrel{def}{=} |y_k|^2 = (y_r^k)^2 + (y_i^k)^2$$

The output can be rewritten as:

$$y_r^k + y_i^k j = g_r u_r^k + g_i u_i^k + d_r^k + j(g_r u_i^k + g_i u_r^k + d_i^k)$$

or in matrix form as:

$$\mathbf{y}_k = \mathbf{G}\mathbf{u}_k + \mathbf{d}_r$$

where

$$\mathbf{y}_k \stackrel{def}{=} \begin{bmatrix} y_r^k \\ y_i^k \end{bmatrix}, \quad \mathbf{u}_k \stackrel{def}{=} \begin{bmatrix} u_r^k \\ u_i^k \end{bmatrix}, \quad \mathbf{d}_k \stackrel{def}{=} \begin{bmatrix} d_r^k \\ d_i^k \end{bmatrix}, \quad \mathbf{G} \stackrel{def}{=} \begin{bmatrix} g_r & -g_i \\ g_i & g_r \end{bmatrix}$$

The control law will be designed to move the control signal $\mathbf{u}_{k+1}$ in the negative gradient direction of the criterion $c_k$. The gradient of the criterion function can be calculated as:

$$\frac{\partial c_k}{\partial u_r^k} = 2\frac{\partial y_r^k}{\partial u_r^k}y_r^k + 2\frac{\partial y_i^k}{\partial u_r^k}y_i^k = 2g_r y_r^k + 2g_i y_i^k$$

$$\frac{\partial c_k}{\partial u_i^k} = 2\frac{\partial y_r^k}{\partial u_i^k}y_r^k + 2\frac{\partial y_i^k}{\partial u_i^k}y_i^k = -2g_i y_r^k + 2g_r y_i^k$$

which finally gives:

$$\nabla c_k = -2\mathbf{G}^T \mathbf{y}_k$$

Note that for the computation of the gradient only an estimate is available as $\mathbf{G}$ is not known. With a $\mu > 0$ step size this will define an adapted control signal as:

$$\mathbf{u}_{k+1} \stackrel{def}{=} \mathbf{u}_k - \mu \hat{\mathbf{G}}^T \mathbf{y}_k \tag{5.9}$$

Let the estimated and the actual transfer matrices of the plant be denoted by:

$$\hat{\mathbf{G}} \stackrel{def}{=} \begin{bmatrix} \hat{g}_r & -\hat{g}_i \\ \hat{g}_i & \hat{g}_r \end{bmatrix} \quad \text{and} \quad \mathbf{G}_0 \stackrel{def}{=} \begin{bmatrix} g_r^0 & -g_i^0 \\ g_i^0 & g_r^0 \end{bmatrix}$$

respectively. The actual output will be obtained from the equation:

$$\mathbf{y}_k = \mathbf{G}_0 \mathbf{u}_k + \mathbf{d} + \mathbf{n}_k \tag{5.10}$$

where $\mathbf{n}_k$ is the measurement noise of the complex amplitude of the output. Substituting eqn. (5.10) into eqn. (5.9) gives that:

$$\mathbf{u}_{k+1} = (\mathbf{I} - \mu\hat{\mathbf{G}}^T \mathbf{G}_0)\mathbf{u}_k + \mu\hat{\mathbf{G}}^T \mathbf{d} + \mu\hat{\mathbf{G}}^T \mathbf{n}_k \tag{5.11}$$

**Lemma 5.4.1** *(a) The adaptive law eqn (5.9) will be stable if the matrix:*

$$\mathbf{I} - \mu \hat{\mathbf{G}}^T \mathbf{G}_0$$

*has eigenvalues with modulus less than 1.*

*(b) If the adaptive law eqn (5.9) is stable and* $\mathbf{n}_k \equiv 0$ *, then*

$$\lim_{k \to \infty} \mathbf{u}_k = \mathbf{G}_0^{-1} \mathbf{d}$$

*where* $\mathbf{u} = \mathbf{G}_0^{-1} \mathbf{d}$ *is the ideal control input to eliminate the vibration of the output.*

*Proof:* (a) is obvious from linear system theory. For (b) the steady state of the given linear stable system can be calculated as:

$$[q\mathbf{I} - (\mathbf{I} - \mu \hat{\mathbf{G}}^T \mathbf{G}_0)]^{-1} \mu \hat{\mathbf{G}}^T (\mathbf{d} + \mathbf{n}_k)_{|q=1} = \mathbf{G}_0^{-1} \mathbf{d} + \mathbf{G}_0^{-1} \mathbf{n}_k$$

which gives the desired result.

The most interesting case, practically, is to find out how small the step size $\mu$ will have to be defined to ensure stability under a given $\hat{\mathbf{G}}$ and its relative error. Let's denote the estimated complex gain associated with $\hat{\mathbf{G}}$ and $\mathbf{G}_0$ by:

$$\hat{\mathbf{g}} \overset{def}{=} \hat{g}_r + j\hat{g}_i \text{ and } \mathbf{g}_0 \overset{def}{=} g_r^0 + jg_i^0$$

respectively. Then the following theorem holds.

**Theorem 5.4.1** *the control law eqn. (5.9) will be stable under any relative error less than* $\delta > 0$ *of the estimate* $\hat{\mathbf{g}}$, *(i.e. for* $|\mathbf{g}_0 - \hat{\mathbf{g}}| \leq \delta |\hat{\mathbf{g}}|$*), if and only if* $\delta < 1$ *and the condition*

$$\mu < \frac{1}{(1+\delta)|\hat{\mathbf{g}}|^2} \tag{5.12}$$

*is satisfied.*

*Proof:* first of all note that:

$$\hat{\mathbf{G}}^T \mathbf{G}_0 = \begin{bmatrix} \hat{g}_r & \hat{g}_i \\ -\hat{g}_i & \hat{g}_r \end{bmatrix} \begin{bmatrix} g_r^0 & -g_i^0 \\ g_i^0 & g_r^0 \end{bmatrix} = \begin{bmatrix} \hat{g}_r g_r^0 + \hat{g}_i g_i^0 & -\hat{g}_r g_i^0 + \hat{g}_i g_r^0 \\ -\hat{g}_i g_r^0 + \hat{g}_r g_i^0 & \hat{g}_i g_i^0 + \hat{g}_r g_r^0 \end{bmatrix}$$

Introducing the notations $a \overset{def}{=} \hat{g}_r g_r^0 + \hat{g}_i g_i^0$ and $b \overset{def}{=} -\hat{g}_r g_i^0 + \hat{g}_i g_r^0$ will give that:

$$\mathbf{I} - \mu \hat{\mathbf{G}}^T \mathbf{G}_0 = \mathbf{I} - \mu \begin{bmatrix} a & b \\ -b & a \end{bmatrix} = \begin{bmatrix} 1 - \mu a & -\mu b \\ \mu b & 1 - \mu a \end{bmatrix}$$

with eigenvalues $1 - \mu(a \pm bj)$. To prove the theorem one has therefore to prove that $|1 - \mu(a \pm bj)| < 1$ for any relative error $< \delta$ if and only if:

$$\mu < \nu(\delta, \hat{\mathbf{g}}) \overset{def}{=} \frac{1}{|\hat{\mathbf{g}}|^2} \min\left\{\frac{2}{1+\delta}, \frac{1}{\delta}\right\} \tag{5.13}$$

By definition it can also be seen that $a + bj = \hat{\mathbf{g}}\overline{\mathbf{g}_0}$ holds. Define the unit disc $C(1) \overset{def}{=} \{z \in \mathbf{C} \mid |1 - z| < 1\}$. In view of the relative error specification the true transfer can be written as $\mathbf{g}_0 = \hat{\mathbf{g}} + \delta|\hat{\mathbf{g}}|re^{j\omega}$ with some $r \in [0,1]$ and $\omega \in [0, 2\pi]$. The uncertainty set of $a + bj$ under maximum relative error $\delta$ of $\hat{g}$ is:

$$D(\delta) \overset{def}{=} \{\mu\hat{\mathbf{g}}(\overline{\hat{\mathbf{g}} + \delta|\hat{\mathbf{g}}|re^{j\omega}}) \mid \omega \in [0, 2\pi], \ r \in [0,1]\} \tag{5.14}$$

Hence $|1 - \mu(a \pm bj)| < 1$ will be satisfied under any relative error $\leq \delta$ of $\hat{g}$ if and only if $D(\delta) \subset C(1)$. It can, however, be easily shown that $D(\delta) \subset C(1)$ if and only if eqn. (5.12) holds. To see this notice that:

$$D(\delta) = \{\mu|\hat{\mathbf{g}}|^2 + \delta\mu|\hat{\mathbf{g}}|^2 re^{j\omega} \mid \omega \in [0, 2\pi], \ r \in [0,1]\} \tag{5.15}$$

and the shaded disk $D(\delta)$ with radius $\delta|\hat{\mathbf{g}}|^2$ will be contained in $C(1)$ if and only if:

$$\mu|\hat{\mathbf{g}}|^2 + \delta\mu|\hat{\mathbf{g}}|^2 < 2 \ \text{ and } \ \delta < 1 \tag{5.16}$$

which is equivalent to eqn. (5.12).

## 5.5 Simulation example

In this Section a combination of the above frequency-domain LMS and time-domain FSF will be used to illustrate the advantages and difficulties which appear in this harmonic control approach.

The simulated plant is a twelfth-order linear dynamical system with pole locations and Bode plots shown in Figure 5.10. Its transfer function is approximately:

$$G(q) = \frac{0.02q^{-1}}{1 - 2.8q^{-1} + 2.62q^{-2} - 0.61q^{-3} + 0.755q^{-4} - 2.9456q^{-5}\dots}$$
$$\frac{\dots}{\dots + 2.9813q^{-6} - 0.8452q^{-7} + 0.0770q^{-8} - 0.8662q^{-9} + 1.0639q^{-10} - 0.4931q^{-11} + 0.0909q^{-12}}$$

where the coefficients are rounded to four decimal points.

The output of this system is affected by a periodic disturbance shown in Figure 5.11 together with its estimated spectrum. The simulated disturbance had harmonic components at the frequencies 1.5Hz, 5Hz, 7Hz, 4Hz, 10Hz, 11.5Hz and an additive white noise of amplitude 0.01.

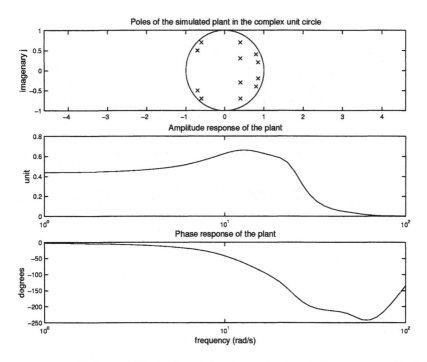

*Figure 5.10    Poles and Bode plot of the plant simulated. Resonant and lightly damped modes can be observed around 2 Hz, 3.5 Hz and 7 Hz*

Disregarding the ripples, the estimate of the spectrum of the disturbance has three high peaks at 1.5022Hz, 5.05Hz, 6.9951Hz and three low peaks at 4.019Hz, 10.077 Hz and 11.554Hz. The control system will aim to cancel only three components of the disturbance, those around 1.5Hz, 5.5Hz and 7Hz. As the precise frequencies may not be known in a practical situation, the control scheme will use an online estimation of the relevant frequencies of the disturbance.

The frequency selective adaptation is based on segmentation of the time axis into equal periods $T(k)$ of $N = 600$ samples. During each period $T(k)$ the output is filtered through three FSFs concentrated at the estimated three most relevant frequencies of the disturbance. The feedback input to the plant is a mix of three sine waves with constant phase and amplitude during each $T(k)$. The frequency selective RLS adaptive system described by the recursive

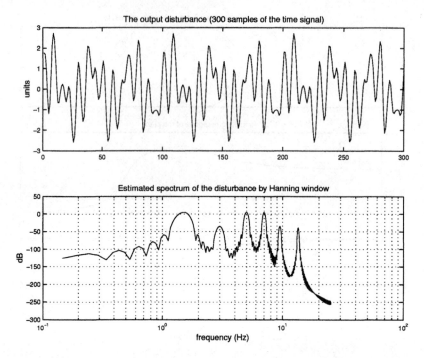

*Figure 5.11    A section of the time signal of the disturbance and the estimate of its spectrum by Hanning window, based on 500 samples*

eqns (5.8) is applied at each frequency to determine suitable amplitudes and phases of the sine waves, based on the adaptation of $\mathbf{w}_k$ by eqns (5.8). As the results of the method are very sensitive to errors in the frequencies of the harmonic components of the disturbance, a frequency estimation is also carried out at the end of each $T(k)$. The frequency of the FSF outputs is estimated by nonlinear optimisation of the squared-sum error function on the basis of the last 100 samples during each period $T(k)$, which ensures that the FSF settled by that time during the previous 500 samples of $T(k)$. To filter out the high-frequency noise of the estimates, a low-pass filter is also applied to each LS frequency estimate $\bar{f}_k^i$. $i = 1, 2, 3$ to obtain smothered estimates by:

$$\hat{f}_{k+1}^i = 0.98\hat{f}_k^i + 0.02\bar{f}_k^i, \quad i = 1, 2, 3 \tag{5.17}$$

$\hat{f}_{k+1}^i$ is then used during time period $T(k+1)$ to set the frequency of the

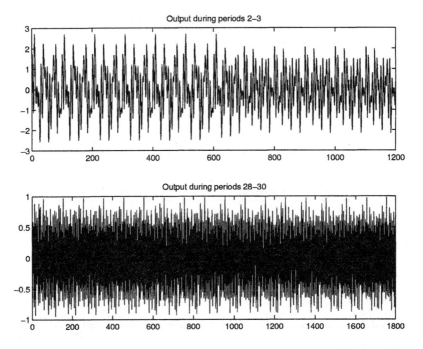

*Figure 5.12* *Time signals of the output during periods $T(2) - T(3)$ and $T(28) - T(30)$*

respective controller sinewave. The length of the time-domain simulation is 29 periods of $T(k)$, $k = 2, 3, ..., 30$. Figure 5.12 shows the reduced vibration of the output during periods $T(28)$, $T(29)$, $T(30)$ compared with the vibration during periods $T(2)$ and $T(3)$ displayed on the top plot. Figure 5.13 displays the complex gains of the control sinewaves during the periods from $T(2)$ to $T(30)$. The bottom plot in Figure 5.13 shows the amplitude of the three output harmonic components, which were obtained by estimation of the amplitude and phase of the sinewave outputs of the three FSF filters concentrated around 1.5Hz, 5.5Hz and 7Hz. This figure shows that the harmonic components of output $y$ at frequencies 1.5Hz, 5.5Hz and 7Hz goes to near zero after 30 periods. The sinewave phase and amplitude estimation during $T(k)$ is carried out by ordinary least-squares fitting of a sinewave to the last 100 output samples of each FSF applied to $y$ during $T(k)$. The top plot in Figure 5.13 shows the convergence of the complex amplitudes of the three sinewave components

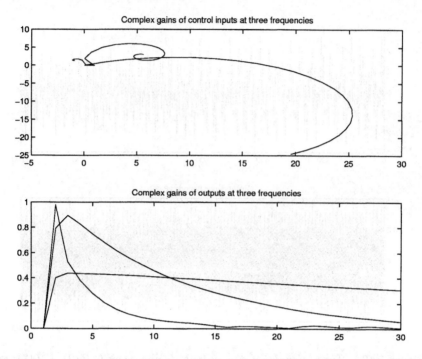

Figure 5.13    *The complex gains of the control inputs and the amplitudes of the estimated output harmonic components during time periods* $T(2) - T(30)$

of the control inputs, the computation of which was based on the FSF-RLS method of eqns (5.8) applied at each main harmonic component of disturbance $d$.

For the sake of completeness, and to illustrate the size of the control input used, Figure 5.14 shows the control input during time periods $T(2) - T(3)$ and $T(28) - T(30)$. Note that in Figure 5.12 the output during periods $T(28) - T(30)$ still contains the harmonic components of the disturbance at frequencies 4Hz, 10Hz and 11Hz which were not intended to be cancelled by the controller. Hence, the controller achieved only about 20 dB reduction in the vibration of the output $y$.

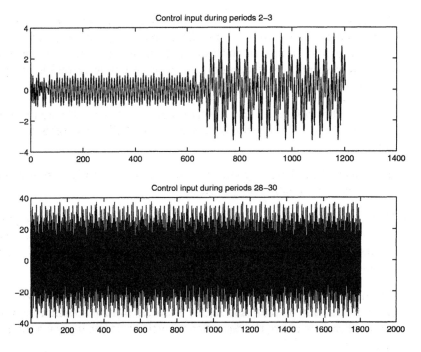

Figure 5.14    *Time signals of the control input during periods* $T(2) - T(3)$ *and* $T(28) - T(30)$

## 5.6    Conclusions

The basic schemes of frequency selective RLS (FS-RLS) and LMS (FS-LMS) methods have been described for periodic disturbance compensation. For the FS-LMS method a stability robustness theorem was given. Advantages of the methods were pointed out as (i) no principle limits of performance as in linear feedback control, (ii) high degree of adaptivity and (iii) high degree of stability robustness. A disadvantage is that the scheme only works for disturbances dominated by a discrete spectrum, although that can be slowly time varying.

This introductory research into the topic has some shortcomings which future research will resolve:

- In the methods presented the discrete spectrum of the disturbances was assumed to be known. Preliminary automatic spectrum analysis of the disturbed output can be used to set the frequencies of interest. Sensitivity to estimation errors in the disturbance frequencies is high. How to use the best estimates and sensitivity analysis should be the topic of future research. Detection of changes in an online system is of great practical interest.

- The effect of white noise disturbances should be analysed more precisely although the simulation shown indicates practically acceptable low sensitivity.

- The effect of nonlinear dynamics added to the basically nonlinear dynamics is a practically relevant question.

- Best numerical implementation can be investigated, parallel computational architectures can be made use of and large numbers of frequency points can be handled while performance is monitored and control action is supervised.

- Extension to multi-input multi-output systems is a future task which is likely to result in a stable and highly adaptive control scheme to numerous applications where the disturbance vector is dominated by a discrete spectrum.

## Chapter 6

# Model-free iterative tuning for periodic noise cancellation

## T Meurers and S M Veres

*School of Engineering Sciences, University of Southampton*
*Southampton, UK*
*Email: t.meurers@soton.ac.uk, s.m.veres@soton.ac.uk*

*In this chapter a model-free time-domain iterative controller tuning method is introduced and extended for periodic noise cancellation using a two degree of freedom (tdf) controller. An assumption is made that a stabilising initial controller is known. In each iteration of the design a controller is to be computed with a better performance than in the previous iteration. The first section will introduce the single-input single-output (SISO) problem, give a description of the iterative design method and explain its advantages compared with model-based approaches. The method is extended to a self-tuning method. Convergence to the optimal solution is improved by introducing frequency selective filters (FSFs). The methodological section is followed by laboratory hardware results to show the usefulness of the different approaches. Finally, conclusions are drawn and future research directions are pointed out.*

## 6.1  Introduction to iterative controller tuning

This section describes an off-line iterative tuning method [205, 323] for a SISO tdf controller to reduce the effect of an output disturbance without modelling secondary path dynamics. A scheme is developed to tune the controller parameters by performing a set of experiments. The collected data is used to calculate the gradient of the cost function of the system. This gradient can be used to change the controller parameters in the negative gradient direction.

The output, which is affected by the noise disturbance, is represented by the signal $y_t$ as in Figure 6.1. The dynamics from the actuator signal $u_t$ to the output $y_t$ are described by an unknown transfer function $G(q)$. There are no bandwidth limitations imposed on the disturbance $d_t$ and hence the output is assumed to be directly excited by the disturbance signal $d_t$ without any further dynamics. It is, however, assumed that the reference signal $r_t$ is correlated to the disturbance signal $d_t$ and this will affect the level of reduction achievable. The reduction of the effect of $d_t$ on $y_t$ will be achieved by a tdf controller which

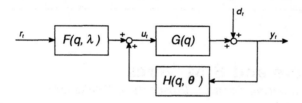

*Figure 6.1     Block diagram of the plant with output disturbance*

consists of a linear feedback and a linear feedforward part denoted by $H(q, \lambda)$ and $F(q, \theta)$, respectively. The controller can be adjusted by its parameter vectors $\theta$ for feedback and $\lambda$ for feedforward. The actuator signal is calculated by:

$$u_t = H(q, \theta)y_t + F(q, \lambda)r_t \tag{6.1}$$

The response of the output $y_t$ to the reference signal $r_t$ and the disturbance signal $d_t$ is:

$$y_t = \frac{1}{1 - G(q)H(q, \theta)}d_t + \frac{F(q, \lambda)G(q)}{1 - G(q)H(q, \theta)}r_t \tag{6.2}$$

where a positive feedback loop is used.

The output signal $y_t$ is measurable and recordable, the disturbance signal cannot be measured. The aim is to tune the controller parameters to minimise the square of the sum for $T$ recorded output signals:

$$J(\theta, \lambda, T) \stackrel{\text{def}}{=} \frac{1}{2}\sum_{t=1}^{T} y_t^2 \tag{6.3}$$

For the tuning of the feedback controller $H(q, \theta)$ the derivative of the cost function with respect to $\theta$ can be calculated as:

$$\frac{\partial J(\theta, \lambda, T)}{\partial \theta} = \sum_{t=1}^{T} y_t \frac{\partial y_t}{\partial \theta} \tag{6.4}$$

where the output sensitivity derivative is of the form:

$$\frac{\partial y_t}{\partial \theta} = \frac{G(q)}{(1 - G(q)H(q, \theta))^2} \frac{\partial H(q, \theta)}{\partial \theta} d_t + \frac{G^2(q)}{(1 - G(q)H(q, \theta))^2} \frac{\partial H(q, \theta)}{\partial \theta} r_t$$

$$= \frac{G(q)}{1 - G(q)H(q, \theta)} \frac{\partial H(q, \theta)}{\partial \theta} \left[ \frac{1}{1 - G(q)H(q, \theta)} d_t + \frac{F(q, \lambda)G(q)}{1 - G(q)H(q, \theta)} r_t \right] \tag{6.5}$$

Eqn. (6.5) can be combined with eqn. (6.2) to obtain:

$$\frac{\partial y_t}{\partial \theta} = \frac{G(q)}{1 - G(q)H(q, \theta)} \frac{\partial H(q, \theta)}{\partial \theta} y_t \tag{6.6}$$

The transfer function $G(q)$ is unknown. The fraction including the unknown transfer function can be eliminated by performing a second experiment. The input signal must be identical with the recorded output signal $y_t$ in the previous experiment. There should be no disturbance signal $d_t$ and no reference signal $r_t$ in this experiment. The system is presented in Figure 6.2, where the output

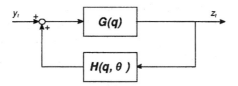

*Figure 6.2    Block diagram for the tuning of the feedback controller*

is denoted by $z_t$. The response of the output $z_t$ to $y_t$ as an input is given by:

$$z_t = \frac{G(q)}{1 - G(q)H(q, \theta)} y_t \tag{6.7}$$

Combining eqn. (6.6) and eqn. (6.7) leads to the output sensitivity derivative without $G(q)$:

$$\frac{\partial y_t}{\partial \theta} = \frac{\partial H(q,\theta)}{\partial \theta} z_t \tag{6.8}$$

With this result, eqn. (6.8), it is possible to calculate the gradient of the cost function $J$ with respect to $\theta$:

$$\frac{\partial J(\theta,\lambda,T)}{\partial \theta} = \sum_{t=1}^{T} y_t \frac{\partial H(q,\theta)}{\partial \theta} z_t \tag{6.9}$$

As shown in Reference [217] the plant itself can be used to generate a gradient estimate. The cost function should be minimised, therefore it is reasonable to change the parameters in the negative direction of the gradient of $J$:

$$\theta_k = \theta_{k-1} - \alpha \frac{\partial J(\theta,\lambda,T)}{\partial \theta} \tag{6.10}$$

Similar is the tuning for the feedforward coefficients. The derivative of the cost function with respect to $\lambda$ is:

$$\frac{\partial J(\theta,\lambda,T)}{\partial \lambda} = \sum_{t=1}^{T} y_t \frac{\partial y_t}{\partial \lambda} \tag{6.11}$$

where the output sensitivity derivative can be written in the form:

$$\frac{\partial y_t}{\partial \lambda} = \frac{G(q)}{1 - G(q)H(q,\theta)} \frac{\partial F(q,\lambda)}{\partial \lambda} r_t \tag{6.12}$$

The fraction including the unknown transfer function can be eliminated by performing another experiment. The input signal must be identical with the reference signal $r_t$ in the first experiment. There should be no disturbance signal $d_t$ and reference signal $r_t$. The block diagram for the experiment is given in Figure 6.3. The output is denoted by $w_t$ and calculated by:

$$w_t = \frac{G(q)}{1 - G(q)H(q,\theta)} r_t \tag{6.13}$$

Eqn. (6.13) can be combined with eqn. (6.12) to obtain:

$$\frac{\partial y_t}{\partial \lambda} = \frac{\partial F(q,\lambda)}{\partial \lambda} w_t \tag{6.14}$$

and therefore for the derivative of the cost function:

$$\frac{\partial J(\theta,\lambda,T)}{\partial \lambda} = \sum_{t=1}^{T} y_t \frac{\partial F(q,\lambda)}{\partial \lambda} w_t \tag{6.15}$$

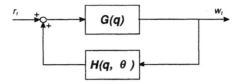

*Figure 6.3     Block diagram for feedforward tuning*

Again it is reasonable to change the parameters in the negative direction of the gradient of $J$:

$$\lambda_k = \lambda_{k-1} - \beta \frac{\partial J(\theta, \lambda, T)}{\partial \lambda} \tag{6.16}$$

**Remark:** The hybrid controller has advantages compared to the feedback controller. The feedforward controller can deal with noise outside the bandwidth of the feedback structure as long as a good correlation exists between the disturbance and the reference signal. The feedback controller provides a short impulse response to the system and enables the feedforward controller to react and converge faster. The feedback controller increases the stability margin of the feedforward system.

## 6.2   The online tuning scheme

The approach described in the previous section is an off-line tuning method for a controller which has fixed coefficients during the control phase. Its performance deteriorates when the dynamics of the plant are changing. To cope with this problem the scheme has to be modified to become a self-tuning controller scheme. In the case of periodic noise, the necessary signals for the calculation of the gradient of the cost function can be extracted out of recorded signals.

We consider the same system as in Figure 6.1. The output signal and the reference signal are recorded as in the off-line approach. This time it is not possible to switch off the disturbance signal and therefore it is not possible to perform experiments like the ones in Figure 6.2 and Figure 6.3, respectivly.

The same gradient of the cost function, eqn. (6.4), for the feedback controller has to be calculated, and the experiment described in Figure 6.4 performed.

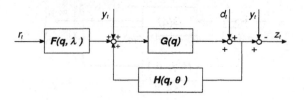

*Figure 6.4    Block diagram for online feedback tuning*

The output is given by:

$$z_t = \frac{G(q)}{1 - G(q)H(q,\theta)}y_t + x_t - y_t \qquad (6.17)$$

where

$$x_t = \frac{1}{1 - G(q)H(q,\theta)}d_t + \frac{F(q,\lambda)G(q)}{1 - G(q)H(q,\theta)}r_t \qquad (6.18)$$

Now we can take advantage of the periodic disturbance signal. The second recording has to start a multiple of disturbance signal periods after the first recording. This means that if one period of the disturbance signal consists of 200 points and three periods should be recorded, the second recording has to start either at sampling step 601, sampling step 801 or so on. Then the value and the gradient of the disturbance signal would be the same as in the first recording and the signals $x_t$ and $y_t$ would be identical. Eqn. (6.17) becomes:

$$z_t = \frac{G(q)}{1 - G(q)H(q,\theta)}y_t \qquad (6.19)$$

which is identical with eqn. (6.7). Now it is possible to calculate the gradient of the cost function, eqn. (6.4), and update the coefficients of the feedback controller with the gradient of $J$:

$$\theta_k = \theta_{k-1} - \alpha\frac{\partial J(\theta,\lambda,T)}{\partial \theta} \qquad (6.20)$$

Similar is the tuning of the feedforward controller. The experiment described in Figure 6.3 cannot be performed because it is not possible to switch off the disturbance signal. Another setup is needed to obtain the data vector $v_t$ to calculate the gradient of the cost function with respect to the feedforward coefficients. A possible way is given in Figure 6.5.

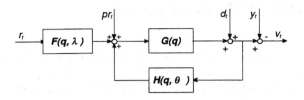

Figure 6.5   Block diagram for online feedforward tuning

The output can be calculated by:

$$v_t = \frac{G(q)}{1 - G(q)H(q,\theta)}pr_t + x_t - y_t \qquad (6.21)$$

where

$$x_t = \frac{1}{1 - G(q)H(q,\theta)}d_t + \frac{F(q,\lambda)G(q)}{1 - G(q)H(q,\theta)}r_t \qquad (6.22)$$

and $pr_t$ is the previously recorded reference signal.

The length of the second recording has to be a multiple of the disturbance signal period. The third recording has to start a multiple of disturbance signal periods after the first recording. Then the signals $x_t$ and $y_t$ are identical owing to the periodic disturbance signal and eqn. (6.21) becomes:

$$v_t = \frac{G(q)}{1 - G(q)H(q,\theta)}pr_t \qquad (6.23)$$

which is identical with eqn. (6.13).

With this vector it is possible to calculate the gradient of the cost function, eqn. (6.11), and update the coefficients of the feedforward controller with the gradient of $J$:

$$\lambda_k = \lambda_{k-1} - \beta\frac{\partial J(\theta,\lambda,T)}{\partial \lambda} \qquad (6.24)$$

With this newly introduced method it is possible to iteratively selftune the coefficients of a tdf controller, the only condition is that the disturbance signal has to be periodic. During one iterative step the plant dynamics $G(q)$ are not allowed to change, otherwise the derivative of the output signal cannot be replaced with the additionally recorded signals.

## 6.3   The online FSF tuning scheme

The previously described approach is a self-tuning scheme for a tdf controller. The controller coefficients are updated after a multiple of periods of the disturbance signal. Therefore the tuning time could be quite long in case of not matching disturbance frequencies. Consider for example disturbance frequencies of 100 Hz, 160 Hz and 250 Hz and a sampling frequency of 4000 Hz. One common period would have 400 points and to eliminate measurement noise at least three periods should be recorded giving a recording length of 1200 points. If there were controllers for each frequency individually the recording length would be 120, 75 and 48 points, respectively, which is in this case ten times faster. A further advantage is that the controller transfer functions are of lower order. A simple way of achieving this would be to split the single problem with multiple disturbance frequencies into a multiple problem with single disturbance frequencies using Butterworth bandpass (frequency selective) filters.

Our initial system is the same as described in Figure 6.1. The disturbance signal $d_t$ consists of different major harmonics and of minor white noise. The disturbance signal frequencies are considered to be known (by spectrum estimation). This single control problem is split up into $N$ control problems, where $N$ is the number of harmonics in the disturbance signal, a possible modification for linear systems. Frequency selective filters (FSF)are inserted into the control system in front of the feedforward and the feedback controller to achieve the separation. For each frequency a tdf controller is tuned. The new control system is given in Figure 6.6. The $N$ FSF split up the reference and the output signal into the different frequency bands; for each band there is a single tdf controller. Each of the controllers is separetly selftuned in parallel. This means that for one frequency band the output signal is recorded but for another the output signal is already used as an input signal. Therefore a possible scenario for tuning the feedback controller is given in Figure 6.7. It is the same situation for the tuning of the feedforward controller. For example, while the previous recorded reference signal of one frequency is used as an input the other frequencies are in the recording phase. This scenario is given in Figure 6.8. The tuning is analogous to the tuning of the unfiltered case, the gradients ($n$ is the number of the harmonic) of the cost functions become:

$$\frac{\partial J^n(\theta^n, \lambda^n, T)}{\partial \theta^n} = \sum_{t=1}^{T} y_t^n \frac{\partial H^n(q, \theta^n)}{\partial \theta^n} z_t^n \tag{6.25}$$

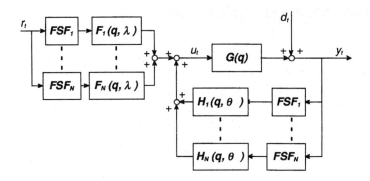

Figure 6.6    *Block diagram for self-tuning FSF system*

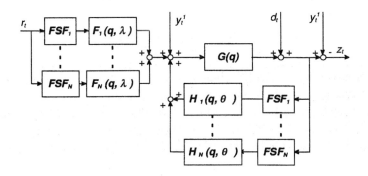

Figure 6.7    *Block diagram for feedback controller tuning first frequency*

and

$$\frac{\partial J^n(\theta^n, \lambda^n, T)}{\partial \lambda^n} = \sum_{t=1}^{T} y_t^n \frac{\partial F^n(q, \lambda^n)}{\partial \lambda^n} v_t^n \qquad (6.26)$$

The coefficients are updated as:

$$\theta_k^n = \theta_{k-1}^n - \alpha \frac{\partial J^n(\theta^n, \lambda^n, T)}{\partial \theta^n} \qquad (6.27)$$

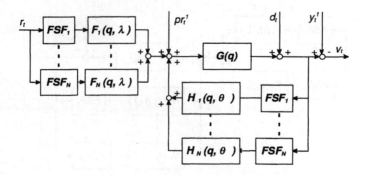

Figure 6.8    *Block diagram for feedforward controller tuning first frequency*

and

$$\lambda_k^n = \lambda_{k-1}^n - \beta \frac{\partial J^n(\theta^n, \lambda^n, T)}{\partial \lambda^n} \qquad (6.28)$$

A further advantage of the FSF-based iterative approach is that noise on the output signal does not influence the tuning of the controller parameters. The noise is filtered out by the filters and only signals in the frequency band tune the coefficients. If the band is small enough then only the disturbance harmonic passes and is used for the adjustment of the parameters.

## 6.4    Simulations

The previous sections introduced different iterative tuning approaches to periodic noise control problems. To illustrate the usefulness of the approaches they were tested in simulation using MATLAB. The results of SISO simulations are shown. Simulation will allow us to gain insight into the working mechanism of the design scheme. The following eighth-order transfer function with a delay of 14 sampling steps was identified to simulate a duct system shown in Figure 6.9.

$$G(q^{-1}) = q^{-14} \frac{0.417q^{-1} - 0.3989q^{-2} + 0.023q^{-3} + \cdots}{1 - 1.6675q^{-1} + 0.8358q^{-2} - 0.0182q^{-3} - \cdots}$$

*Figure 6.9    Photograph of the simulated duct*

$$\frac{+0.0232q^{-4} + 0.0201q^{-5} + 0.0178q^{-6} - 0.3608q^{-7} + 0.5728q^{-8}}{0.0165q^{-4} - 0.0133q^{-5} - 0.0076q^{-6} - 0.0003q^{-7} - 0.0525q^{-8}} \qquad (6.29)$$

The schematic of the sytem is given in Figure 6.10. The sampling frequency

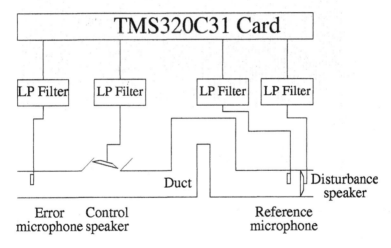

*Figure 6.10    Schematic of the duct system*

was set for all different approaches to 4 kHz, and the feedback controller was realised in a tenth-order FIR structure and the feedforward controller in a sixtieth-order FIR structure for the non FSF-based approaches. The FSF-based iterative controller had for each frequency an eleventh-order FIR feedforward controller structure and a third-order feedback controller structure. These simple controller structures were chosen to show that the performance for a more complex plant can be improved with the iterative tuning method. The disturbance signal was a mix of three sinewaves with frequencies of 100 Hz, 160 Hz and 250 Hz and a white noise signal $w_t$ with variance 0.01, leading to:

$$d_t = \frac{1}{3}(sin(2\pi 100t) + sin(2\pi 160t) + sin(2\pi 250t)) + w_t \qquad (6.30)$$

shown in Figure 6.11. The square cost function for 1200 samples was here $J = 200$. The reference signal $r_t$ was obtained from $d_t$ by:

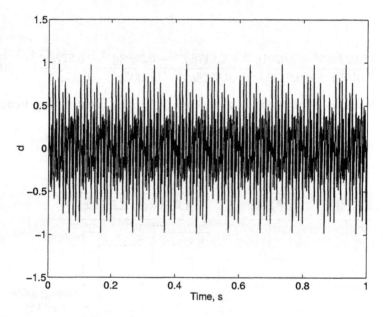

*Figure 6.11    Disturbance signal*

$$r_t = 0.4d_{t-10} + e_t \qquad (6.31)$$

where $e_t$ was a white noise with variance 0.001 .

The first approach to be examined is the off-line approach as it nicely shows the improvement of the controller during the iterative steps. The initial controller coefficients were set to zero. For each iterative step 1200 points of the reference and the output signal were recorded. Then the recorded output signal was used as an input signal to record another set of output data. Finally, the recorded reference signal was used as an input signal to record the last needed set of output data. With these recorded data it was possible to update the controller coefficients using an adaption gain of $\alpha = 0.0002$ and $\beta = 0.001$. The controller was tuned for ten seconds. In the curve of the cost function, which is shown in Figure 6.12, it can be seen that the initial improvement is relatively

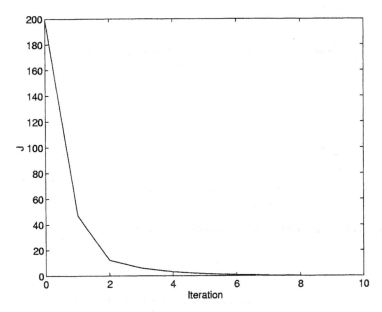

*Figure 6.12*    Cost function value of the output signal

large compared with the end of the tuning. This means that there is a plateau around the optimal controller parameters. If the controller parameters do not exactly match the optimal solution a very high reduction is still achieved. The final value of the cost function after ten seconds was $J = 0.0275$. This was a reduction of 39 dB compared to an uncontrolled disturbance signal. The controlled output is given in Figure 6.13. The same system, described by eqn. (6.29), was assumed for the self-tuning iterative controller design. The same disturbance signal, defined in eqn. (6.30), was exciting the output signal and a tdf controller was to be tuned. The initial controller coefficients were all set to zero. The self-tuning phase for five seconds is shown in Figure 6.14. It can be clearly seen that each iterative tuning step consists of three different stages. The initial recording of the output and reference signal was followed by using

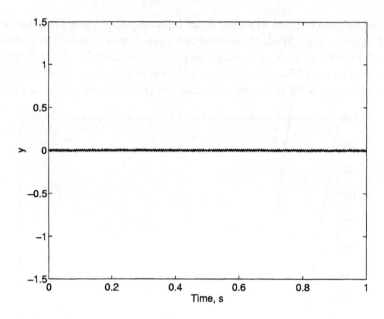

*Figure 6.13    Controlled output signal*

the recorded output signal as an input. The current output signal becomes larger as the output signal has a larger amplitude than the reference signal. The low-amplitude reference signal used as an input has not much influence on the output. It can also be seen that the output signals have initially a larger amplitude compared with the end phase of the tuning. This shows clearly that the noise is reduced with the readjusted controller.

The self-tuning phase was decided to be five seconds; the controlled output signal with the iteratively tuned controller is given in Figure 6.15. The value of the cost function for 1200 recorded samples came down to $J = 1.64$, a value that can also be found in Figure 6.12 at iteration step 5. The reduction was 21 dB.

One problem with this tuning method is the need to match as exactly as possible the starting points for the second and third recording, which have to be a multiple of the disturbance signal period. If this is not the case, for example owing to the sample time it is not possible to match this point, the recordings of the second and third experiment will not be the derivative of the output signal. However, simulations have shown that even if the starting point

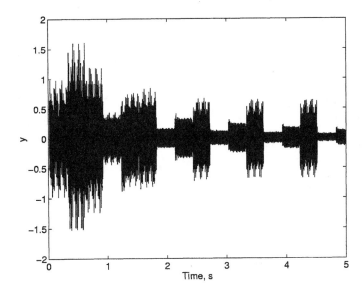

*Figure 6.14    Self-tuning phase*

is missed by five points it is still possible to tune the controller parameters. The adaption gain has to be lowered and a controller close to the optimal one would be found. The performance is still very good as in most cases there is a plateau in the cost function around the optimal controller.

The final case to look at for the system given in eqn. (6.29) was the FSF iterative tuning approach. The disturbance signal consisted of three different major harmonics and therefore there were three tdf controllers to be tuned. Each of these had the initial controller coefficients set to zero. Further, it was assumed that these frequencies were known. The output was split up into three different frequency-dependent signals using third-order Butterworth bandpass filters. The bandwidths of each of these filters were given by the disturbance frequency ±10 per cent which also eliminates unwanted white noise in the tuning. The self-tuning phase was decided to have a length of 2.5 seconds as it was assumed that the tuning in this approach is much faster. The step sizes were $\alpha = \frac{1}{300}$ for the feedback controller and $\beta = \frac{1}{30}$ for the feedforward controller. In Figure 6.16 the tuning phase is shown. It was not possible to detect the different stages of the tuning process in this case as the signals involved had smaller amplitudes. The output attenuation is

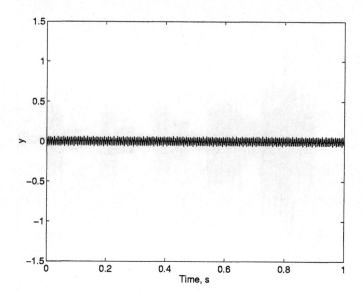

*Figure 6.15    Controlled output signal*

significantly improved during the first second. After 2.5 seconds the tuning was switched off leading to the controlled output signal given in Figure 6.17. The value of the cost function was $J = 0.1192$. The reduction at the output was 32 dB, a very good reduction considering that the tuning phase was shorter and the controllers had less coefficients than the controllers in the unfiltered case.

## 6.5  Conclusions

The iterative controller tuning approach was introduced for SISO systems, and tuning a controller efficiently with or without a permanent existing disturbance signal was demonstrated. How to speed up convergence against an optimal control solution by introducing frequency selective filtering into the control system was also shown. All presented methods were tested in simulation. The results show a very high noise reduction of more than 30 dB at the output with an easy tuning mechanism but without any knowledge of the plant behaviour.

As the basic scheme was outlined and illustrated in simulation there was no consideration of the robustness of the controller obtained. Although optimum

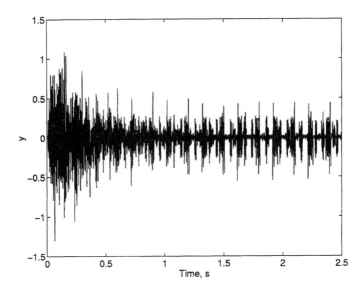

*Figure 6.16    Self-tuning phase*

performance is achievable for a plant with stationary dynamics, it is questionable whether this good performance would be retained under perturbation of the dynamics. Future research will need to aim at some kind of robustification of the controller obtained.

Application to practical problems is straightforward and tuning and testing of robustness can be based on a sequence of experiments. As many applications allow for a large number of experiments to be performed, and since there is no need for extensive modelling, the methodology is relatively simple and practicable.

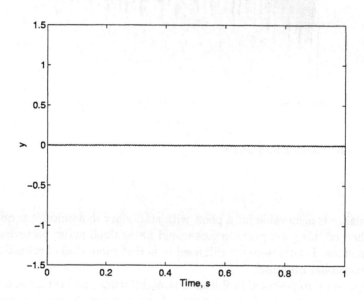

*Figure 6.17    Controlled output signal*

## Chapter 7

# Model-based control design for AVC

## S. M. Veres

*School of Engineering Sciences, University of Southampton*
*Southampton, UK*
*Email:s.m.veres@soton.ac.uk*

*A commonly used modern approach to the design of controllers is the $H_\infty$ design method; the purpose of this chapter is to present a tutorial on the method and how it can be used for the design of active vibration controllers. The essence of $H_\infty$ design is that it accounts for the dynamical uncertainty of the model used for control. As both physical and empirical models tend to be approximate descriptions of the dynamics of the plant, control design tends to be based on inaccurate models. To achieve good performance this implies that control design should account for possible uncertainties. Design for uncertainty can be done by suitable implementations of $H_\infty$ controllers. This chapter gives a tutorial introduction into $H_\infty$ control for active vibration control. The presentation is much simplified so that the reader can appreciate the approach without extensive reading of the literature.*

## 7.1 Introduction

The purpose of vibration control is to dampen the response of a flexible structure to external excitation. Examples of this are bridges, high-rise buildings, still tables, beams, robot arms, enclosures etc. In all cases there are the alternatives of passive or active control solutions.

Active control solutions will be defined as those which will require external energy to produce control signals within the controlled system. On the other hand, passive control systems do not need external energy to produce damping: examples are car suspension and resonators to counteract engine vibrational modes. Adding structural elements to a flexible structure in order to improve vibrational characteristics can also be considered as passive vibration control. It is clear, however, that passive techniques provide fewer degrees of freedom for control design than do active techniques. Also, they are more likely to increase the total weight of the equipment, especially below 1 kHz. They are, however, cheaper and more reliable in many industrial applications. Although in principle the ideas of this Chapter for vibration control can be made use of in the context of passive designs, the main area of their applicability will clearly lie in active control techniques with higher degrees of freedom. One would not, however, like to exclude the possibility of passive designs derived from ideas borne out of the investigation of active techniques.

In active vibration control by feedback, the crude starting point is the given maximum complexity of the controller that we can afford. This complexity will determine the maximum model orders which we can realistically handle in an online implementation. As the real plant order might be much higher, fitting a lower-order model is necessary which can then only be described by use of an uncertainty band of the frequency response. The objective of $H_\infty$ design for robust performance is to obtain a controller which performs well for all possible dynamics of the plant within the uncertainty band.

The first section (7.2) will outline the problem setup for one measured output and one control input, i.e. for the SISO case. For clear presentation of the principles the whole Chapter will only study the SISO case. Complete generalisation can, however, be made to the multicontrol multimeasurements case, although the discussion length would be much longer and more involved [114].

Section 7.3 derives a solution to the $H_\infty$ control problem by optimisation for tolerance of modelling inaccuracies under performance constraints. Section 7.4 describes the algorithmic steps of the design for robust performance. Section 7.5 reviews and gives references to the literature on empirical modelling of plant dynamics suitable for the design method described.

**Notations**

C    field of complex numbers

D    complex open right half-plane $D \stackrel{def}{=} \{s \in C \mid Real(s) > 0$

item[C]   complex open unit disk $C \overset{def}{=} \{s \in \mathbf{C} \mid |s| < 1$

$\|\cdot\|_\infty$   norm of a transfer function $G(s)$ is defined by $\|G\|_\infty \overset{def}{=} \sup_{s \in D} |G(s)|$

A   algebraic domain of proper and stable transfer function $G(s)$, i.e. $G(s)$ the ratio of real coefficient polynomials in $s$ so that the denominator polynomial is nonzero and stable (i.e. all of its zeros have negative real parts) and the order of the numerator polynomial is not greater than that of the denominator polynomial. The linear operator associated with a transfer function $G(s)$ will be denoted by $G_{op}$.

## 7.2   Problem description

A schematic description of the control system is displayed in Figure 7.1. Here

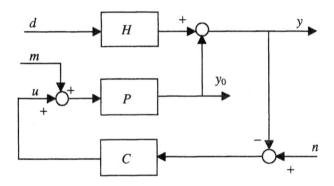

Figure 7.1    *Block diagram of a linear feedback system for disturbance cancellation*

$d$ is the external excitation and $y$ is the measured output signal, the vibration of which has to be dampened in some sense to be formulated later. The control input is denoted by $u$. $y_0$ is also available in some cases, for instance when the external excitation $d$ is 0, i.e. the plant is without external excitation. In either case it will be assumed that measurement noise $n$ and actuator noise $m$ are present to some degree. We will assume that the plant dynamics is basically linear and therefore its response to disturbance and control inputs can be describe via transfer functions by the equation:

$$y = Hd + P(m + u) = Hd + P(m + C(n - y)) \tag{7.1}$$

which then implies:

$$y = \frac{1}{1 + PC}\tilde{d} + \frac{P}{1 + PC}m + \frac{PC}{1 + PC}n, \quad \tilde{d} = Hd \tag{7.2}$$

As $d$ is rarely measurable directly, the point of view will be here that the spectrum of $d$ is known plus some approximate model $\hat{H}$ of the transfer function $H$ is available. Given a conservative bound $\Phi_d(\omega)$ of the spectrum of $d$, the spectrum of $\tilde{d}$ can be bounded by $\tilde{\Phi}_d(\omega) = |\hat{H}(j\omega)|^2 \Phi_d(\omega)$.

The objective function will measure the maximum of the harmonics of $y$. This will be named the $H_\infty$ objective. The notations:

$$S(s) \overset{def}{=} \frac{1}{1 + P(s)C(s)}, \quad R(s) = \frac{P(s)}{1 + P(s)C(s)}, \quad T(s) \overset{def}{=} \frac{P(s)C(s)}{1 + P(s)C(s)} \tag{7.3}$$

are for a nominal plant model $P(s)$ and designed controller $C(s)$. $S(s)$ is called the *sensitivity* function and $T(s)$ the *complementary sensitivity* function of the feedback loop.

The objective of limiting the harmonics of $y$ (due to $d$) below a given $\varepsilon > 0$ will be reformulated in the form:

$$\|SW_1\|_\infty \leq 1 \tag{7.4}$$

where $W_1$ is a weighting function such that:

$$|W_1(j\omega)|^{-1}|H(j\omega)|\Phi(\omega)_d^{1/2} \leq \varepsilon, \quad \omega \in [0, \infty) \tag{7.5}$$

The component of $y$ due to disturbance $d$ is $y_d = SHd$. The harmonic component of $y_d$ at frequency $\omega$ can be obtained by the application of a frequency selective filter $F^\omega(s)$ which is a narrowband filter with gain 1 at $\omega$ and well attenuating away from $\omega$.

By eqn. (7.5) $|H(j\omega)|\Phi^{1/2}(\omega) \leq \varepsilon|W_1(j\omega)|$ which can be multiplied by $|S(j\omega)|$ to obtain:

$$|S(j\omega)H(j\omega)\Phi^{1/2}| \leq \varepsilon|W_1(j\omega)S(j\omega)|.$$

By eqn. (7.4) the latter implies $|S(j\omega)H(j\omega)|\Phi^{1/2}(\omega) \leq \varepsilon$. This means that if $W_1(s)$ is chosen to satisfy eqn. (7.4) then for any frequency selective filter $F^\omega(s)$ centred to any $\omega \in [0,\infty)$ the amplitude of the harmonic content of $y_d$ at omega is:

$$\limsup_{t \to \infty} |F^\omega_{op} y_t| \leq \|F^\omega(j\omega)S(j\omega)H(j\omega)\|\Phi_d(\omega)^{1/2} \leq \varepsilon \qquad (7.6)$$

As $d$ is assumed to be the dominant disturbance, eqn. (7.4) will be our performance objective. The effects of $m, n$ will also be looked at later.

More importantly, we have to first look at the effect of possible mismodelling of $P$. Uncertainty description of $P$ will here be given in terms of relative accuracy bounds given at all frequencies. It will be assumed that the actual transfer function $P_0(j\omega)$ of the plant satisfies:

$$\left| \frac{P_0(j\omega)}{P(j\omega)} - 1 \right| \leq |W_2(j\omega)|, \ \forall \omega \qquad (7.7)$$

where $W_2$ is an uncertainty weighting function which is roughly increasing with frequency. $W_2$ is normally given by the designer. Eqn. (7.7) can be reformulated so that:

$$P_0 = (1 + \Delta W_2)P \qquad (7.8)$$

holds for some $\|\Delta\|_\infty \leq 1$.

A basic requirement of the functioning control system is its internal stability. Internal stability means that all signals within the system remain bounded if the inputs are bounded. The transfer matrix from $[d\ n]^T$ to $[y\ y_0]$ is:

$$\begin{bmatrix} y \\ y_0 \end{bmatrix} = \begin{bmatrix} \frac{H}{1+PC} & \frac{PC}{1+PC} \\ \frac{-HPC}{1+PC} & \frac{PC}{1+PC} \end{bmatrix} \begin{bmatrix} d \\ n \end{bmatrix} = \begin{bmatrix} HS & T \\ -HT & T \end{bmatrix} \begin{bmatrix} d \\ n \end{bmatrix} \qquad (7.9)$$

With stable $P$ the stability of $S$ is required for internal stability, which we will simply refer to as stability from now on. Stability for the nominal model $P$ does not, however, imply that for the real plant dynamics $P_0 = (1 + \Delta W_2)P$ the system is stable. An important result of robust control theory [69, 114] is the following lemma.

**Theorem 7.2.1** *$C$ provides robust stability for all plants $P_0 = (1 + \Delta W_2)P$ with $\|\Delta\|_\infty \leq 1$ if and only if $\|W_2T\|_\infty < 1$.*

The problem remains however that simultaneous satisfaction of performance $\|SW_1\|_\infty \leq 1$, and $\|W_2T\|_\infty < 1$ for stability robustness, are not sufficient for *robust performance* in the sense that the performance requirement is kept under model uncertainty. If $P$ is perturbed to $(1+\Delta W_2)P$ then $S$ is perturbed to:

$$\frac{1}{1+(1+\Delta W_2)PC} = \frac{S}{1+\Delta W_2 T} \tag{7.10}$$

then the robust performance condition is:

$$\|W_2 T\|_\infty < 1 \text{ and } \left\|\frac{W_1 S}{1+\Delta W_2 T}\right\|_\infty < 1, \ \forall\Delta : \|\Delta\|_\infty \le 1 \tag{7.11}$$

This condition is relatively easy to convert into a condition on a single $H_\infty$ norm.

**Theorem 7.2.2 ([69])** *Robust performance is satisfied if and only (7.11) if and only if*

$$\||W_1 S| + |W_2 T|\|_\infty < 1 \tag{7.12}$$

This test is important and can be nicely interpreted graphically. Let $L(j\omega) = PC(j\omega)$ denote the nominal loop gain of the control system at frequency $\omega$. Figure 7.11 shows circles drawn around the points $-1$ and $L(j\omega)$ with radius $|W_1(j\omega)|$ and $|W_2 L(j\omega)|$ for a given frequency $\omega \ge 0$. Note that the distance of $-1$ and $L(j\omega)$ is $|1 + L(j\omega)|$. If the circles do not intersect for any $\omega$ then:

$$|W_1(j\omega)| + |W_2 L(j\omega)| \le |1 + L(j\omega)| \tag{7.13}$$

for any $\omega$, which implies $\||W_1 S| + |W_2 T|\|_\infty < 1$.

Robust performance is now achieved if and only if these two circles remain disjoint for any $\omega \ge 0$. This is the condition for keeping the performance defined by $W_1$ for any perturbations of the dynamics as defined by $W_2$ via eqn. (7.8).

The criterion:

$$\gamma(C) = \||W_1 S| + |W_2 T|\|_\infty < 1$$

is difficult to handle in design and therefore the criterion:

$$\beta(C) = \||W_1 S|^2 + |W_2 T|^2\|_\infty^{1/2} < 1 \tag{7.14}$$

will be used. As the relation:

$$\gamma(C) \le \sqrt{(2)}\beta(C) \tag{7.15}$$

holds, minimisation of $\beta(C)$ results in good control design. Satisfaction of the inequality:

$$\||W_1 S|^2 + |W_2 T|^2\|_\infty < \frac{1}{2} \tag{7.16}$$

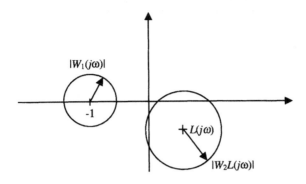

Figure 7.2    *Non-intersecting disks for robust performance condition*

will then give that:

$$\||W_1 S| + |W_2 T|\|_\infty < 1 \tag{7.17}$$

There are necessary conditions on $W_1$ and $W_2$ in order that the controller $C(s)$ satisfying eqn. (7.16) exists. Next, a procedure will be described to compute a $C(s)$ for proper transfer functions $W_1(s)$ and $W_2(s)$ under the conditions that:

$$\left\|\frac{|W_1|^2 |W_2|^2}{|W_1|^2 + |W_2|^2}\right\|_\infty < \frac{1}{2} \tag{7.18}$$

The necessary condition eqn. (7.18) means that:

(i) For a given performance requirement defined by $W_1$ the model has to be known with an accuracy defined by $W_2$ so that eqn. (7.18) must hold. For every $\omega$ eqn. (7.16) is equivalent to:

$$|W_2(\omega)| \leq \frac{|W_1(\omega)|}{\sqrt{2|W_1(\omega)|^2 - 1}} \quad \text{if } |W_1(\omega)|^2 > 1/2 \tag{7.19}$$

as the relevant case for disturbance attenuation is when $|W_1(\omega)| > 1 > 1/\sqrt{2}$ because one does not want to allow for amplification of disturbance harmonics of $d$ in $y$.

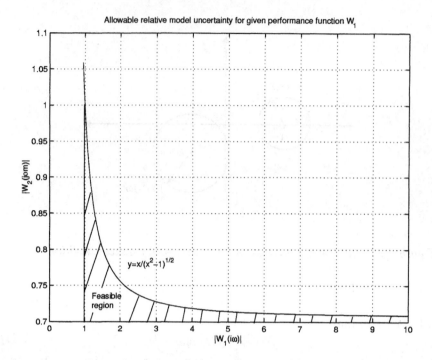

Figure 7.3    *For each frequency the pair of $|W_1(j\omega)|$, $|W_2(j\omega)|$ must be in the feasible region in order for robust performance to be achievable with a proper controller $C(s)$. Note that there is an additional condition for solvability which is Step 7 of the design procedure in the next section*

(ii) Figure 7.3 shows the feasible gains of attenuation $|W_1(j\omega)|$ for a given relative uncertainty of the model, for which a controller $C(s)$ with robust performance as defined by eqn. (7.11) may exist.

While one is focusing on optimising the attenuation of $d$ on output $y$, the transfer from $n$ to $y$ will be $T$. Based on that $\beta(C)$ keeps $W_2T(j\omega)$ low, the attenuation of $n$ will be inversely proportional to $|W_2(j\omega)|$. This is favourable in practice as $n$ tends to be high frequency measurements noise. Similarly, $m$ will be attenuated proportionally with $|P(j\omega)|/|W_1(j\omega)|$. For this not to be too high at high frequencies $W_1(s)$ must be defined so that it has a relative degree not greater than the plant. If the plant relative degree is not known then it is safe to set the relative degree of $W_1(s)$ to 0.

## 7.3 $H_\infty$ controller optimisation under model uncertainty

The design steps are relatively simple and in this section they will be listed and further references provided for reading. The objective is to compute a controller $C(s)$ which satisfies the design criterion $\beta(C) < 1$ for a pair of $W_1$, $W_2$ weighting functions given to define the performance requirement and bound of uncertain dynamics, respectively.

Step 1 Compute:
$$V(s) \overset{def}{=} \frac{W_1(s)W_1(-s)W_2(s)W_2(-s)}{W_1(s)W_1(-s) + W_2(s)W_2(-s)} \qquad (7.20)$$

If $\|V\|_\infty < 1/2$ then the problem may be solvable and proceed with the next step. If $\|V\|_\infty \geq 1/2$ then the problem is not solvable and either $W_1$ and $W_2$ will have to be similarly reduced or redesigned, depending on the practical assessment and the results of empirical system identification as it is described a few sections later.

Step 2 A coprime factorisation is computed for the plant model in the form:
$$P = \frac{N}{M}, \quad NX + MY = 1, \quad N, M \in \mathbf{A} \qquad (7.21)$$

Step 3 The following auxiliary rational functions are defined and computed:
$$\begin{aligned} R_1 &\overset{def}{=} W_1MY \quad S_1 \overset{def}{=} W_2NX \\ R_2 &\overset{def}{=} W_1MN \quad S_2 \overset{def}{=} -W_2MN \end{aligned} \qquad (7.22)$$

Step 4 By spectral factorisation of the rational functions $V_1(s), \in \mathbf{A}$ and $V_2(s), \in \mathbf{A}$ are computed so that:
$$\begin{aligned} V_2(s)V_2(-s) &= R_2(s)R_2(-s) + S_2(s)S_2(-s) \\ V_1(s)V_1(-s) &= R_1(s)R_1(-s) + S_1(s)S_1(-s) - V(s) \end{aligned} \qquad (7.23)$$

Step 5 A spectral factor $F(s) \in \mathbf{A}$ is obtained from:
$$F(s)F(-s) = \frac{1}{2} - V(s) \qquad (7.24)$$

Step 6 Define and compute $T_1(s) \overset{def}{=} F^{-1}(s)V_1(s) \in \mathbf{A}$ and $T_2(s) \overset{def}{=} F^{-1}(s)V_2(s) \in \mathbf{A}$

Step 7 Compute $\zeta_{opt}$, the minimal matching error for the minimisation of

$$\zeta_{opt} \overset{def}{=} \min_{Q \in \mathbf{A}} \|T_1 - T_2 Q\|_\infty \qquad (7.25)$$

If $\zeta_{opt} < 1$ then compute the minimising $Q(s)$ in eqn. (7.25). If $\zeta_{opt} \geq 1$ then the robust performance problem is not solvable. Under reliable relative error bound $|W_2(j\omega)|$ a new and milder $W_1(s)$ has to be defined or the plant is to be remodelled with a different order of model and uncertainty bounds. Then the design algorithm has to be repeated from Step 1. This gives the design an iterative character.

Step 8 If $Q(s)$ is not proper then roll it off by multiplying it with a suitable:

$$R_{lp}(s) = (1 + \tau s)^{-k}, \quad \tau > 0, \quad k \geq 0 \qquad (7.26)$$

with sufficiently small $\tau > 0$.

Step 9 Finally compute the controller as:

$$C(s) \overset{def}{=} \frac{X + M Q R_{lp}}{Y - N Q R_{lp}} \qquad (7.27)$$

Note that the above design algorithm provides solutions for continuous transfer functions. In practice, one frequently obtains discrete-time models, i.e. model-fitted periodically sampled input-output data. These models are described by discrete-time transfer functions which are rational functions of the open complex unit disk:

$$\mathbf{C} \overset{def}{=} \{z \mid |z| < 1 \, , z \in \mathbf{C}\}$$

For instance, if $\hat{P}(z)$ is an estimated discrete-time model of a plant then the $H_\infty$ design procedure can be transformed into an $H_\infty$ problem in continuous time by using the conformal mapping:

$$z \mapsto \frac{1 - s}{1 + s} \qquad (7.28)$$

which maps the open unit disk onto the open right half plane:

$$\mathbf{D} \overset{def}{=} \{s \mid Re(s) > 0, \, s \in \mathbf{C}\}$$

The conformal mapping has an algebraically identical inverse:

$$s \mapsto \frac{1 - z}{1 + z} \qquad (7.29)$$

and therefore it is one-to-one between C and D. Hence, a discrete-time control design problem can be converted into a continuous-time one and vice versa. For a given $\hat{P}(z)$ the continuous-time design problem is solved for $\hat{P}((1-s)/(1+s))$. The solution can be converted back to the discrete-time form using the inverse mapping of eqn. (7.29). Also, note that the application of this conformal mapping is not for the purpose of obtaining equivalent discrete and continuous-time models for periodic sampling for which, for instance, Tustin's formula can be used. There is no sampling assumed here, the aim is only to find the continuous-time $H_\infty$ problem formulation which is equivalent to the discrete-time problem. This can be done without regard to the sampling period.

Finally, note that direct implementation of the above design procedure can lead to numerically ill-conditioned computations. For more complex plant dynamics state-space implementation is needed as described and coded in the *Robust Control Toolbox* [193].

## 7.4  Examples

### 7.4.1  Glass plate attenuation

Consider the glass vibration control problem as schematically described in Figure 7.4. Assume that the response of the glass plate at low frequencies

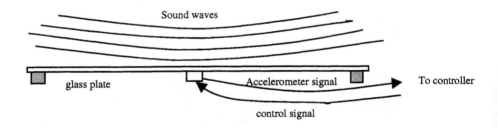

Figure 7.4    *Generic schematic diagram of glass vibration control system when the disturbance is sound waves*

has been measured as having one resonance frequency. A frequency-response model has been produced with uncertainty bands as indicated in Figure 7.5 where the frequency is in $rad/s$. An approximate transfer function model can be measured when the disturbance is turned off. The transfer function model will be here:

$$P(s) = \frac{64480}{193440 + 40s + s^2} \tag{7.30}$$

with associated relative uncertainty bound defined by:

$$|W_2(j\omega)| = \left| \frac{1}{600 + j\omega} \right| \tag{7.31}$$

The disturbance of the sound waves is assumed to have a profile as shown in

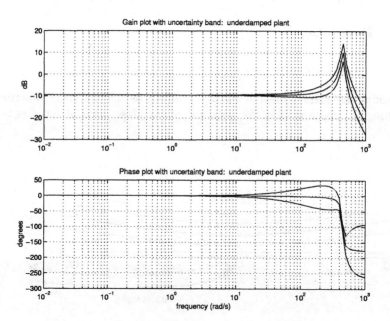

Figure 7.5    *Measured response of the glass plate to the actuator is within the band indicated. The tope line is for $|P(j\omega)|(1 + |W_2(j\omega)|)$, the middle line for $|P(j\omega)|$ and the bottom line for $|P(j\omega)|(1 - |W_2(j\omega)|)$. The resonant peak is around $430 rad/s \sim 72Hz$*

Figure 7.14. This profile was here defined by:

$$W_1(s) = \frac{290170}{48360 + 90s + s^2} \tag{7.32}$$

to provide an approximate envelope and it is shown in the top part of Figure 7.14. The bottom diagram shows the envelope function $W_2(s)$ of the uncertainty band in Figure 7.6. The design of a feedback controller $C(s)$ was carried

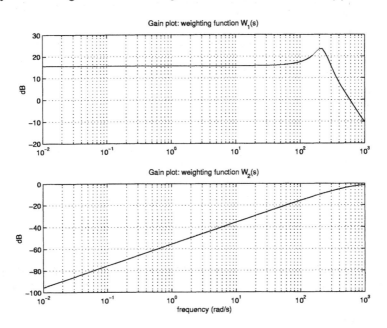

*Figure 7.6   The profile $|W_1(j\omega)|$ of the soundwave disturbances and the model uncertainty bound $|W_2(j\omega)|$*

out using the design procedure described in the previous section with final low-pass filtering by $(1/(1+0.001s)^3$. The controller has the Bode plot shown in Figure 7.7. The attenuated disturbances can be described by $|W_1(j\omega)S(j\omega)|$ and robust performance is kept as long as the absolute error of modelling is smaller then $|W_2(j\omega)T(j\omega)|$ as shown in Figure 7.8. The attenuation level of the disturbance is described by $|S(j\omega)|$ as shown in the top graph of Figure 7.9. The bottom graph indicates that even the actuator noise will be attenuated to a minor degree by $|T(j\omega)|$. As $\|V\|_\infty = 0.3827$ is near to $1/2$ the

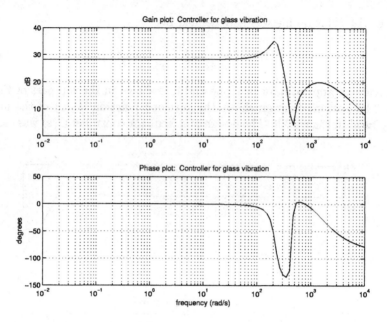

*Figure 7.7    Bode plots of $C(s)$ for glass vibration control*

design computed is nearly the best achievable for robust performance given the relative uncertainty level by $W_2$ and performance requirement by $W_1$.

### 7.4.2   Active mounts

An important application is the vibration isolation of the base plate of an instrument from bench/ground vibrations as illustrated in Figure 7.10. The response of the base plate to the actuator is assumed to be measurable and is shown with uncertainty bands in Figure 7.11. This transfer function model will be here:

$$P(s) = \frac{645}{1934 + 4.4s + s^2} \tag{7.33}$$

with relative uncertainty bound defined by:

$$|W_2(j\omega)| = \left| \frac{0.1}{60 + j\omega} \right| \tag{7.34}$$

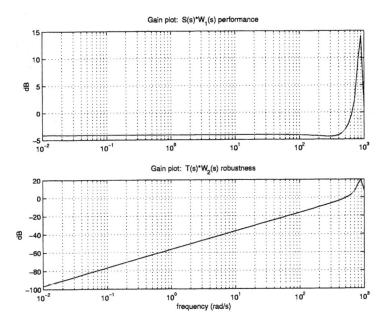

*Figure 7.8*    *Profile of the attenuated disturbance and the allowable maximum absolute modelling error to retain performance for the glassplate attenuation problem*

The usefulness of the design approach can be appreciated by pointing out that the choice of $W_1(s) = H(s)$ as the transfer function from the foundation to the instrument base plate leads to a design of an active mount. This can be used to insulate the instrument from vibration of the bench top. The response $H(s)$ of the base plate to excitation from the bench has been measured and is shown in Figure 7.11. This approximate transfer function model is measured when there is no feedback and in our case it is chosen as:

$$H(s) = \frac{386.9}{1934.4 + 17.6s + s^2} \qquad (7.35)$$

The design of a feedback controller $C(s)$ was carried out as previously with final lowpass filtering by $(1/(1+0.001s))^3$. Bode diagrams of the controller are shown in Figure 7.12.

The attenuated disturbances can be described by $|H(j\omega)S(j\omega)|$ and robust performance is kept as long as the absolute error of modelling is smaller then

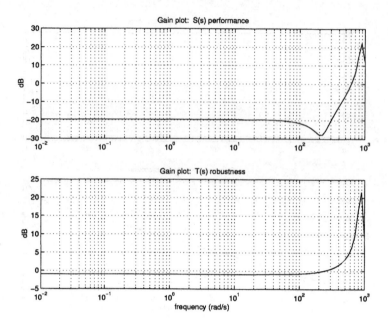

Figure 7.9    *Attenuation of the disturbances by $|S(j\omega)|$ and attenuation of actuator noise by $|T(j\omega)|$ for the glass plate*

$|W_2(j\omega)T(j\omega)|$, as shown in Figure 7.13.

### 7.4.3    Instrument base plate/enclosure attenuation

It sometimes happens that the machine generates vibration itself and one wants to reduce the vibration of the base plate and the attached enclosure. This is realistic when the machine and the actuator rest on well attenuating mounts and there is little side effect of active control. The disturbance exerted on the plant from parts inside the instrument is assumed to have a profile as shown in Figure 7.14. This profile was here defined by:

$$W_1(s) = \frac{46427}{7738 + 35s + s^2} \tag{7.36}$$

The transfer function model will be as in the previous example:

$$P(s) = \frac{645}{1934 + 4.4s + s^2} \tag{7.37}$$

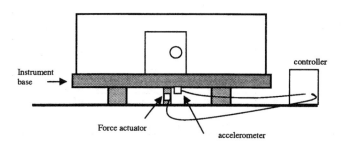

Figure 7.10 Generic schematic diagram of instrument vibration control system used as an active mount

with relative uncertainty bound defined by:

$$|W_2(j\omega)| = \left|\frac{0.1}{60 + j\omega}\right|.$$ (7.38)

The bottom plot shows $W_2(s)$ of the uncertainty band in Figure 7.14. The design of a feedback controller $C(s)$ was carried out with final lowpass filtering by $(1/(1 + 0.001s))^3$. The controller is shown in Figure 7.15. The attenuated disturbances can be described by $|W_1(j\omega)S(j\omega)|$ and robust performance is kept as long as the absolute error of modelling is smaller then $|W_2(j\omega)T(j\omega)|$, as shown in Figure 7.16. The attenuation level of the disturbance is described by $|S(j\omega)|$ as shown in the top graph of Figure 7.17. The bottom graph indicates that even the actuator noise will be attenuated to a minor degree by $|T(j\omega)|$.

## 7.5 Identification of empirical models for control

System identification has by now an extensive literature and in this section we outline the main points and give references.

There are methods of estimation:

1. Discrete-time models to periodically sampled input-output data:

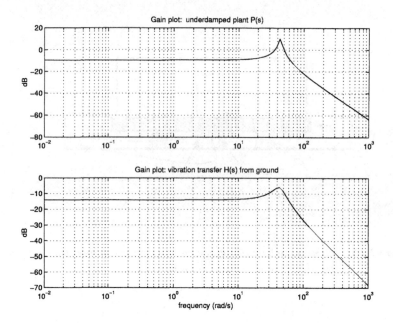

Figure 7.11    *Measured response of the base plate of the instrument to the actuator and to ground vibrations*

(1.1) by parameter estimation of transfer function models

(1.2) by fitting parametric models to frequency response measurements

2. Continuous-time models:

(2.1) by parameter estimation of transfer function models

(2.2) by fitting parametric models to frequency response measurements

3. To estimate models in alternating iterations of modelling and control redesign based on repeated experiments:

(3.1) via a stochastic modelling framework

(3.2) via a model unfalsification framework

A good introduction into the approach of (1.1) can be found in References [224] and [177]. A recent survey of (1.2)and (2.2) is provided in the Reference [195]. $H_\infty$ identification for (2.2) has been well investigated and is

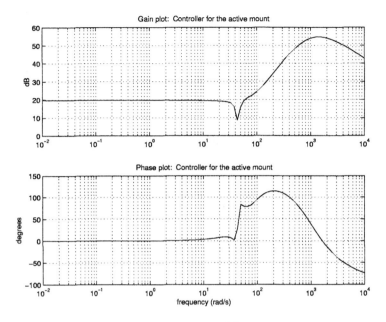

*Figure 7.12    Bode plots of the controller $C(s)$ for the active mount*

summarised in Reference [323] and in papers [131], [347] and [223]. Identification of continuous-time parametric models is surveyed in References [317] and [318].

During the last decade it has been recognised that modelling for control can only be done iteratively as the optimal model depends on some knowledge of the optimal controller [258, 270]. In some extreme cases it can be shown that an open-loop model, which is very good for prediction, can be completely wrong for feedback control design, giving even an unstable closed-loop system. On the other hand a relatively bad open-loop model can be good for control [270].

The recently developed methodology of iterative identification and control redesign has therefore a great potential for active vibration control applications and is relatively unexploited so far. The Reference [323] gives a summary of iterative approaches and related papers are References [12],[28],[31],[270],[271] and [248].

A novel approach based on iterative model unfalsification and control re-

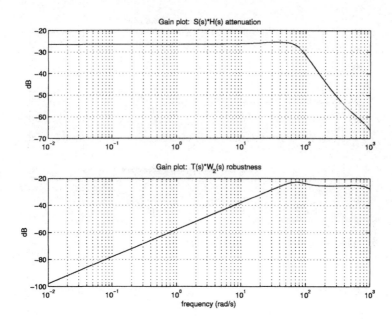

*Figure 7.13    Frequency profile of the attenuated disturbance and the allowable maximum absolute modelling error to retain performance for the active mount*

design is initiated in the References [322] and [341] and in the Reference [323] and is largely unexploited in the area of active vibration control, although the expected benefits in performance and reliability are considerable in this approach. The main point of an unfalsification-based approach is that it retains most benefits and improves upon all other iterative approaches.

## 7.6    Conclusions

This Chapter gives a gentle introduction to model based robust control design via the $H_\infty$ approach. Robust performance is defined as a performance which satisfies the requirements within the model uncertainty. The main points are:

(i) Model uncertainty limits the robust performance achievable.

(ii) A complete method is presented for the computation of a robust controller under given model uncertainty defined by a weighted $H_\infty$-norm.

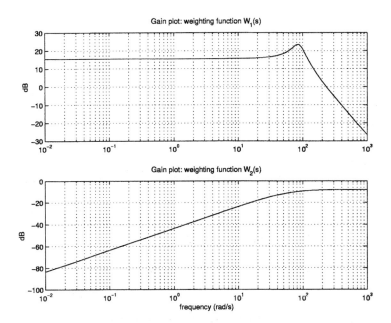

Figure 7.14    *The profile* $|W_1(j\omega)|$ *of the disturbance and the model uncertainty bound* $W_2(j\omega)$

For complicated cases state-space versions of computation are necessary to keep numerical stability.

Finally, it is pointed out that iterations of remodelling and control redesign are necessary in order to achieve the potentially best solutions. Computing models with uncertainty bounds are not discussed here, a comprehensive list of references is provided instead.

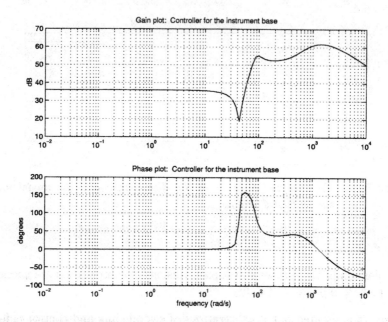

Figure 7.15    *Bode plots of the controller $C(s)$ for instrument vibration*

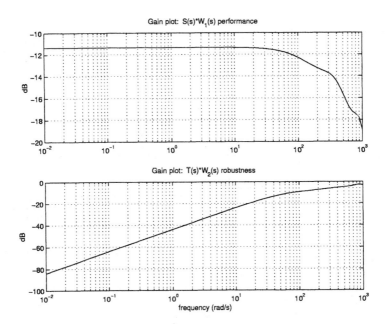

Figure 7.16    *Profile of the attenuated disturbance and the allowable maximum absolute modelling error to retain performance for base plate/enclosure attenuation*

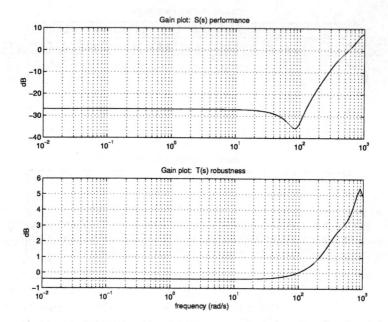

Figure 7.17    *The attenuation of the instrument base plate vibration by $|S(j\omega)|$ and the attenuation of actuator noise by $|T(j\omega)|$*

*Chapter 8*

# Active control of noise and vibration using neural networks

## M. O. Tokhi and R. Wood

*Department of Automatic Control and Systems Engineering*
*The University of Sheffield, Sheffield, UK, Email: o.tokhi@sheffield.ac.uk*

*Neuro-adaptive active control schemes for noise cancellation and vibration sup-*
*pression are developed and presented in this chapter. Multilayered perceptron*
*and radial basis function neural networks are considered in both the modelling*
*and control contexts. A feedforward ANC structure is considered for optimum*
*cancellation of broadband noise in a three-dimensional propagation medium.*
*Online adaptation and training mechanisms allowing the neural network ar-*
*chitecture to characterise the optimal controller within the ANC system are de-*
*veloped. The neuro-adaptive ANC algorithms thus developed are implemented*
*within a free-field noise environment and simulation results verifying their per-*
*formances in the cancellation of broadband noise are presented and discussed.*

## 8.1 Introduction

Active control is realised by detecting and processing the noise/vibration (disturbances) by a suitable electronic controller so that, when superimposed on
the disturbances, cancellation occurs. Owing to the broadband nature of
these disturbances, it is required that the control mechanism realises suitable
frequency-dependent characteristics so that cancellation over a broad range of
frequencies is achieved [172, 303]. In practice, the spectral contents of these

disturbances as well as the characteristics of system components are in general subject to variation, giving rise to time-varying phenomena. This implies that the control mechanism is further required to be intelligent enough to track these variations, so that the desired level of performance is achieved and maintained.

Through his experiments on reducing transformer noise, Conover realised the need for a black box controller, which would adjust the cancelling signal in accordance with information gathered at a remote distance from the transformer [58]. Later it has been realised by numerous researchers that an essential requirement for an active noise/vibration control system to be practically successful is to have an adaptive capability. Implementing an adaptive control algorithm within an active control system will allow the controller characteristics to be adjusted in accordance with the changes in the system.

Much of the work reported on ANC has centred on conventional adaptive controllers [82, 87, 100, 121, 250, 303]. These do not, however, have the long-term memory of intelligent systems to remember the optimal control parameters corresponding to different plant configurations. Intelligent controllers have the ability to sense environment changes and execute the control actions required. Neural networks, on the other hand, have had great success in areas such as speech, natural language processing, pattern recognition and system modelling [10, 165, 216, 261]. Not much work has, however, been reported in the area of intelligent neural network control applied to ANC systems [308, 309]. It is attempted in this chapter to develop intelligent ANC systems incorporating neural networks.

Various types of neural network architecture have been represented in the literature. Among these the radial basis function (RBF) and the multilayered perceptron (MLP) neural networks have commonly been used in modelling and control of dynamic systems. The RBF and MLP networks are similar in some respects [38]. The RBF networks, however, can be considered to be superior in many ways to the MLP type. The MLP models are highly nonlinear in the unknown parameters and parameter estimation must be based on nonlinear optimisation techniques, which require intensive computation and which can result in problems associated with local minima. The RBF models do not suffer from such problems. One of the main advantages of the RBF networks is that they are linear in the unknown parameters and consequently global minimum in the error surface is achieved. Such capabilities of neural networks are utilised in the development of several neuro-adaptive active control algorithms in this chapter, using both the MLP and RBF networks.

## 8.2 Neural networks

This section presents a brief description of the MLP and RBF neural networks with their training algorithms, as considered throughout this chapter.

### 8.2.1 Neural network models

Neural network models can be expressed in the general form of

$$y(t) = f\left(y(t-1), \ldots, y(t-n_y), u(t), \ldots, u(t-n_u), \varepsilon(t-1), \ldots, \varepsilon(t-n_\varepsilon)\right) + \varepsilon(t) \tag{8.1}$$

where $f(\cdot)$ is a nonlinear function, $y(t)$ is the output, $u(t)$ is the input, $\varepsilon(t)$ is the residual and $n_y$ and $n_u$ are the maximum lags in the output and input, respectively. The above is known as the nonlinear autoregressive moving average with exogenous inputs (NARMAX) model [173]. The NARMAX model can also be expressed by the regression equation:

$$y(t) = \sum_{i=1}^{M} p_i(t)\theta_i + \varepsilon(t) \quad , \quad t = 1, \ldots, N \tag{8.2}$$

where $p_i(t)$ are the regressors, $y(t)$ represents the dependent variables, $\varepsilon(t)$ is some modelling error which is uncorrelated with the regressors, $\theta_i$ represent the unknown parameters to be estimated, $M$ represents the number of parameters of the regressors and $N$ is the data length. An MLP or RBF network can be represented by using the extended model set representation in eqn. (8.2). The input to the neural network is thus given by:

$$X(t) = [y(t-1), \ldots, y(t-n_y); u(t), u(t-1), \ldots, u(t-n_u)]^T \tag{8.3}$$

Using the neural network it is possible to realise or approximate the underlying dynamics of the nonlinear function $f(\cdot)$ in eqn. (8.1).

### 8.2.2 Multilayered perceptron networks

An MLP neural network is made up of layers of neurons between the input and output layers. The output of each neuron can be expressed as:

$$x_i^l(t) = F\left(\sum_{j=1}^{n_l-1} w_{ij}^l x_j^{l-1}(t) + b_i^l\right)$$

where, $w_{ij}^l$ is the weight connection between the $j$th neuron of the $(l-1)$th layer and the $i$th neuron of the $l$th layer, $b_i^l$ is the threshold of the neuron and $n_{l-1}$ is the number of neurons of the $(l-1)$th layer and $F(\cdot)$ is the nonlinear transformation or activation function. The functionality of the network is determined by specifying the strength of the connection paths (weights) and the threshold parameter of each neuron. The input layer usually acts as an input data holder and distributes inputs to the first hidden layer. The inputs then propagate forward through the network and each neuron computes its output according to the learning rule chosen.

The backpropagation training algorithm is commonly used to adapt the connection weights of an MLP network. A set of training data inputs is given to the network which propagate through the neuron layers to give an output prediction. The error between this prediction and the required output is used to adjust the gradient information, which is backpropagated through the network to change the weights connecting the neurons. The backpropagation training algorithm is a gradient search (steepest descent) method which adjusts the weights so that application of a set of inputs produces the desired outputs. It is also dependent on the user selectable parameters to the extent that if an inappropriate combination of learning rates and momentum constants are chosen the algorithm performs badly. However, with the use of better initial conditions and the adaptive learning rate epoch the run time can be reduced drastically.

An advanced backpropagation algorithm is utilised here. The algorithm uses a better initialisation of the weights and biases, which drastically reduces the training time [222]. Moreover, an adaptive learning rate is employed which helps the network avoid local error minima. As a result of using both of these methods together the training time is reduced by a factor of over 60. The $k$th correction of the weights and biases in this algorithm are described as:

$$\Delta w(k) = -\eta_w \frac{\partial E}{\partial w}(k-1) + \alpha_w \Delta w(k-1)$$

$$\Delta b(k) = -\eta_b \frac{\partial E}{\partial b}(k-1) + \alpha_b \Delta b(k-1)$$

(8.4)

where $E$ is the total error between the required output (training data) and the actual output and $\eta$ and $\alpha$ are the learning and momentum constants. The first term in eqn. (8.4) indicates that the weights and biases are corrected in proportion to the gradients $\partial E/\partial w$ and $\partial E/\partial b$, respectively. The second term is added to accelerate the training.

### 8.2.3 Radial basis function networks

An RBF expansion with $n$ inputs and a scalar output implements a mapping according to:

$$y(t) = w_o + \sum_{i=1}^{n_r} w_i G\left(\|\mathbf{x} - c_i\|\right)$$

where $\mathbf{x}$ is the $n$-dimensional input vector, $G(\cdot)$ is the basis function, $\| \cdot \|$ denotes the Euclidean norm, $w_i$ represent the weights, $c_i$ represent the centres and $n_r$ represents the number of centres. The above mapping can be implemented in a two-layered neural network structure where, given fixed centres, the first layer performs a fixed nonlinear transformation, which maps the input space onto a new space. The output layer implements a linear combiner on this new space. Thus, the RBF expansion can be viewed as a two-layered neural network, which has the important property that it is linear in the unknown parameters. Therefore, the problem of determining the parameter values is reduced to one of a linear least-squares optimisation.

Since RBF expansions are linearly dependent on the weights, a globally optimum least-squares interpolation of nonlinear maps can be achieved. However, it is important to emphasise that the RBF performance is critically dependent upon the given centres. Although it is known that the fixed centres should suitably sample the input domain, most published results simply assume that some mechanism exists to select centres from data points and do not offer any real means for choosing the centres. The orthogonal forward regression (OFR) algorithm [158] has been proposed as the mechanism to select the centres and, using the idea of generalisation [38] coupled with the model validity techniques, it has been shown that a parsimonious RBF model can be obtained. The OFR algorithm is used in this chapter to train the RBF network and also select the appropriate centres. This is briefly described below.

Eqn. (8.2) can be written as:

$$\mathbf{z} = \mathbf{P}\Theta + \Xi \qquad (8.5)$$

where

$$\mathbf{z} = \begin{bmatrix} z(1) \\ \vdots \\ z(N) \end{bmatrix}, \quad \mathbf{P} = \begin{bmatrix} \mathbf{p}_1 & \cdots & \mathbf{p}_M \end{bmatrix}, \quad \Theta = \begin{bmatrix} \theta_1 \\ \vdots \\ \theta_M \end{bmatrix}, \quad \Xi = \begin{bmatrix} \xi(1) \\ \vdots \\ \xi(N) \end{bmatrix}$$

$$\mathbf{p}_j = \begin{bmatrix} p_j(1) & \cdots & p_j(N) \end{bmatrix}^T \qquad j = 1, \ldots, M$$

An orthogonal decomposition of $\mathbf{P}$ can be obtained as:

$$\mathbf{P} = \mathbf{W}\mathbf{A} \qquad (8.6)$$

where $\mathbf{A} = \{\alpha_{ij}\}$; $i = 1, \ldots, M - 1$, $j = 2, \ldots, M$, is an $M \times M$ unit upper triangular matrix and $\mathbf{W} = \{\mathbf{w}_j\}$; $j = 1, \ldots, M$, is an $N \times M$ matrix with orthogonal columns, which satisfy $\mathbf{W}^T\mathbf{W} = \mathbf{D}$ and $\mathbf{D}$ is a positive diagonal matrix:

$$\mathbf{D} = \operatorname{diag}\{d_j\}, \quad d_j = \langle \mathbf{w}_j, \mathbf{w}_j \rangle \quad , \quad j = 1, \ldots, M$$

where $\langle \cdot, \cdot \rangle$ denotes the inner product, that is:

$$\langle \mathbf{w}_i, \mathbf{w}_j \rangle = \mathbf{w}_i^T \mathbf{w}_j = \sum_{t=1}^{N} w_i(t) w_j(t)$$

Eqn. (8.5) can be rearranged as:

$$\mathbf{z} = \left(\mathbf{P}\mathbf{A}^{-1}\right)(\mathbf{A}\Theta) + \Xi = \mathbf{W}\mathbf{g} + \Xi$$

where $\mathbf{A}\Theta = \mathbf{g}$.

Since $\xi(t)$ is uncorrelated with the regressors:

$$\mathbf{g} = \mathbf{D}^{-1}\mathbf{W}^T\mathbf{z}, \quad g_j = \frac{\langle \mathbf{w}_j, \mathbf{z} \rangle}{\langle \mathbf{w}_j, \mathbf{w}_j \rangle} \quad , \quad j = 1, \ldots, M$$

The number of all the candidate regressors can be very large even though adequate modelling may only require $M_s$ ($\ll M$) significant regressors. The OFR procedure identifies the significant regressors.

The sum of squares of the dependent variable is:

$$\langle \mathbf{z}, \mathbf{z} \rangle = \sum_{i=1}^{M} g_i^2 \langle \mathbf{w}_i, \mathbf{w}_i \rangle + \langle \Xi, \Xi \rangle$$

The error reduction ratio due to $\mathbf{w}_i$ is thus expressed as the proportion of the dependent variable variance expressed in terms of $\mathbf{w}_i$:

$$[err]_i = \frac{g_i^2 \langle \mathbf{w}_i, \mathbf{w}_i \rangle}{\langle \mathbf{z}, \mathbf{z} \rangle} \quad , \quad 1 \le i \le M$$

Thus, using eqn. (8.6), $\mathbf{W}_s$ is computed and hence $\mathbf{P}_s$ from $\mathbf{P}$ using the classical Gram-Schmidt procedure. From the $i$th stage, by interchanging the

$i$ to $M$ columns of $\mathbf{P}$, a $\mathbf{p}_i$ is selected which gives the largest $[err]_i$ when orthogonalised into $\mathbf{w}_i$. The selection procedure is as outlined below.

Denote $\mathbf{w}_1^{(i} = \mathbf{p}_i;\ i = 1, \ldots, M$, compute:

$$g_i^{(i} = \frac{\langle \mathbf{w}_1^{(i}, \mathbf{z} \rangle}{\langle \mathbf{w}_1^{(i}, \mathbf{w}_1^{(i} \rangle}, \qquad [err]_1^{(i} = \frac{\left(g_1^{(i}\right)^2 \langle \mathbf{w}_1^{(i}, \mathbf{w}_1^{(i} \rangle}{\langle \mathbf{z}, \mathbf{z} \rangle}$$

If it is assumed that $[err]_1^{(i} = \max\left\{[err]_1^{(i}, 1 \le i \le M\right\}$, then $\mathbf{w}_1 = \mathbf{w}_1^{(i} (= \mathbf{p}_i)$ is selected as the first column of $\mathbf{W}_s$ together with the first element of $\mathbf{g}_s$, $g_1 = g_1^{(i}$, and $[err]_1 = [err]_1^{(i}$.

For the second stage, $i = 1, \ldots, M$ and $i \ne j$, compute:

$$\alpha_{12}^{(i} = \frac{\langle \mathbf{w}_1^{(i}, \mathbf{z} \rangle}{\langle \mathbf{w}_1^{(i}, \mathbf{w}_1 \rangle}, \qquad \mathbf{w}_2^{(i} = p_i - \alpha_{12}^{(i} \mathbf{w}_1$$

$$g_2^{(i} = \frac{\langle \mathbf{w}_2^{(i}, \mathbf{z} \rangle}{\langle \mathbf{w}_2^{(i}, \mathbf{w}_2^{(i} \rangle}, \qquad [err]_2^{(i} = \frac{\left(g_2^{(i}\right)^2 \langle \mathbf{w}_2^{(i}, \mathbf{w}_2^{(i} \rangle}{\langle \mathbf{z}, \mathbf{z} \rangle}$$

If it is assumed that $[err]_2^{(k} = \max\left\{[err]_2^{(i}, 1 \le j \le M \text{ and } i \ne j\right\}$, then $\mathbf{w}_2 = \mathbf{w}_2^{(k}(= \mathbf{p}_k - \alpha_{12}\mathbf{w}_1)$ is selected as the second column of $\mathbf{W}_s$ together with the second column of $\mathbf{A}_s$, $\alpha_{12} = \alpha_{12}^{(k}$, the second element of $\mathbf{g}_s$, $g_2 = g_2^{(k}$, and $[err]_2 = [err]_2^{(k}$.

The error reduction ratio offers a simple and effective means of selecting a subset of significant regressors from a large number of candidates in a forward regression manner. At the $i$th step a regressor is selected if it produces the largest value of $[err]_i$ among the number of candidates. The selection procedure is continued until the $M_s$th stage when:

$$1 - \sum_{i=1}^{M_s} [err]_i < \rho$$

where $0 < \rho < 1$ is a desired tolerance. This leads to a subset of $M_s$ regressors. The parameter estimate $\hat{\Theta}_s$ is then computed from the subset model:

$$\mathbf{A}_s \Theta_s = \mathbf{g}_s$$

where $\mathbf{A}_s$ is an upper triangular matrix.

Several basis functions have been proposed for the RBF networks [242]. Among these, the thin-plate-spline function has been demonstrated to give the best performance [148]. This function is, thus, used here.

## 8.2.4   Structure validation

To ensure that a fitted neural network model adequately represents the underlying mechanism, which produced the data set, a process of model validation can be employed. Model validity tests are procedures designed to detect the inadequacy of a fitted model.

Common measures of predictive accuracy used in control and system identification are to compute the one-step-ahead (OSA) output prediction and model predicted output (MPO) of the system. The OSA prediction of the system output is expressed as:

$$\hat{y}(t) = f\left(u(t), u(t-1), \ldots, u(t-n_u), y(t-1), \ldots, y(t-n_y)\right)$$

where $f(\cdot)$ is a nonlinear function and $u$ and $y$ are the inputs and outputs respectively. The residual or prediction is given by:

$$\varepsilon(t-1) = y(t) - \hat{y}(t) \tag{8.7}$$

Often $\hat{y}(t)$ will be a relatively good prediction of $y(t)$ over the estimation set even if the model is biased because the model was estimated by minimising the prediction errors.

The MPO is defined by:

$$\hat{y}(t) = f\left(u(t), u(t-1), \ldots, u(t-n_u), \hat{y}(t-1), \ldots, \hat{y}(t-n_y)\right)$$

and the deterministic error or deterministic residual is given as in eqn. (8.7). If the fitted model behaves well for the OSA and the MPO this does not necessarily imply that the model is unbiased. The prediction over a different set of data often reveals that the model could be significantly biased. One way to overcome this problem is by splitting the data into two sets, namely, the estimation set and the test set (prediction set). The estimation set is used to train the neural network. The test set is presented as new inputs to the trained network and the predicted output is observed. If the fitted model is correct, that is, correct assignment of lagged $us$ and $ys$, then the network will predict well for the prediction set. In this case the model will capture the underlying dynamics of the system. If both the OSA and the MPO of a fitted model are good over the estimation and the prediction data sets then most likely the model is unbiased.

A more convincing method of model validation is to use correlation tests [27]. If a model of a system is adequate then the residuals or prediction errors $\varepsilon(t)$ should be unpredictable from all linear and nonlinear combinations of past inputs and outputs. It has been shown that for an estimated model to be reasonably acceptable and accurate, the following conditions should hold [27].

$$\phi_{\varepsilon\varepsilon}(\tau) = E\left[\varepsilon(t-\tau)\varepsilon(t)\right] = \delta(\tau)$$
$$\phi_{u\varepsilon}(\tau) = E\left[u(t-\tau)\varepsilon(t)\right] = 0 \;;\; \forall\tau$$
$$\phi_{u^2\varepsilon}(\tau) = E\left[\left(u^2(t-\tau)-\overline{u}^2(t)\right)\varepsilon(t)\right] = 0 \;;\; \forall\tau$$
$$\phi_{u^2\varepsilon^2}(\tau) = E\left[\left(u^2(t-\tau)-\overline{u}^2(t)\right)\varepsilon^2(t)\right] = 0 \;;\; \forall\tau$$
$$\phi_{\varepsilon(\varepsilon u)}(\tau) = E\left[\varepsilon(t)\varepsilon(t-1-\tau)u(t-1-\tau)\right] = 0 \;;\; \tau \geq 0$$

where $\phi_{u\varepsilon}(\tau)$ indicates the crosscorrelation function between $u(t)$ and $\varepsilon(t)$, $\varepsilon u(t) = \varepsilon(t+1)u(t+1)$ and $\delta(\tau)$ is an impulse function. The OSA prediction and MPO along with correlation tests are implemented and used in the work presented in this chapter to validate the trained neural network models.

## 8.3 Neuro-active noise control

A schematic diagram of the geometrical arrangement of an SISO feedforward ANC structure is shown in Figure 8.1a. This structure has previously been considered in various noise and vibration control applications [58, 87, 172, 180, 215, 218, 250, 251, 301, 303]. The unwanted (primary) point source emits broadband noise into the propagation medium. This is detected by a detector located at a distance $r_e$ relative to the primary source, processed by a controller of suitable transfer characteristics and fed to a cancelling (secondary) point source located at a distance $d$ relative to the primary source and a distance $r_f$ relative to the detector. The secondary signal thus generated is superimposed on the primary signal so as to achieve cancellation of the noise at and in the vicinity of an observation point located at distances $r_g$ and $r_h$ relative to the primary and secondary sources, respectively.

A frequency-domain equivalent block diagram of the ANC structure is shown in Figure 8.1b, where $E$, $F$, $G$ and $H$ are transfer functions of the acoustic paths through the distances $r_e$, $r_f$, $r_g$ and $r_h$, respectively. $M$, $C$ and $L$ are transfer characteristics of the detector, the controller and the secondary source, respectively. $U_D$ and $U_C$ are the primary and secondary signals at the source locations whereas $Y_{OD}$ and $Y_{OC}$ are the corresponding signals at the observation point, respectively. $U_M$ is the detected signal and $Y_O$ is the observed signal. The block diagram in Figure 8.1b may be considered either in the continuous frequency ($s$) domain or the discrete frequency ($z$) domain.

The objective in Figure 8.1 is to force the observed signal, $Y_O$, to zero.

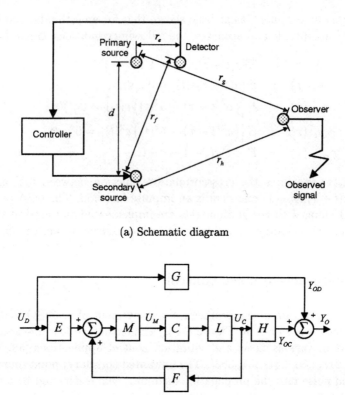

(a) Schematic diagram

(b) Block diagram

**Figure 8.1**    *Active noise control structure*

This is equivalent to the minimum variance design criterion in a stochastic environment. This requires the primary and secondary signals at the observation point to be equal in amplitudes and have a phase difference of 180° relative to each other. Thus, synthesising the controller within the block diagram of Figure 8.1*b* on the basis of this objective yields [310]:

$$C = \frac{G}{ML(FG - EH)} \tag{8.8}$$

This is the required controller transfer function for optimum cancellation of broadband noise at the observation point.

In practice, the characteristics of sources of noise vary due to operating conditions leading to time-varying spectra. Moreover, the characteristics of transducers, sensors and other electronic equipment used are subject to variation due to environmental effects, ageing etc. To design an ANC system so that the controller characteristics are updated in accordance with these changes in the system such that the required performance is achieved and maintained, a self-tuning control strategy, allowing online design and implementation of the controller, can be utilised.

For online design of the controller, the system can be considered with $U_M$ as input and $Y_O$ as output. Thus, owing to the state of the secondary source let the system behaviour be characterised by two subsystems, namely, when the secondary source is off, with an equivalent transfer function denoted by $Q_0$, and when the secondary source is on, with an equivalent transfer function denoted by $Q_1$. Thus, synthesising the controller within the system and using $Q_0$ and $Q_1$ yields [310]:

$$C = \left[1 - \frac{Q_1}{Q_0}\right]^{-1} \tag{8.9}$$

This is the required optimal controller design rule given in terms of the transfer characteristics $Q_0$ and $Q_1$ which can be measured/estimated online.

### 8.3.1    Frequency-response measurement scheme

In this Section a frequency-response measurement (FRM) scheme is adopted to devise a neuro-active control strategy. Eqn. (8.9) is the required controller design rule given in terms of the transfer characteristics $Q_0$ and $Q_1$ which can be measured/estimated online. To ensure that variations in the characteristics of system components are accounted for an online design and implementation of a neuro-controller can be devised. This can be achieved with the following algorithm.

*Algorithm 8.1 FRM scheme:*

(*i*) Obtain the actual frequency responses of $Q_0$ and $Q_1$ using online measurement of input/output signals.

(*ii*) Use eqn. (8.9) to obtain the corresponding frequency-response of the
ideal (optimal) controller.

(*iii*) Train a neural network structure to characterise the ideal controller.

(*iv*) Implement the neuro-controller on a digital processor.

Moreover, to monitor system performance and update the neuro-controller
upon changes in the system a supervisory level control [301] can be utilised.


### 8.3.2    Decoupled linear/nonlinear system scheme

The control strategy introduced in the previous Section is extended and developed here within a decoupled linear/nonlinear system framework, on the basis
of exploiting the capabilities offered by neural networks. It is evidenced in previous studies that in an ANC system the characteristics of the transducers and
electronic components used dominantly contribute to the nonlinear dynamics
of the system. This allows the explicit identification of linear and nonlinear
components within the ANC structure and development of the corresponding
neuro-control strategy. Two alternative methods are proposed and verified on
the basis of this strategy in this section.

The dominant nonlinear dynamics in an ANC system can be thought of
as those present within the characteristics of transducers and electronic components used. These characteristics in general take the form of (amplitude)
limiting transformation; i.e. the input/output transformation is linear up to a
certain input signal level and reaches saturation (nonlinear behaviour) beyond
this level.

Note in the ANC structure shown in Figure 8.1 that the detector, secondary source and associated electronics are all in cascade with one another.
This allows the nonlinear dynamics present in these components to be lumped
together as a single nonlinear function in cascade with the controller. In this
manner, the equivalent block diagram of the ANC structure can be represented
as in Figure 8.2, where $f_n$ is the nonlinear function representing the nonlinear
dynamics of the detector, secondary source and their associated electronics.

The controller in an ANC system is principally required to compensate for
the characteristics of the system components in the secondary path so as to
result in 180° phase difference of the secondary signal relative to the primary
signal at the observation point. This compensation for the detector, secondary
source and their associated electronics, as noted in the design relation in eqn.

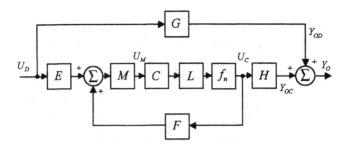

*Figure 8.2    The decoupled linear/nonlinear ANC system.*

(8.8), appears inversely within the required controller transfer function. This implies that, for optimum cancellation to be achieved at the observation point with the structure in Figure 8.2, the controller is additionally required to compensate for the nonlinear function $f_n$. Thus, to extend the neuro-adaptive ANC strategy presented in Section 8.3.1, to include the nonlinear function $f_n$, two alternative schemes namely direct function learning (DFL) and inverse function learning (IFL) are proposed. These are schematically outlined in Figure 8.3, where the ideal controller represents the characteristics in eqn. (8.8) corresponding to the linear dynamic characteristics of the system.

It follows from Figure 8.3a that realisation of the DFL requires a characterisation of the nonlinear function $f_n$. This can be achieved by treating the detector (microphone) and secondary source (loudspeaker) with an acoustic separation between them in cascade as a unit, driving the unit by a signal of large enough amplitude to excite the nonlinear dynamics of the unit, and training a neural network to characterise the unit. This will result in a neural network emulator characterising the nonlinear function $f_n$. Note that the characteristics of the acoustic path between the loudspeaker and the microphone will not dominantly affect the characteristic behaviour of the nonlinear dynamics of the unit. In this manner, the direct nonlinear function emulator (DNFE) thus obtained can be used to represent the nonlinear function block in Figure 8.3a and thus train the neuro-controller accordingly.

Note in Figure 8.3a that the DFL scheme can conceptually be considered as training the neuro-controller to the remainder of the ideal controller after extracting a component equivalent to the nonlinear function $f_n$ [340]. The DFL scheme of training the neuro-controller can accordingly be formulated by

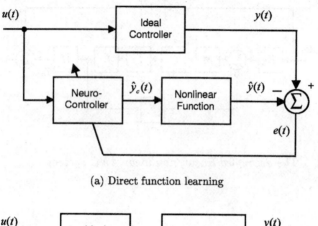

(a) Direct function learning

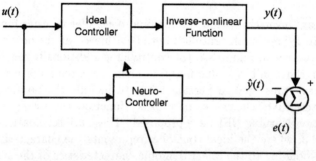

(b) Inverse function learning

Figure 8.3    *Neuro-controller training with decoupled linear/nonlinear system scheme*

the following algorithm.

*Algorithm 8.2 DFL scheme:*

(*i*) Train a neural network emulator to characterise the nonlinear dynamics of the system. This results in the DNFE.

(*ii*) Train the neuro-controller through the process shown in Figure 8.3*a* with the DNFE, representing the nonlinear function block.

(*iii*) Implement the trained neuro-controller within the ANC system of Figure 8.2.

Moreover, to monitor system performance and update the neuro-controller upon changes in the system a supervisory level control [301] can be utilised.

It follows from Figure 8.3*b* that realisation of the IFL scheme requires a suitable characterisation of the inverse nonlinear function $f_n^{-1}$. Such a characterisation can be achieved through training a neural network to the inverse of the nonlinear dynamics of the system. This will result in an inverse nonlinear function emulator (INFE). The INFE thus obtained can be used within the IFL scheme of Figure 8.3*b*, in place of the inverse nonlinear function block, to train the required neuro-controller for use within the ANC system in Figure 8.2 for broadband cancellation of noise at the observation point. This accordingly allows formulation of the following algorithm for the IFL scheme.

*Algorithm 8.3 IFL scheme:*

(*i*) Train a neural network emulator to characterise the inverse nonlinear dynamics of the system. This results in the INFE.

(*ii*) Train the neuro-controller according to the process described in Figure 8.3*b* with the inverse nonlinear function block represented by the INFE.

(*iii*) Implement the trained neuro-controller within the ANC system in Figure 8.2.

Moreover, to monitor system performance and update the neuro-controller upon changes in the system a supervisory level control [301] can be utilised.

### 8.3.3   Direct neuro-modelling and control scheme

In this Section a control strategy for the general case of an ANC system, where nonlinear dynamics may also be present due to the characteristics of the propagation medium, is presented. This involves direct neuro-modelling of system characteristics and online design and implementation of the neuro-controller.

The approach presented in Section 8.3.2 assumes explicit decoupling of the system into linear and nonlinear parts. This is achieved at the modelling stage by exciting the linear and nonlinear dynamics of the system, in turn, with suitable signals and extracting the corresponding system dynamics accordingly. To allow both the linear and nonlinear dynamics of the system be incorporated together within the design, a formulation of the modelling and control of the system, using the controller design rule in eqn. (8.9), is assessed in this Section.

Consider the ANC system in Figure 8.1. For optimum cancellation of the noise at the observation point, the controller characteristics are given by eqn. (8.9). This can be restated as:

$$C = \left[1 - Q_1 Q_0^{-1}\right]^{-1}$$

which can be realised by adopting an inverse modelling approach. To allow nonlinear dynamics of the system to be incorporated within the design, it is proposed to train suitable neural networks to characterise the system models $Q_0^{-1}$ and $Q_1$. In this manner, the corresponding neuro-controller, for optimum cancellation of the noise, can be trained as shown in Figure 8.4, with a suitable input signal covering the dynamic range of interest of the system. An ANC system design and implementation approach can accordingly be formulated as follows.

*Algorithm 8.4 Direct neuro-modelling and control scheme:*

(*i*) With $C = 0$, train a neural network to characterise the inverse of the system between the detection and observation points. This gives characterisation of $Q_0^{-1}$.

(*ii*) With $C = 1$, train a neural network to characterise the system between the detection and observation points. This gives characterisation of $Q_1$.

(*iii*) Train a neural network according to Figure 8.4 to characterise the design rule in eqn. (8.9). This gives the required neuro-controller.

(*iv*) Implement the neuro-controller within the ANC system.

Moreover, to monitor system performance and update the neuro-controller upon changes in the system a supervisory level control [301] can be utilised.

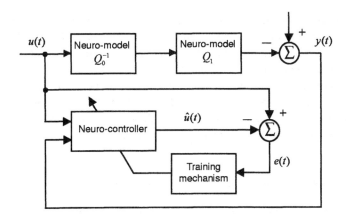

Figure 8.4    *Neuro-controller training with direct neuro-modelling and control scheme*

## 8.4    Implementations and results

To test and verify the neuro-control strategies described in the previous section a simulation environment characterising the ANC system of Figure 8.1 within a free-field medium was developed using practical data [311]. A tan-sigmoid function, representing the nonlinear dynamics, $f_n$, of the system in Figure 8.2 was also incorporated into the simulation environment. The simulation algorithm was coded within MATLAB and implemented on a 486 PC. A pseudorandom binary sequence (PRBS) simulating a broadband signal in the range $0 \sim 500$ Hz was utilised as the primary noise source. The amplitude of the signal was varied to excite the various dynamic ranges of the system. The system components were arranged with the observer at 1.5 metres away and equidistant from the primary and secondary sources.

In these experiments the backpropagation and the orthogonal forward regression algorithms were used for training the MLP and RBF networks, respectively. The networks incorporated tansigmoid neurons, and the thin-plate-spline basis function was used with the RBF networks. The input data format of eqn. (8.3) was utilised for the neural networks throughout these experiments. The networks were trained with 500 data points using a varied-amplitude PRBS signal as input. Half of the data was used as the training set and the remaining half as the test set. The OSA prediction, MPO and correlation tests were used in each case to validate the neural network structure. In these tests all the correlation tests were found within the 95 per cent confidence interval. The full data set was then used in each case to train the validated neural network for subsequent use.

To determine the amount of cancellation achieved with the system over a broad frequency range of the noise, the autopower spectral densities of the noise before and after cancellation are obtained and the corresponding difference, that is, the cancelled spectrum, is evaluated.

### 8.4.1    Frequency-response measurement scheme

To implement the MLP neuro-controller a network with two hidden layers each having five neurons, input vector with $n_u = n_y = 8$ and learning rate (at start) of 0.0002 was set up. The input and the corresponding output of the ideal controller were used as the training pair. Figure 8.5a shows the predicted output superimposed on the ideal controller output. It is noted that good output prediction with an error index of 0.0073 was achieved.

The performance of the ANC system was monitored at the observation point with the MLP neuro-controller. The cancelled spectrum thus achieved is shown in Figure 8.5b. It is noted that an average level of around 20 dB cancellation was achieved with the MLP neuro-controller. This was similar to the performance achieved with the ideal controller, demonstrating that a performance as good as that of the optimal controller is achieved with the MLP neuro-adaptive ANC system.

To implement the RBF neuro-controller a network with 65 centres and input vector with $n_u = n_y = 15$ was set up. The input and the corresponding output of the ideal controller were used as the training pair. Figure 8.6a shows the predicted output superimposed on the ideal controller output. It is noted that good output prediction with an error index of 0.0002 was achieved.

The performance of the ANC system was monitored at the observation point with the RBF neuro-controller. The cancelled spectrum thus achieved is shown in Figure 8.6b. It is noted that an average level of around 20 dB cancellation was achieved with the RBF neuro-controller. This was similar to the performance achieved with the ideal controller.

### 8.4.2    Decoupled linear/nonlinear system scheme

To realise the neuro-controller using the DFL scheme an MLP neural network emulator incorporating one hidden layer with 14 neurons and $n_u = n_y = 5$ was trained with a learning rate (at start) of 0.00002 to characterise the non-

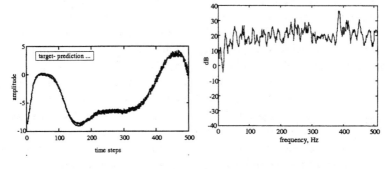

(a) Output prediction by neuro-controller

(b) Cancelled spectrum at observation point

Figure 8.5    Performance with MLP neuro-controller; FRM scheme

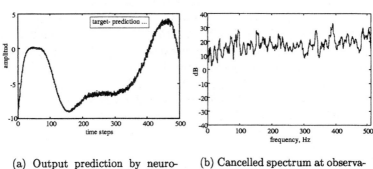

(a) Output prediction by neuro-controller

(b) Cancelled spectrum at observation point

Figure 8.6    Performance with RBF neuro-controller; FRM scheme

linear function $f_n$. It was noted that the network gave good output prediction with an error index of 0.00001. The DNFE thus obtained was utilised in Figure 8.3a, in place of the nonlinear function block $f_n$, with the ideal controller characteristics, to train the neuro-controller.

An MLP neural network incorporating two hidden layers each with eight neurons, $n_u = n_y = 16$, and a learning rate (at start) of 0.00004 was accordingly trained as the neuro-controller. Figure 8.7a shows the predicted output superimposed on the ideal controller output. It is noted that the network achieved good output prediction with an error index of 0.00038. The neuro-controller thus trained was implemented within the ANC system in Figure 8.2 and the performance of the system was monitored at the observation point. Figure 8.7b shows the cancelled spectrum thus obtained. It is noted that an average level of around 35 dB cancellation was achieved with the system. To investigate the effect of input noise level on the performance of the system, the primary signal was varied, in turn, in amplitude so as to excite the non-linear function $f_n$ at various regions. These included the linear region, upper (positive) nonlinear region and lower (negative) nonlinear region. It was noted that the system performed very well in these cases [340].

To realise the neuro-control strategy using the IFL scheme an MLP neural network emulator incorporating one hidden layer with 14 neurons, $n_u = n_y = 5$ and a learning rate (at start) of 0.00002 was trained to characterise the inverse nonlinear function $f_n^{-1}$. This was utilised in Figure 8.3b accordingly with the ideal controller characteristics. In this case an MLP and an RBF neuro-controller were trained according to the scheme in Figure 8.3b.

An MLP neural network incorporating two hidden layers each with five neurons and $n_u = n_y = 9$ was first trained to characterise the neuro-controller. Figure 8.8a shows the predicted neuro-controller output superimposed on the actual output. It is noted that the network achieved good output prediction with an error index of 0.0008. The neuro-controller thus obtained was implemented in the ANC system of Figure 8.2 and the performance of the system was monitored at the observation point. This is shown in Figure 8.8b. It is noted that an average level of around 35 dB cancellation was achieved with the MLP neuro-controller.

An RBF neural network incorporating 54 centres and $n_u = n_y = 18$ was then trained to characterise the neuro-controller. Figure 8.9a shows the predicted neuro-controller output superimposed on the actual output. It is noted that the network achieved good output prediction with an error index of 0.00024. The neuro-controller thus obtained was implemented in the ANC system of Figure 8.2 and the performance of the system was monitored at the observation point. This is shown in Figure 8.9b. It is noted that an average level of around 40 dB cancellation was achieved with the RBF neuro-controller. This is slightly better than that with the MLP neuro-controller. Tests were also carried out with input signals exciting various regions of the nonlinear function. It was noted that the system achieved performances similar to those presented above in each case [340].

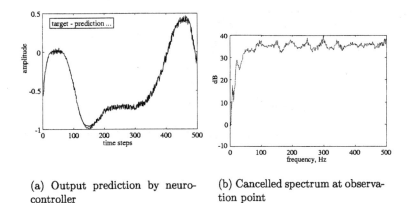

(a) Output prediction by neuro-controller

(b) Cancelled spectrum at observation point

**Figure 8.7** *Performance with MLP neuro-controller; DFL scheme*

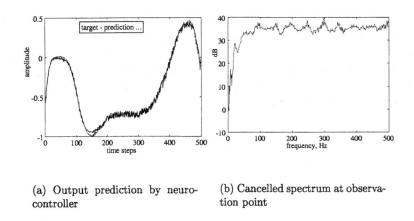

(a) Output prediction by neuro-controller

(b) Cancelled spectrum at observation point

**Figure 8.8** *Performance with MLP neuro-controller; IFL scheme*

### 8.4.3 Direct neuro-modelling and control scheme

To assess and verify the direct neuro-modelling and control strategy the decoupled linear/nonlinear structure of Figure 8.2 is used. In this process both MLP and RBF neural networks are utilised.

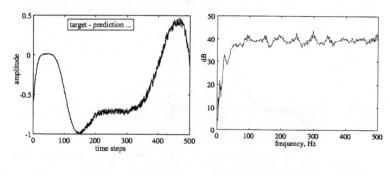

(a) Output prediction by neuro-controller

(b) Cancelled spectrum at observation point

*Figure 8.9     Performance with RBF neuro-controller; IFL scheme*

An MLP network incorporating two hidden layers each with nine neurons and $n_u = n_y = 9$ was trained with a learning rate (at start) of 0.00001 to characterise $Q_0^{-1}$. A further MLP network incorporating two hidden layers each with four neurons and $n_u = n_y = 14$ was trained with a learning rate (at start) of 0.00005 to characterise $Q_1$. Figure 8.10 shows the output prediction achieved with these networks.

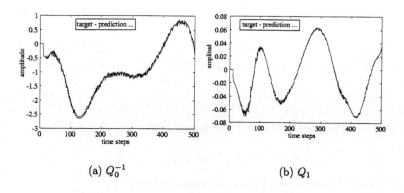

(a) $Q_0^{-1}$

(b) $Q_1$

*Figure 8.10     Output prediction by the MLP neural network characterisation*

An MLP network incorporating two hidden layers with 25 neurons in each layer and $n_u = n_y = 18$ was trained with a learning rate (at start) of 0.000001 to characterise the optimal controller according to the scheme in Figure 8.4 using the $Q_0^{-1}$ and $Q_1$ networks obtained above. The MLP neuro-controller thus obtained was implemented within the ANC system of Figure 8.2 and its performance assessed. The amplitude of the input signal was varied so as to excite the full dynamic range of the nonlinear function $f_n$. The performance of the system as monitored at the observation point is shown in Figure 8.11. It is noted that an average level of cancellation of around 40 dB was achieved with the system.

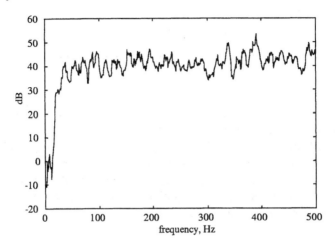

Figure 8.11    *Cancelled spectrum with MLP neuro-controller; (direct neuro-modelling and control scheme)*

To assess the direct neuro-modelling and control strategy with RBF networks, a network incorporating 56 centres and $n_u = n_y = 14$ was trained to characterise $Q_0^{-1}$. A further RBF network incorporating 52 centres and $n_u = n_y = 16$ was trained to characterise $Q_1$. Figure 8.12 shows the output prediction achieved by the networks. It is noted that the networks gave good output prediction with error indices of 0.0001 and 0.00002 for $Q_0^{-1}$ and $Q_1$, respectively.

An RBF network incorporating 67 centres and $n_u = n_y = 22$ was trained according to the scheme in Figure 8.4 to characterise the optimal controller, using the $Q_0^{-1}$ and $Q_1$ networks obtained above. The RBF neuro-controller

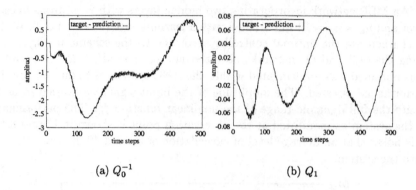

$$\text{(a) } Q_0^{-1} \qquad\qquad\qquad \text{(b) } Q_1$$

*Figure 8.12     Output prediction by the RBF neural network characterisation*

thus obtained was implemented within the ANC system of Figure 8.2 and its performance assessed. The amplitude of the input signal was varied so as to excite the full range of the nonlinear function $f_n$. The corresponding performance of the system as monitored at the observation point is shown in Figure 8.13. It is noted that an average level of cancellation of around 40 dB was achieved with the system. This is similar to the performance achieved with the MLP neuro-controller, see Figure 8.11.

## 8.5    Conclusions

The development of several neuro-adaptive ANC systems has been presented and verified in the cancellation of broadband noise in a three-dimensional propagation medium.

The MLP neural network, with backpropagation training algorithm, and the RBF network, with OFR training algorithm, have been introduced. The capability of the networks in characterising dynamic systems has been investigated. Both the OSA prediction and the MPO have been used as the training methods. Model validity tests using correlation tests have been carried out. It has been shown that with a suitable choice of the input data structure the system data can faithfully be predicted with an acceptable prediction error minimum.

An SISO feedforward ANC structure has been considered for optimum cancellation of broadband noise at an observation point in a three-dimensional propagation medium. The controller design relations have been formulated such as to allow online design and implementation and hence an adaptive control algorithm.

The MLP and RBF neural networks have been incorporated within the adaptive control algorithm to characterise the optimal controller. In this manner, a neuro-adaptive ANC system based on the FRM scheme has been developed for cancellation of broadband noise. The FRM neuro-adaptive ANC algorithm thus developed has been tested within an acoustic environment characterising a non-dispersive propagation medium and its performance verified in the cancellation of broadband noise in comparison with the optimal controller. It has been shown that the FRM neuro-ANC system achieves as good a performance as the system incorporating the optimal controller.

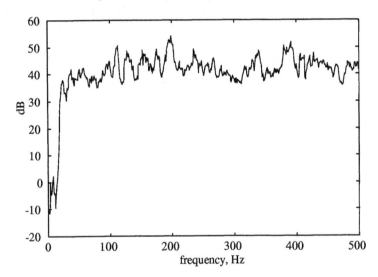

*Figure 8.13    Cancelled spectrum with RBF neuro-controller; (direct neuro-modelling and control scheme)*

The DFL and IFL neuro-adaptive ANC strategies have been developed, within a decoupled linear/nonlinear system framework, and verified in the cancellation of broadband noise in a free-field environment. Both the MLP and RBF networks have been utilised to realise the DFL and IFL schemes. It has been demonstrated through simulation tests that the system performances with both MLP and RBF neuro-controllers are significant and comparable to one another. It has also been demonstrated through extended (amplitude) primary signals that the neuro-controllers give significant levels of noise cancellation over dynamic ranges of the system for which the controller has not been trained.

A neuro-adaptive ANC strategy based on a direct neuro-modelling and control scheme has been presented. The online design and implementation of the ANC system has been achieved through neuro-modelling of the two subsystems, $Q_1$ and $Q_0$, and design of the neuro-controller. The ANC system has accordingly achieved similar results with MLP and RBF neuro-controllers. The control strategy has been tested within the decoupled ANC system and significant levels of cancellation of broadband noise in each case have been achieved.

The significance of the neuro-control strategies has been demonstrated through the level of performance achieved in the cancellation of broadband noise. This has established the basis and potential for further developments of noise cancellation of vibration suppression schemes in practical applications.

*Chapter 9*

# Genetic algorithms for optimising ASVC systems

## C. H. Hansen, M. T. Simpson and B. S. Cazzolato

*Department of Mechanical Engineering, University of Adelaide*
*Adelaide, South Australia*
*Email: chansen@watt.mecheng.adelaide.edu.au*

*In feedforward active noise control problems, it is always necessary to opti-
mise both the physical system and the electronic controller. Optimisation of
the physical system is principally concerned with determining the optimum location of the control sources and error sensors. Here, genetic algorithms are
investigated for this purpose and modifications to the standard algorithm that
are necessary for this application are discussed. Optimising the electronic controller is principally concerned with finding the optimal control filter weights
that will produce the most noise reduction when the reference signal is filtered
and then input to the control sources. Normally, gradient descent algorithms
are used for this purpose. However, for nonlinear systems, such as control
sources (loudspeakers or shakers) with significant harmonic distortion, the gradient descent algorithm is unsatisfactory. Here a genetic algorithm is developed
specifically for control filter weight optimisation. It is able to handle nonlinear
filters and requires no cancellation path identification. The disadvantage is
that it is relatively slow to converge.*

## 9.1   Introduction

Genetic algorithms may be applied to active sound control systems in two quite different ways. First, they may be used to optimise the locations of the control sources. Second, they may be used to adapt the weights of the digital filters which generate the signals to drive the control sources that cancel the unwanted sound. Both of these applications are discussed in detail here. Previous work and limitations of genetic algorithms as well as their advantages in certain applications will be discussed.

The active control of sound differs from the active control of electronic or radiofrequency noise in that it is not just a signal processing problem. The level of control that is achievable is dependent on a number of additional parameters that are unrelated to the signal processing aspects of the problem. Once the cost function which must be minimised has been chosen (e.g. sound power, sound pressure at a number of locations, acoustic energy density or acoustic potential energy), these parameters must be optimised in a strict sequence to achieve an overall optimised system, as illustrated in Figure 9.1.

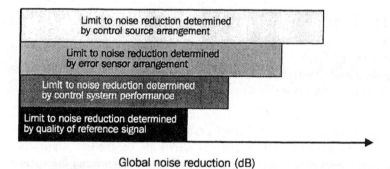

Global noise reduction (dB)

*Figure 9.1   Performance hierarchy of an active control system*

The first parameter that must be optimised is the control source arrangement (number and location), as this will determine the maximum amount of cost function control achievable with an ideal error sensor arrangement and an ideal electronic controller. The next parameter is the error sensor arrangement (number and locations) which will determine how close it will be possible to get with an ideal controller to the maximum achievable control set by the control source arrangement. Next the electronic controller must be optimised and this includes optimisation of the algorithm type, algorithm coefficients (such as convergence coefficient and leakage coefficient), filter size and type (linear IIR or FIR, nonlinear   neural network or polynomial) and filter weight values. The final parameter to be optimised is the quality of the reference signal. For gradient descent algorithms to be effective in optimising the control filter weights and to minimise extraneous unwanted noise being introduced into the system through the control sources, it is necessary that the reference signal be well correlated with the error signals and with the sound to be minimised.

Here will be discussed two aspects of the optimisation problem for which genetic algorithms may play an important role. The first is the optimisation of control source location. In this Chapter an example of optimising the location of a number of control actuators on a stiffened cylinder with floor structure will be considered and results using four different algorithms will be compared. Once the control actuator locations have been optimised it is a relatively straightforward procedure to optimise the locations of the error sensors which should be at locations where there is greatest difference between the existing primary sound field and the theoretically optimally controlled field. This procedure, which involves the use of multiple regression, is outside the scope of this Chapter and is discussed at length in References [122] and [252].

The second aspect of control system optimisation to be discussed involves optimisation of the controller. There are some special cases where a genetic algorithm may be used in place of a gradient descent algorithm to optimise the control filter weight values and so obtain improved performance. An example of the nonlinear control of a harmonically vibrating beam will be discussed. The one remaining aspect not yet considered is the optimisation of the reference signal. This generally means that the reference signal should be obtained by nonacoustic means if possible (such as with a tachometer) and, if a microphone is used, care must be exercised to ensure that the reference signal is not influenced by flow noise or by the control signal. However, these considerations are discussed in Reference [122] and will not be discussed further here.

## 9.2   The genetic algorithm

Genetic algorithms are well known in many disciplines for their efficient optimisation capabilities [111] and it is the purpose of this Chapter to demonstrate their usefulness in active control applications for optimising two of the above parameters, the control source arrangement and the control filter weight values.

The complex multimodal nature of most practical active sound and vibration control applications involving large structures and multiple control sources, means that an exhaustive search of all possible control source configurations is usually impractical. Even gradient-based search methods which move from point to point in the direction of maximum sound reduction (commonly referred to as a hill-climb search), are ineffective for searching for a global optimum arrangement of control source locations across many local optima. Here it will be shown that the use of a genetic algorithm can reduce the search space of a complex system to a manageable size and consistently result in an optimum control source arrangement which provides sound reductions close to the theoretical optima.

Recently, the genetic algorithm search technique has been applied to the problem of loudspeaker placement in aircraft interiors to minimise the interior sound levels [22, 23, 155, 156, 189, 316]. Other interior noise problems involving the use of genetic algorithms for optimising loudspeaker placement have also been investigated [241]. In addition, related work involving the optimisation of vibration actuator placement on truss structures [244, 349] along with sensor placement for modal identification [342] has been undertaken. Reference [167] investigated the optimal placement of four actuators on a rectangular panel to minimise the sound transmitted into an enclosure for which the panel was one of the walls. The genetic algorithm was also used to find the optimum locations of the 45 error microphones in the enclosure. Reference [64] investigated the use of a genetic algorithm for the optimisation of sensor and actuator locations for the active control of sound transmission through a double panel partition. Reference [190] used a genetic algorithm to optimise the error sensor location for the active control of sound radiated by an electrical transformer. Reference [326] used the genetic algorithm for the optimisation of piezoelectric actuator locations to control the sound radiated by a beam. The author also used the genetic algorithm to optimise the error microphone locations, although he could have used the difference between the radiated primary and predicted optimally controlled sound fields [122].

Here, an example will be shown of the use of a genetic algorithm for optimisation of vibration control actuator locations on a stiffened cylinder to minimise the sound radiated into the interior space. Although this problem has been reported previously by the authors [265], it is used here to illustrate the differences and similarities between genetic algorithms that are suitable for control source optimisation and those that are suitable for control filter weight optimisation. Here a number of different algorithms are described, the advantages of each are discussed and the relative performance of each when applied to the problem of sound radiated into the interior of a stiffened cylinder is compared.

For optimisation of control filter weights in a feedforward control system, a genetic algorithm [145, 290, 291] is especially useful in problems involving a nonlinear acoustic or vibration system [328], a nonlinear cancellation path transfer function (often as a result of nonlinear control sources or sources with nonnegligible harmonic distortion), or when there is acoustic or vibration feedback from the control source to the reference sensor. In all cases the system to be controlled must be stationary or varying very slowly. Two important advantages of a genetic algorithm (rather than the conventional filtered-$x$ or filtered-$u$ gradient descent algorithms) for adaptation of control filter weights are that the error signal need not be fully correlated with the reference signal (so there can be noise in the reference signal) and filters other than the standard FIR and IIR filters may be used. The relaxation of the requirement for correlation between the reference and error signals means that knowledge of the cancellation path (control source input to error sensor output) transfer function, which is essential for a conventional gradient descent-type algorithm to remain stable, is unnecessary. The removal of the restriction of the control filter type allows the use of complex and nonlinear filter structures to treat sound problems which may not be treatable using conventional filter structures and gradient descent algorithms. In addition, the genetic algorithm does not suffer from the same instability problems associated with a filtered-$u$ algorithm and IIR filters used in systems characterised by significant acoustic or vibration feedback from the control source to the reference sensor. A typical control system arrangement in which a genetic algorithm is used for control filter weight adjustment is shown in Figure 9.2. Here, the genetic algorithm evaluates the error signal against specified performance criteria for a particular arrangement of digital filter weights, adjusts the filter weights then reevaluates the error signal, gradually directing the filter weights to produce the optimum error signal.

The genetic algorithm (GA) may be regarded as a guided random search which begins with the arbitrary specification of a population, typically consisting of between 40 and 100 solutions. Finding the optimal solutions involves breeding new solutions from the original population which involves three processes: fitness evaluation, selection of appropriate parents and use of the parents to breed new population members (children). Fitness evaluation involves the assessment of the particular individual solution in terms of its ability to minimise the cost function in the active noise control system. For optimisation of control source locations, a particular solution is one configuration of all the control sources and for the control filter weight optimisation, one solution is a particular set of filter weight values.

Selection (used only for the filter weight optimisation process) involves re-

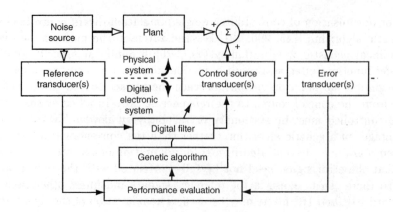

*Figure 9.2    Control system arrangement with a genetic algorithm*

moving a proportion of the population based on their lack of fitness (usually about 30 per cent of the total). Breeding is the use of the remainder of the population to generate a new population for fitness testing. The choice of parents for breeding is based on probability considerations with fitter individuals more likely to be chosen; this will be discussed in more detail later. The characteristics of the two individuals are combined randomly to produce a third individual (child) which becomes a member of the next generation. The breeding process continues until there are enough new individuals created to completely replace all members of the breeding population (except the fittest one when an elitist model is used). The breeding cycle then repeats until it is clear that further new high-fitness solutions are unlikely to be found.

To achieve optimal convergence behaviour, each genetic algorithm must be tuned to suit the particular problem at hand although, with the new approach formulated in this Chapter, the algorithm for actuator location on the stiffened cylinder is now more generally applicable to other actuator placement problems. The search behaviour can be altered significantly by varying the parameters controlling the algorithm such as crossover probability, mutation probability, selection probability distribution and population size. The influence that these parameters have may be described in terms of their effect on selective pressure and population diversity [332]. Selective pressure is defined as the bias which exists towards the high-fitness members of the breeding population during parent selection. Population diversity is a measure of the degree to which strings in the breeding population are distributed throughout the solution space. Selective pressure and population diversity are inversely related. For example, as selective pressure increases, the focus of the search shifts towards the high-fitness members of the breeding population, often at the expense of excluding low-fitness solutions from the breeding process. This represents a hill-climb procedure in which population diversity is commonly lost resulting in a narrow band of similar string types. A low selective pressure means that more members of the breeding population are included in the search process, resulting in a wider search with more diverse string combinations produced. However, in this case the search slows down due to lack of focus on the high-fitness individuals.

The key to a successful global optimisation search is maintaining the correct balance between selective pressure and population diversity. In the following sections, various methods of controlling selection pressure and population diversity are described, with the aim of improving search behaviour such that optimum performance is achieved. Consideration will be given to what choices from the many alternatives should be made in the context of developing appropriate genetic algorithms for application to control source placement and control filter weight selection. Questions such as the use of an individual more than once in generating the next generation of children will also be addressed. The breeding cycle can continue indefinitely for the case of the control filter weight updates, but for the control source optimisation it is usually stopped at a point when it becomes clear that no significant improvement in the best solution is likely.

### 9.2.1  Coding selection for search variables

Genetic algorithms are sufficiently robust to exhibit favourable convergence characteristics for a range of different coding alternatives. However, there may be significant benefit in selecting one coding type over others in a given problem configuration due to improved convergence and performance characteristics. Here, various coding alternatives are reviewed with the appropriateness of each discussed with a view to obtaining best empirical search performance in a generalised multimodal optimisation problem. It is shown that the problems of control source location optimisation and control filter weight optimisation are not necessarily best served by the same coding method.

The first coding type considered is that of a single number (integer or floating point) coding scheme [139, 155, 316, 342] so that each control source location or each filter weight value is represented by a single number proportional to its coordinate values or weight value, respectively. In this case, each position at which actuators may be located (in the problem) is assigned a unique integer location number. This approach lends itself directly to finite-element modelling with each node in the model represented by a unique integer. The coding is then formed by concatenating the location numbers representing each actuator position (or filter weights in the case of the control filter) into a single numerical string, of length equal to the number of control sources (or number of filter weights). Although maintaining a constant number of control sources is inherent to this coding method, duplicate actuator positions within individual strings are possible, requiring additional constraints to ensure that all actuator locations (in a single string) are distinct. Clearly this is not a problem for the control filter weights as it is acceptable for different weights in the same filter to have the same value.

The application of the single number (numerical) coding method is not constrained to locations represented by integer numbers. The possibility arises for the real floating-point values of coordinate locations (representing each control source position) to be used directly in the string structure. However, to ensure that an even distribution of control source locations is achieved (throughout the problem space), the continuous variable range defining the control source coordinate locations must be discretised into a finite number of equidistant values suitable for finite difference models. The real values of these discretised variables could then be used directly in the string structure in a similar manner to that used for the integer values. This is the approach adopted here for optimisation of the control filter weights.

A second coding scheme which has been used for actuator placement optimisation consists in using a binary string to trace the status of each possible actuator placement location in the problem space. With each location assigned a unique position in the binary string, an actuator present is assigned a bit value of '1' in the string, with the remaining (empty) positions assigned a bit value of '0'. This scheme is not really practical for the problems of interest here because large binary string lengths would be required to trace a relatively low number of actuators among a large number of potential actuator locations in the placement problems to be considered and a similar argument applies to its application to filter weight optimisation. However, the approach may be useful for the actuator placement problem when the potential number of actuator locations is relatively small.

A third coding scheme has been proposed in which a multivariable binary string (MVBS) is used to represent the positions of each actuator configuration. In this coding scheme, each individual actuator location is mapped to an $n$-bit binary string. To represent $M$ actuator positions, the length of the binary string $n$, must satisfy the requirement: $2^n \geq M$. Once actuators are assigned a unique binary number representing their location, the binary numbers themselves are concatenated into a single (combined) binary string. To represent continuous variables in this manner, the variable range is discretised into $2^n$ separate equidistant values, enabling each variable value to be mapped to a distinct $n$-bit binary number.

On the basis of the results obtained using a multimodal test example used by others to test genetic algorithm performance in other applications, Reference [265] suggested that multivariable binary string (MVBS) coding would work best for the control source placement problem. Although the multimodal example was representative of the problem of sound radiation into a stiffened cylinder, it was not the same and it is shown here that for this particular problem, MVBS coding does not produce such good results as does integer coding. Reference [328] showed that for the control filter weight optimisation problem, the MVBS coding scheme was impractical because of the large jumps in the actual weight values which occurred when the more significant bits in the string were changed. These large jumps in filter weight values cause corresponding large jumps in control filter output which appear as annoying pops when the control source is a loudspeaker.

### 9.2.2 Parent selection

Two selection processes are carried out during implementation of the genetic algorithm: the choice of individuals to be removed or kept and the choice of parents to be used for breeding the next generation. Both processes are usually implemented using a simulated (and biased) roulette wheel where each segment on the roulette wheel corresponds to an individual with the size of the segment proportional to the probability of the individual being chosen (selection probability). Selection probabilities are assigned such that low-fitness individuals are more likely to be removed and such that high-fitness individuals (those which result in greatest reduction of the cost function – noise or vibration) are more likely to be chosen as parents for breeding. Selection without replacement is used for removal, where once an individual is chosen it is removed from the roulette wheel. For parent selection, replacement is used so that the selected parent is not removed from the roulette wheel after being selected. This allows the entire population to be available for breeding of each new individual in the next generation. In addition, when an elitist model is used, the implication is that the parent with the best fitness is carried into the next generation population unchanged.

The selection probability of member $i$ for removal (or killing) is:

$$P_r(i) = \frac{1.0 - f_i}{\sum_{j=1}^{n_{pop}} f_j} \qquad (9.1)$$

where $f_i$ is the assigned string fitness. The selection probability of member $i$ for breeding is:

$$P_s(i) = \frac{f_i}{\sum_{j=1}^{n_{pop}} f_j} \qquad (9.2)$$

Concerns over stochastic errors involved with the roulette wheel selection procedure have led to proposals of many alternative parent selection schemes [111]. The parent selection scheme known as "stochastic remainder selection without replacement" has been widely accepted as providing an adequate, more accurate alternative to roulette wheel selection. This method ensures that the expected number of breeding opportunities allocated to strings in each generation correspond as closely as possible to the breeding probability distribution. All parent strings required for one generation are preselected, in proportion to the breeding probability distribution, into a separate parent selection population (of size $2n_{pop}$). To do this, an expected value parameter $e_i$, for member $i$, is defined as:

$$e_i = \frac{f_i}{\bar{f}} \qquad (9.3)$$

where $\bar{f}$ is the mean fitness value of the breeding population. The truncated integer value of $e_i$ (if $e_i > 1$) indicates the number of whole copies that each string is allocated in the selection population.

Once completed, the fractional value of $e_i$ (regardless of integer value) is used as a success probability in a weighted coin toss procedure. If successful, a string is allocated a whole copy in the selection population, and removed from further contention. This process continues until all available places in the selection population have been filled. Then, when choosing two parent strings for breeding, all that is required is the random selection of two parent candidates from the selection population (without replacement).

It is clear from eqns (9.1) to (9.3) that the numerical value of each probability and hence the difference between the fitness of various individuals will depend on the fitness measure. If there is not much numerical difference between fitness values in the population, the search may stagnate in its ability to find new solutions, owing to a general lack of direction in allocating reproductive opportunities. Second, an unusually high-fitness individual may be found early in a search, causing a disproportionate number of reproductive opportunities to be allocated to it. This usually results in that individual dominating the breeding population within a few generations, leading to search stagnation due to the lack of population diversity, which in turn could result in the search terminating in a local optimum.

To overcome these problems, more control is required over the way in which the selection probability distribution is allocated. A common method used for this purpose is linear scaling [111]. The principle of this scaling procedure is to scale each string fitness value such that the differences in all fitness values are enhanced (or reduced) to maintain an appropriate bias towards high fitness individuals in a population (selective pressure). Therefore, the relatively minor differences in fitness values of a near-homogeneous population may be scaled up such that adequate selective pressure is maintained throughout a search.

Even with fitness scaling procedures, the selection probability distribution may not always be appropriate for the desired search behaviour. A more direct way of controlling selective pressure has been proposed [332] where the selection probability is assigned to members of the breeding population according to the rank position of their fitness values rather than in proportion to their absolute fitness values. Of course, different probability distributions are used for removal selection and breeding selection as shown in Figure 9.3. The use of rank selection had the effect of removing problems associated with either high-fitness string domination or search stagnation as a result of using fitness proportional schemes and was found to provide the best results for both of the applications addressed in this Chapter.

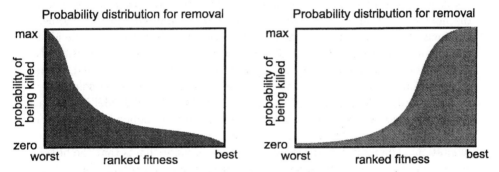

Figure 9.3    *Typical probability distributions for killing and breeding*

Altering the distribution of the selection probability determines the nature of the search performed. A linear probability distribution ensures that the median assigned fitness corresponds to the median rank position of the population. A nonlinear probability distribution creates a bias towards higher fitness individuals as shown in Figure 9.3 and will tend to display hill-climbing characteristics. It is generally advantageous to use moderate selective pressure at the start of the search to enable a wide search for high-fitness solutions.

However, when the search is converging on a final solution, a hill-climb solution is usually more appropriate to speed up the process by allocating a higher selective pressure.

For the case of control filter weight adaptation, maximisation of the convergence rate of the genetic algorithm is of prime importance in terms of maximising the tracking performance of the control system. Convergence speed can be increased by the removal of low-fitness individuals from the population prior to breeding as discussed above. However, the associated disadvantage of reducing population diversity resulted in the removal process being unsuitable for optimisation of control source locations.

One of the problems encountered with the standard implementation of the genetic algorithm is that individuals with higher fitness values are not automatically guaranteed survival into the next generation. As the composition of the breeding population dynamically changes throughout the duration of the search, it is likely that some good solutions may not be reproduced. Therefore, the potential exists for best-fitness solutions found in a search to be lost during the generational transition, requiring several more generations to recover. To overcome this problem, an elitist model [111] was implemented (in both applications discussed here) to ensure survival of the previous best string into the next generation, if that string was not already present. In this manner, the focus of the search was maintained on the best performance individual throughout the search process. One drawback, however, is that occasionally the search may focus exclusively on the best performance solution found, at the expense of undertaking a wider search for other potentially better solutions.

The steady-state genetic algorithm (SSGA) [288, 332, 333, 349] is a variation of the standard genetic algorithm and has been developed in an attempt to minimise the potential loss of vital search information from generation to generation. The main difference with this type of algorithm is that the breeding population is sorted in order of fitness magnitude and retains the highest fitness solutions found from the entire search undertaken so far. Reproduction of new child strings (i.e. crossover and mutation) is the same as in the previous generic algorithm (GA) case. However, after each child string is created and assigned a fitness value (by the objective function), its relative performance is immediately compared with those in the existing breeding population. If the fitness exceeds that of the lowest fitness string in the population, then the newer child string is inserted at the correct rank position in the population (displacing the low fitness solution from the bottom) and becomes an immediate active member of the breeding process. This contrasts significantly with the original genetic algorithm concept, where each child produced must wait until the next generation before it is used as a parent. Also, the SSGA eliminates the need to apply the elitist model (discussed above) as the retention of high-fitness solutions is inherent.

The steady-state genetic algorithm was found to be the best approach for the control source optimisation problem, but was unsuitable for ensuring good continuous performance for the online filter weight optimisation application. This is because, in the latter application, better results are achieved if the best performing individual from a particular population is left in the control filter while the next generation is being calculated. After all individuals in the next generation are determined, the fitness of each of them is tested sequentially and the best individual from the new population is then left in the control filter while the next generation is calculated.

It is of interest to discuss how the original population of individuals (usually 100 in number for the control source optimisation and 40 for the control filter weight optimisation) was established. In the case of control source location optimisation, the population was selected at random from using the available locations and the specified number of control sources. However, in the case of control filter weight adaptation, the requirement to maintain good online performance meant that it was best to begin with very small values of the weights for all members of the population and use a high mutation rate with the mutation amplitude decreasing as the filter weights converged towards their optimum values. At the beginning of the optimisation process for the control filter weights, the population was 40, the mutation rate was 30 per cent of all filter weights and the mutation amplitude was 3 per cent of the maximum possible filter weight value. As the filter weights converged these values were changes in five stages to final values of 3, 20 per cent and 0.01 per cent, respectively.

### 9.2.3   Crossover

Crossover is a process where a new individual (child) string is created by the random copying of information from two parent strings, with either parent string equally likely to be the information source for each element of the child string. A biased coin toss (with a success probability, $P_c$) is used in determining whether the crossover will occur, or whether one parent string (with highest fitness value) is copied directly to the child population. For the examples discussed here, it was found advantageous for $P_c$ to be always set equal to one.

The random uniform crossover method [288] simply involves randomly copying string information from either parent (with equal probability) to fill each string position in the child string as shown in Figures 9.4a. and 9.4c. The exact form in which crossover is performed remains fundamentally unchanged, despite the fact that the string may be made up of either separate integer numbers, or bit values from a binary string. However, the effect that crossover has on the variable values represented in the child string varies significantly between coding types.

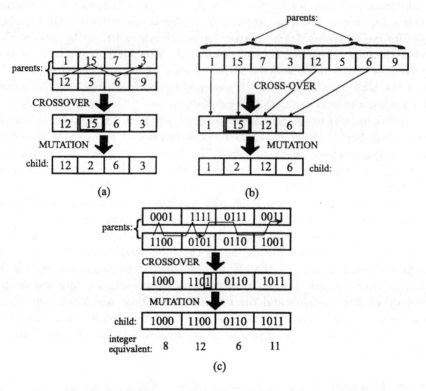

Figure 9.4    *Illustration for crossover and mutation (a) numerical string coding, (b) numerical string coding, (c) multivariable binary string coding*

In the case of numerical strings, each position in the string (consisting of either an integer value or a discretised real number) represents the whole value of an individual search variable. When crossover occurs, the newly created child string will contain only a random mix of values which exist in either parent string (Figure 9.4a). This contrasts with the crossover of multivariable binary strings, where partial exchanges of the binary information representing each search variable can occur (Figure 9.4*b*). The result of partial exchanges of binary information is that the values represented by the child string may not necessarily be the same as those in the corresponding positions of either parent (in contrast to the case of the numerical string). Thus, an inherent diversity is observed when two variable values represented as binary numbers undergo crossover.

The net effect of this increased diversity during MVBS crossover remains unclear. Although increased diversity in the search population is desirable in sustaining a meaningful genetic algorithm search for longer periods of time [332], it may also cause the search to lose focus and slow down the convergence rate. Therefore, as suggested above, it remains necessary to investigate the benefits of either coding method empirically.

In a traditional crossover operation, the new child is generated by randomly copying string information from one or other of the parents as shown in Figures 9.4*a* or 9.4*c*. However, it can be seen that the method illustrated in Figure 9.4*a* eliminates the possibility of the two values corresponding to any particular string location for the two parents ever appearing together in the child. Thus it would seem that a better technique (at least for the control source location optimisation problem) would be not to use crossover in the strict sense but instead allow any of the string values for both parents to appear in any of the string locations for the child; that is, use a meta-string representation as shown in Figure 9.4*b*. When an element is selected from the meta-string it is removed to avoid duplication through reselection. If a duplicate is selected, which can happen when there are multiple entries in the meta-string, particularly near convergence, it is rejected and the selection is made again. This ensures uniqueness of all actuator locations in the final solution. Clearly this approach is not appropriate for the control filter weights as in that case, position in the string is also important. Thus the scheme illustrated in Figure 9.4a is used for optimisation of the control filter weights and the example used here is the optimisation of the weights in a number of different nonlinear filters for minimising vibration in a beam using a nonlinear control source.

The technique of Figure 9.4*b*, which is appropriate for the control source optimisation problem, differs subtly from the traditional crossover technique used for breeding of integer coded strings in the number of unique combinations of actuator locations that can be produced. The simplest crossover scheme involves randomly copying string information from either parent to fill each string of the child and for a string length of $n$ in each parent there are $2^n$ possible unique children. The technique represented in Figure 9.4*b* produces $(2n)!/(n!)^2$ possible unique children. As an example, for a five-actuator system, the crossover method can result in 32 possible different children whereas the meta-string method can result in 252 different children; thus the latter method is a more exhaustive search of the solution space. All three types of crossover illustrated in Figure 9.4 are used here for the example problem involving optimisation of control source location and the performance of each approach is compared.

### 9.2.4   Mutation

It is important to have a mechanism by which the population can be prevented from becoming too uniform; that is, population diversity is essential to ensure that local optima do not become the final solution. The maintenance of population diversity is achieved using a process known as mutation, where one of the string values is occasionally changed on a random basis in an attempt to prevent the irreversible loss of string information which may be vital to the progress of the search. In addition to mutation, diversity can be encouraged by penalising solutions that are similar to other solutions in the population.

The mutation operation itself is an occasional (with small probability) random alteration of string values. For integer strings, this involves replacing selected string values with a randomly selected integer value (Figure 9.4a or 9.4b). For binary strings, this simply means flipping the value of a single-bit position (Figure 9.4c). By itself, mutation represents a random walk through the parameter space. However, in the context of genetic algorithms, mutation plays an important secondary role in reintroducing lost information from overzealous exploitation by the parent selection and crossover operations. In practice, mutation minimises the risk of the search terminating at a local optimum, which is far removed from the global optimum.

For the optimisation of control source location when binary strings are used, mutation involves the random flipping of bits (for example five or six in one thousand bit transfers) during the crossover operation. When numerical strings were used, on average, 1 in 50 strings were affected and one number in that string would be replaced by a number selected randomly within the search space.

For the control filter weight optimisation (for which numerical strings were used), the optimum mutation rate was found to be similar in terms of the percentage of weights in each string which were mutated in a given population. Once a weight in a particular string was selected (approximately 30 per cent probability) for mutation, it was changed by a random amount limited to a specific small range. The optimum range was found to be problem specific and is a compromise between the need to ensure convergence to a global optimum and the need to maximise the online performance of the controller. In fact, for the cases considered here, the mutation amplitude was probably too small to ensure that the entire search space was covered, resulting in control filter weight combinations which may not be globally optimal.

Another way of encouraging population diversity is to prevent the existence of duplicate solutions in the breeding population. To do this, forced mutation is used, which consists of a simple mutation procedure with a high probability, such that duplicate strings are significantly altered from their original state, by introducing randomly selected values at selected string locations. By allowing some values of that string to undergo a random mutation and be replaced by a different (valid) string value, a new (distinct) search solution is created and included back in the search in the normal way. For the case of numerical strings such as those used for control filter weight optimisation, all filter strings were mutated anyway to some extent so forced mutation did not need to be applied.

The forced mutation concept may also be used in the control source location optimisation problem to take care of solutions in which more than one control source occupies the same location or if two control sources do not satisfy minimum clearance requirements. Clearly, this is not a problem for the control filter weight optimisation because more than one weight may have the same value. However, in the actuator placement cases considered here, the forced mutation operator is also used to deal with circumstances where a newly-created string already exists in the breeding population to be used next.

In addition to preventing duplicate strings and string values, forced mutation allows the reintroduction of new genetic material into the search. This occurs at a much faster rate than natural mutation, without significant detriment to the overall search progress. The result is potentially significantly higher diversity levels being sustained throughout the search procedure, meaning that for multimodal search problems, the genetic algorithm search is much less prone to focussing exclusively on a local optimum. Thus, forced mutation enables a more diverse range of strings to be reproduced, allowing the search to span a wider proportion of the solution space when searching for the global optimum solution.

### 9.2.5 Sharing

In natural genetics, the concepts of niche and species are fundamental to the simultaneous coexistence of many different living species. In each case, stable sub-populations of a particular species are formed when there exists increased competition for finite resources (which are vital for survival) limiting population growth and size. A similar concept to this has been developed in genetic algorithms [112]. In this method, a practical sharing scheme was developed so that similar string solutions (in the breeding population) competed for breeding opportunities and continued survival. Sharing is a method which achieved this by applying a penalty (based on the value returned by a summary measurement function) to the parent selection probabilities of breeding population members that are similar in the string values they contain. In this way, the uncontrolled growth of a single string type amongst all population members is limited according to the number of similar solutions which coexist around it, thus preventing the search from stagnating due to homogeneity. In multimodal problems, it was reported previously that as a result of sharing, stable subpopulations were formed around each optimum, where the number of strings around each occurred in relative proportion to the magnitude of the locally optimum fitness value.

Although the concept of sharing is too time consuming to implement in the control filter weight optimisation process, it is easily applied to the control source location optimisation problem and is used here as an effective mechanism for maintaining adequate levels of diversity amongst members of the breeding population. To do this, a sharing distance $d_{ij}$ is defined which determines the neighbourhood and degree of sharing that exists between strings $i$ and $j$ of the breeding population. A sharing function $F_s$, is defined such that:

$$F_s = F_s(d_{ij}) \tag{9.4}$$

and is characterised by a linear inverse relationship with the sharing distance $d_{ij}$, illustrated in Figure 9.5.

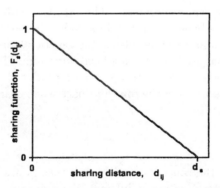

Figure 9.5    *Sharing function versus distance between adjacent locations*

A constant $d_s$ is defined, which represents the upper bound for sharing distance values, below which sharing is considered significant. The sharing function values range from 0 (strings have no similarity) to 1 (strings are exactly the same). The niche count $M_i$ is defined (for string $i$) as the sum of the sharing function values for each string comparisons such that:

$$M_i = \sum_{j=1}^{n_{pop}} F_s(d_{ij}) \qquad (9.5)$$

The fitness value $f_i$ (of string $i$) is then derated in direct proportion to its niche count $M_i$ such that:

$$f_i' = \frac{f_i}{M_i} \qquad (9.6)$$

Altering the fitness values in this manner has the effect of increasing the focus of the search towards high-fitness solutions which are under-represented within the breeding population, thus maintaining sustainable levels of diversity.

## 9.3   Control source location optimisation example

### 9.3.1   Analytical model

In this Section, optimisation of the control source locations is compared for each of the breeding schemes shown in Figure 9.4 for some physical examples involving sound transmission into an air-filled rib stiffened cylinder with a stiffened floor. The stiffeners were constructed on 1 mm thick sheet steel. The model, illustrated in Figure 9.6, has the following dimensions: length, $L = 3$ m; radius, $R = 0.45$ m and floor angle $\theta_f = 40°$. The thickness of the shell and floor material was: $h_f = h_s = 0.00086$ m, with each having the following material properties: Young's modulus $E = 209$ GPa, Poisson's ratio $\nu = 0.3$ and material density $\rho = 7930$ kg/m$^3$. The fluid within the enclosure was modelled with density $\rho_0 = 1.19$ kg/m$^3$ and the speed of sound $c_0 = 343$ m/s.

Thirty longitudinal stiffeners were attached to the external cylinder surface, with six stiffeners attached to the internal floor structure. Each stiffener crosssection resembled an uneven C section, with a height of 0.027 m and with upper and lower flange widths of 0.018 m and 0.0065 m, respectively.

A finite-element model of the structure was developed using the Ansys finite-element analysis software. A total of 1824 shell elements were used to represent both the cylinder and floor structures; a total of 864 beam elements were used for representing the longitudinal stiffeners attached to the shell

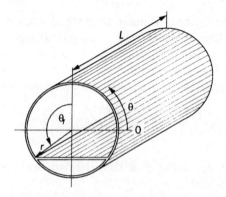

*Figure 9.6    Stiffened cylinder model*

surfaces. In addition, a separate finite element model was developed for the interior acoustic volume. The entire enclosed volume was represented by 4464 acoustic volume elements. The structure and acoustic models were solved separately, to determine the resonance frequencies and mode shapes. Modal coupling theory [239, 272, 273] was used to predict the system response for specified dynamic loading conditions (in this case, point forces on the cylinder surface). Quadratic optimisation theory [220] was then used (for specified primary and control force locations) to determine the optimum control forces for minimisation of the interior acoustic potential energy (cost function).

To calculate the cost function for a particular control source configuration on the structure, the following calculation method was used. The total pressure $p(\vec{y})$ at an arbitrary location $\vec{y} = (r, \theta, z)$ in the enclosure is equal to the sum of the pressure field due to the action of the primary excitation, $p_p(\vec{y})$, and the influence of the applied control forces, $p_c(\vec{y})$, such that:

$$p(\vec{y}) = p_p(\vec{y}) + p_c(\vec{y}) \qquad (9.7)$$

where the subscripts $p$ and $c$ denote primary and control, respectively.

The acoustic potential energy inside the (bounded) enclosure is defined as [40]:

$$E_p = \frac{1}{4\rho_0 c_o^2} \int_V |p(\vec{y})|^2 \, \partial \vec{y} \qquad (9.8)$$

which can be expressed as:

$$E_p = \frac{1}{4\rho_0 c_o^2} \int_V [p(\vec{y})]^{\mathrm{H}} \, [p(\vec{y})] \, \partial \vec{y} \qquad (9.9)$$

Using the orthogonal properties of the rigid-walled acoustic modes in the present model allows eqn. (9.9) to be expressed as [292]:

$$E_p = \frac{1}{4\rho_0 c_o^2} [P]^{\mathrm{H}} [P] \tag{9.10}$$

where $[P]$ is the $(n \times 1)$ vector of acoustic modal amplitudes given by:

$$[P] = [Z_a] [Z_i]^{-1} [\Gamma] \tag{9.11}$$

where $[Z_a]$ is the $(n \times m)$ matrix of acoustic modal radiation transfer functions, $[Z_i]$ is the $(m \times m)$ modal structural impedance matrix and $[\Gamma]$ is the $(m \times 1)$ vector of modal generalised forces which may be expressed as:

$$[\Gamma] = [\Psi] [F] \tag{9.12}$$

where $[\Psi]$ is the $(m \times i)$ structural mode shape matrix and $[F]$ is the $(i \times 1)$ vector of force inputs.

For $i$ complex control forces modelled as point forces, eqn. (9.10) may be expressed as [346]:

$$E_p = \frac{1}{4\rho_0 c_o^2} \left( [F_c]^{\mathrm{H}} [a] [F_c] + [F_c]^{\mathrm{H}} [b] + [b]^{\mathrm{H}} [F_c] + c \right) \tag{9.13}$$

where

$$[a] = [\Psi_c]^{\mathrm{H}} \left( [Z_i]^{-1} \right)^{\mathrm{H}} [Z_a]^{\mathrm{H}} [Z_a] [Z_i]^{-1} [\Psi_c] \tag{9.14}$$

$$[b] = [\Psi_c]^{\mathrm{H}} \left( [Z_i]^{-1} \right)^{\mathrm{H}} [Z_a]^{\mathrm{H}} [Z_a] [Z_i]^{-1} [\Psi_p] \tag{9.15}$$

$$c = [\Psi_p]^{\mathrm{H}} \left( [Z_i]^{-1} \right)^{\mathrm{H}} [Z_a]^{\mathrm{H}} [Z_a] [Z_i]^{-1} [\Psi_p] \tag{9.16}$$

The optimal control force vector, $[F_c]_{opt}$, to minimise the acoustic potential energy inside the enclosure is found from quadratic optimisation to be:

$$[F_c]_{opt} = - [a]^{-1} [b] \tag{9.17}$$

For the present example, the use of the above cost function formulation to guide the genetic algorithm in the search for optimum control source locations requires that the reduction in the total acoustic potential energy be calculated. This is defined as the difference in the acoustic potential energy levels both before and after active control is applied. Therefore, the point at which the total acoustic potential energy reduction is maximised corresponds to the same point where the total internal acoustic potential energy level is minimised. For the comparisons reported here an excitation frequency of 85 Hz was used.

## 9.3.2   Genetic algorithm formulation

The advantage of a cost function calculated using finite-element data rather than classical analysis is that the genetic algorithm once developed is more generic to a wide variety of different practical optimisation placement problems. In the finite-element model, each node on the structure and the acoustic interior was assigned a unique node number which could be used to identify a unique actuator location for the genetic algorithm. Actuator locations on the floor structure were not used for the results discussed here. The search grid size was 23 (axial) by 62 (circumferential) locations. The genetic algorithm was then used to find the four control actuator locations which resulted in the least acoustic potential energy in the enclosed space for a particular primary excitation (which for the cases considered here was ten actuators placed at random over the surface of the cylinder). Coding actuator locations using node numbers from a finite-element model has the advantage of being blind to any changes in the crosssection shape of the model. For all search cases, actuators were not permitted to be located at the same node position as a primary source, although they were allowed to approach as close as the next adjacent node position. Also, the node locations at the edge of the structure were not used as the modal displacements at these node positions were zero, so actuators at these locations would have no effect.

For reasons of higher search efficiency as explained previously, a steady-state genetic algorithm was used for the test comparisons performed. Two different coding schemes were implemented. An integer string coding and a multivariable binary string coding. The three different crossover methods discussed previously were implemented and the results achieved using each are compared for a population size of 100 individuals. For each case results were averaged for ten separate runs, beginning with random control actuator locations. In all cases, "Stochastic remainder selection without replacement" was used rather than simple roulette wheel selection for the selection of parent strings for breeding. Rank probability distributions were utilised. The fitness was not allocated according to the magnitude of the fitness value, but rather the rank position of the string in the breeding population.

To simplify initial comparisons between different crossover methods in the cases outlined below, the crossover probability, $P_c$, was set to 1. This means that each child produced is only as a direct result of the crossover that occurs between two parents. For the case of the modified integer coding method, it was possible for all of the information in a child to come from only one parent. However, for the integer and binary string coding methods, there were always two parents involved in producing a child except for the case where the crossover probability was set to a value less than one, in which case there is always a chance that either parent would be selected and copied directly to the child. However, a constraint imposed on the searching process prevents two identical strings from coexisting in the same breeding population (as this duplication would represent a loss of population diversity). Thus, any new solution to be inserted into the breeding population that is identical to one already existing there, will undergo a forced mutation. This means that the values of the string have a high probability ($P_m = 0.3$) of undergoing a mutation process. A crossover probability of less than unity represents a mechanism where the random variation of strings already existing in the breeding population can occur. At the beginning of a search, this process can assist in performing a wider search for good solutions but, if used too liberally, can detract from the effective recombination and exploitation of string information already existing in the breeding population.

At the end of a search, the principle of preventing string duplication in the breeding population has a different role in the continued search for optimum solutions. Lower levels of population diversity will ultimately occur near the end of a search, due to the eventual continued selection and retention of the best solutions from the entire search process. This often means that only a few key string combinations (consisting of one, maybe two actuator locations that occur in combination) are common to many strings in the population. With increasing levels of population uniformity, the likelihood of duplicate strings being produced as a result of crossover is significantly increased. When this occurs, forced mutation is enacted to potentially reintroduce population diversity without detracting from the existing search *via* crossover. Setting the crossover probability, $P_c$, to a value less than 1.0 is a way of enacting this random searching mechanism (*via* forced mutation) at an earlier stage in a search than is likely to otherwise naturally occur.

Comparisons are to be made in this Chapter regarding the effectiveness of different crossover methods in the context of a realistic actuator placement optimisation problem. Any value of crossover probability less than unity could have the effect of clouding eventual comparison results between different crossover methods, which is why it was set to a value of unity. Also, to aid the comparison of different crossover methods, the elitist model, fitness scaling and sharing operations were not used. The elitist model is a method that ensures the continued survival of the best solution found so far in a search from generation to generation. However, this work is based on the use of a steady-state genetic algorithm, where the principles of the elitist model are inherent to the algorithm design. A fitness proportionate probability distribution and fitness scaling operator were not used in these comparisons, owing to the direct controllability of the selection pressure that was afforded by the use of a rank probability distribution method in combination with the steady-state genetic algorithm. The use of a sharing function (to maintain population diversity near the end of a search) was not adopted for the comparisons performed in this Chapter. This is due to the need to calculate a string fitness penalty function that is dependent on the degree of location similarity that each new string has to other existing members the breeding population. The genetic algorithm used for the present comparisons uses node numbers (from the finite element model) to specify the actuator locations on the structure. However, adjacent node numbers do not always have a direct correlation to adjacent locations on the structure. For a sharing function to be most effective, the location of each node on the structure would need to be known to the genetic algorithm, so that the correct penalty can be assigned to each string. This was not undertaken for the present comparisons.

To compare the performances of the different crossover methods, a search problem more representative of the complexity of a propeller pressure distribution on a fuselage structure is developed. This involves randomly locating ten primary point forces (of random amplitude and phase) over the exterior surface of the stiffened cylinder structure (excluding the floor). The task performed by the genetic algorithm is to optimally locate four control sources on the surface of the structure (excluding the floor) to minimise the total acoustic potential energy levels inside.

The web publishing forum is used here to demonstrate the relative effectiveness of the different alternative approaches in converging to an optimal solution.

### 9.3.3 Results

For each coding scheme shown in Figure 9.4, ten searches were conducted for the same arrangement of primary sources and for random starting arrangements of the four control actuators. In addition, a random search was conducted to show how it compared with the genetic search approach. Results showing interior potential energy reduction as a function of iteration number are shown in Figure 9.7 for each of the coding schemes and for the random search. For each scheme, the best result out of a total of ten searches is shown. It is clear from Figure 9.7 that the use of a genetic algorithm yields better overall search results than a random search for optimum actuator locations. In addition, the superior performance of the integer string crossover methods compared with multivariable binary string crossover can be clearly seen. The modified integer crossover method (Figure 9.4 *b*) was found to perform the best, both in terms of the rate of convergence and the final fitness value of the solutions achieved. The reason for this is that the genetic information containing individual actuator locations is no longer restricted to occupying just the one string position in the breeding population. The recombination of the parent string values in a different random order when formulating the child string appears to aid the overall progress and convergence rate of the search. This creates a greater opportunity for diversity in string recombinations, as the pool of genetic information available is now potentially the whole breeding population.

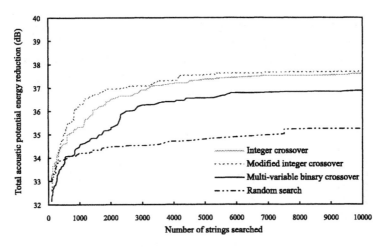

Figure 9.7    *Average of ten search performance for the different coding schemes* $(P_c = 1)$

Another reason for the improved performance of the modified integer string coding method displays a much greater potential to create similar string solutions (that is, strings with the same string values rearranged in a different order) in the breeding population. As a constraint only exists for preventing identical string solutions (that is, strings with the same string values arranged in the same order), this creates the opportunity for greater selective pressure to be applied to particular solution types in the breeding population. This promotes further reproduction and recombination opportunities to these solutions. This, combined with the use of forced mutation, appears beneficial to the overall task of finding near-global optimum solutions in the present example.

In the context of the present example, the diversity advantages of using the multivariable binary string (discussed previously in Reference [265]) were not realised in the context of the present actuator placement problem. It appears that the more focussed exploitative nature of the integer coding methods (in conjunction with using the FEA structure node location numbers to represent actuator locations) is sufficient for finding the optimum control source positions for the present example. This was clearly demonstrated by the continued convergence of the MIS crossover method, well beyond the stage where other algorithm types have stagnated in their ability to find improved search solutions.

In Figure 9.8 are shown the optimum actuator locations corresponding to the different curves of Figure 9.7. It is clear that substantial levels of noise reduction are possible with a number of different actuator configurations. It is interesting to note that for the integer crossover and binary string crossover coding methods, two of the actuator optimal locations are identical.

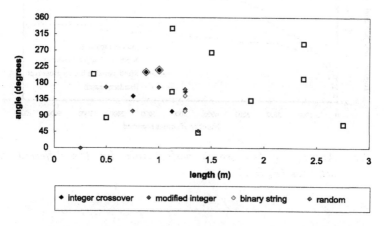

Figure 9.8    *Optimal control actuator locations using various genetic algorithm crossover methods of Figure 9.4 and a random search. The squares represent the locations of the control sources (10) and the diamonds represent the control actuator locations (4 for each crossover scheme)*

To test the effect of changing the crossover probability to something other than 1.0, the searches were repeated for crossover probabilities of 0.6 and 0.8 and the results are shown in Figures 9.9, 9.10 and 9.11. The results for all of the cases tested are summarised in Table 9.1. It can be seen that reducing the crossover probability to a value less than unity generally resulted in superior search performance. This is apparent for the integer string and modified integer string crossover methods, where a crossover probability of $P_c = 0.8$ was found to provide the optimum average acoustic potential energy reduction for ten separate run cases. For the multivariable binary string (MVBS) crossover method, there was no improvement in the average search results achieved over ten separate runs with decreasing crossover probability. However, significant improvement was observed for the initial convergence rate of the MVBS genetic algorithm search when the crossover probability was lowered. The best initial convergence rate for the MVBS method was observed for the case where $P_c = 0.6$. The MVBS search was found to converge comparably with the two integer string crossover methods (for $P_c = 0.6$) until the stage where approximately 2000 strings were searched. After this, the MVBS crossover method lagged in its ability to seek further improvement, and generally did no better that for the case when $P_c = 1.0$.

| | Maximum | | | Minimum | | | Average | | |
|---|---|---|---|---|---|---|---|---|---|
| Value of $P_c$ | 1.0 | 0.8 | 0.6 | 1.0 | 0.8 | 0.6 | 1.0 | 0.8 | 0.6 |
| Modified integer coding (Figure 9.4b) | 38.7 | 39.2 | 38.9 | 37.0 | 37.6 | 37.0 | 37.7 | 38.3 | 38.0 |
| Integer coding (Figure 9.4a) | 38.6 | 39.1 | 38.3 | 37.2 | 37.1 | 37.3 | 37.6 | 37.9 | 37.7 |
| Binary coding (Figure 9.4c) | 37.9 | 38.2 | 37.9 | 35.9 | 36.0 | 36.2 | 36.9 | 36.9 | 36.9 |
| Random search ($P_c$ is undefined) | 36.6 | 36.6 | 36.6 | 34.5 | 34.5 | 34.5 | 35.2 | 35.2 | 35.2 |

Table 9.1
  *Variability in interior potential energy reduction (dB) over a total of ten searches*

Figure 9.11 shows the improvement in average search performance for the case of modified integer string (MIS) crossover for decreasing crossover probability. Here, the initial convergence rates for the average search results decrease with decreasing crossover probability. However, for the case of $P_c = 0.8$, the search continued on to seek further improvement longer than for any other search case. These results indicate two main findings in the context of using genetic algorithms to seek optimum vibration control source locations on a

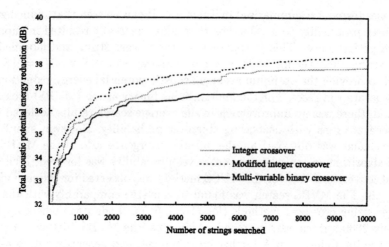

Figure 9.9    *Average of ten search performance for the different coding schemes* $(P_c = 0.8)$

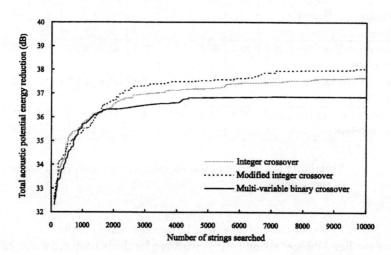

Figure 9.10    *Average of ten search performance for the different coding schemes* $(P_c = 0.6)$

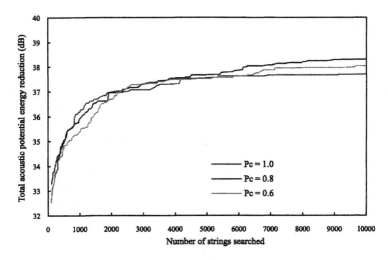

*Figure 9.11*    *Search performance for the modified integer string coding for various values of $P_c$*

structure to minimise the noise transmitted inside. First, the value of allowing string values from either parent to occupy any child string position is demonstrated with the superior performance of the modified integer string crossover method. Second, the value of allowing some forced mutations on parent strings (for the cases where crossover is not performed), to introduce new genetic information to the search, without significant detriment to the overall search performance. Allowing forced mutation of an unchanged parent string provided a mechanism for new genetic information to be introduced to the search process at a much higher rate than might be possible *via* natural mutation alone. For the case of a steady-state genetic algorithm, this provides a mechanism for the random variation of the strings selected for breeding. Thus, two search types coexisted (a randomised variation of string values and the genetic recombination two different strings) in a manner which was beneficial to the overall search for improvement in control source locations.

## 9.4    Example of control filter weight optimisation

A single channel example involving vibration control of a simple aluminum beam, supported at both ends, will be used to demonstrate the effectiveness of the genetic algorithm with three different types of filter structure. The beam crosssection was 25 mm by 50 mm and it was driven on the 50 mm face by two electrodynamic shakers, one providing the primary excitation sinusoid at 133 Hz and the other providing the control force. Higher harmonics were introduced into the control force by inserting a rubber pad between the shaker and the beam as shown in Figure 9.12, which should be looked at with Figure 9.2 to show how the genetic algorithm is involved in adapting the controller.

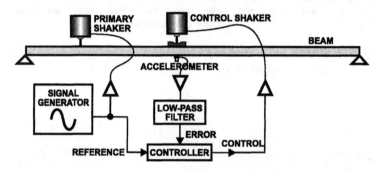

*Figure 9.12    Arrangement for nonlinear control of beam vibration*

The error signal was measured using an accelerometer (with a 1 kHz low-pass filtered output) attached to the beam directly opposite the shaker. The controller sampling rate was 2.5 kHz. The performance measurement (mean-square acceleration) was averaged over 50 samples with a 100 sample delay prior to each measurement to allow any transient vibration from the previous filter weights to subside. This means that it took 2.5 s to test the performance of one generation of the 40 individuals. Note that the genetic algorithm could have handled power or intensity as an error signal just as easily.

During convergence to the final filter weights, the genetic algorithm parameters were manually changed in five stages as indicated in Table 9.2. In the first four stages, parents were not removed from the population when selected for breeding, thus allowing the possibility of a child being generated with both parents being the same. In the last stage, identical parents were not allowed (breeding copies = N), although, after each child string was created, the parents were returned to the mating pool. The term uniform crossover probability refers to the probability of using a weight from the least fit parent in the crossover operation. The rank probability distributions used for killing selection and parent breeding selection are shown in Figures 9.13 and 9.14.

| PARAMETER | SET 1 | SET 2 | SET 3 | SET 4 | SET 5 |
|---|---|---|---|---|---|
| Title | initial | intermediate | standard | fast tracking | smallest population |
| Population size | 40 | 30 | 20 | 10 | 3.00 |
| Killing or removal % | 60 | 65 | 65 | 70 | 70 |
| Uniform crossover probability % | 45 | 50 | 50 | 50 | 50 |
| Mutation probability % | 30 | 30 | 30 | 30 | 30 |
| Mutation amplitude (% max) | 3 | 1.50 | 0.30 | 0.03 | 0.01 |
| Breeding copies? | Y | Y | Y | Y | N |
| Copy mutation probability % | 20 | 20 | 20 | 25 | - |
| Copy mutation amplitude % | 15 | 2.5 | 1.2 | 0.06 | - |

Table 9.2
*Genetic algorithm parameter settings for filter weight optimisation*

Three different filter structures were tested. The first was a linear FIR filter which was only capable of producing frequencies present in the reference signal, in this case a 133 Hz sinusoid. The second filter structure used was a nonlinear polynomial filter consisted of two 50-tap FIR filters, the inputs of which were the reference signal raised to the fourth and fifth powers, respectively, with the control signal obtained by adding the filter outputs (Figure 9.15). Raising a signal to the fourth power gives a signal consisting of the second and fourth-order harmonics of the initial signal content. Similarly, raising a signal to the fifth power gives a signal with first (fundamental), third, and fifth-order harmonic content. Hence, this polynomial filter, referred to as a P4P5 filter, can only produce harmonics (including the fundamental) of the reference signal up to the fifth order.

The third filter tested was a neural-network-based filter structure which is shown in Figure 9.16, and has 50 taps, one hidden layer with 20 nodes, and one (linear) output layer node (designated 50 x 20 x 1). Four different transfer functions were utilised simultaneously in the hidden layer, as shown in Figure 9.16, with equal numbers (that is, five) of each type being used.

Final converged vibration levels obtained using each of the three types of filter structure are shown in Figures 9.17-9.19. The genetic algorithm adapted

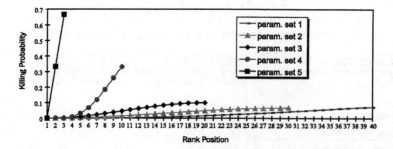

Figure 9.13    *Killing probability as a function of rank order in the population for selecting parents for removing from the population prior to breeding for the control filter weight genetic algorithm*

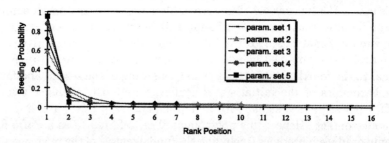

Figure 9.14    *Breeding probability as a function of rank order in the population for selecting parents for breeding for the control filter weight genetic algorithm*

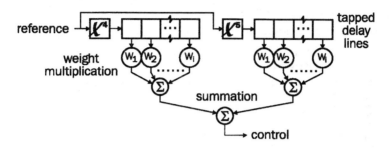

*Figure 9.15    P4P5 filter configuration*

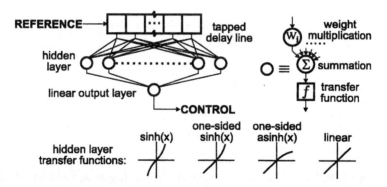

*Figure 9.16    Neural network filter structure*

FIR filter gave a maximum of 12 dB mean-square error (MSE) reduction within 40 s. The P4P5 filter gave 12 dB at 50 s, and a maximum of 36 dB within three min. The 50 x 20 x 1 neural network filter gave 24 dB at 50 s, 30 dB in 6 min, and a maximum of 32 dB MSE reduction within 15 min. A summary of the attenuation achieved at the harmonic peaks is given in Table 9.3.

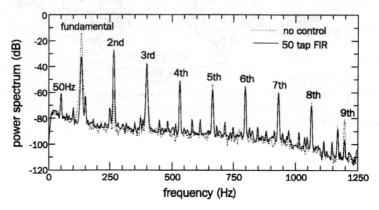

*Figure 9.17    Control results of a 50 tap FIR control filter*

| | 1st | 2nd | 3rd | 4th | 5th | 6th | 7th | 8th | 9th |
|---|---|---|---|---|---|---|---|---|---|
| FIR | 18 | -32 | -8 | -5 | 4 | -4 | -3 | 3 | 16 |
| P4P5 | 40 | 8 | 7 | 13 | 10 | 15 | 26 | 37 | 29 |
| NN | 47 | -2 | 12 | -7 | 17 | 14 | 27 | 33 | 26 |

Table 9.3

*Error signal power spectrum attenuation (dB) at each harmonic for the FIR, P4P5 and neural network filters*

For the FIR filter case, attenuation of the fundamental peak at 133 Hz is limited owing to the introduction of the higher-order harmonics by the nonlinear control source. The P4P5 filter achieved the best overall reduction, with all harmonic peaks being attenuated. In comparison, the neural network filter structure has given greater control of the first, third, and fifth-order components, but has caused the second and fourth-order components to increase.

Note that the presence of small quantities of higher-order harmonics in the reference signal seen by the controller (due to harmonic distortion in the signal

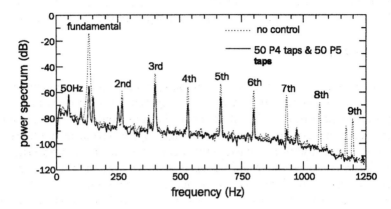

Figure 9.18    *Control results with P4P5 control filter*

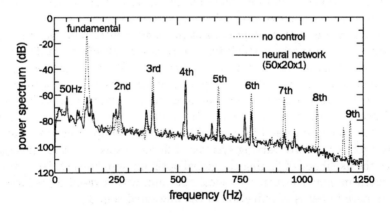

Figure 9.19    *Control results with a neural network control filter*

generator) has allowed the attenuation of higher-order harmonics which would not normally be possible for the FIR (eighth and ninth harmonics) and P4P5 (sixth to ninth) filter structures when given a purely sinusoidal reference.

## 9.5   Conclusions

It has been shown how two previously seemingly intractable problems in feed-forward active noise control can be solved by the use of genetic algorithms. The problems are optimisation of the control source configuration in complex, multisource systems and control filter weight optimisation in nonlinear systems or in systems with nonlinear control actuators (such as piezoceramics or magnetostrictive devices). In both cases it is shown that the genetic algorithm exhibits consistent performance (in terms of the minimum value of the cost function achieved) over a series of tests, even though the final solutions (in terms of string values) were different for each test. For the case of optimisation of control source locations, it is shown that it was necessary to remove a large number of poorly performing individuals from the population prior to breeding in order to speed up the convergence to optimum. It was also found that the convergence speed was dependent on the type of breeding and coding used (see Figure 9.4). The optimum solution was also found to be quite frequency dependent, making the usefulness of control source location optimisation questionable when control of sound transmission through a complex structure must be achieved over a band of frequencies.

For the case of online adaptation of the control filter weights, the genetic algorithm showed excellent stability, mainly as a result of not needing knowledge of the cancellation path transfer function. Other advantages included independence of the type of filter structure on the performance of the genetic algorithm and relaxation of the requirement for the error signal to be linearly correlated with the reference signal. However, the genetic algorithm is relatively slow to converge and for control filter weight adaptation it is thus only suitable for very slowly varying systems. Efforts to speed up the convergence reduce the likelihood of finding the global optimum weight coefficients; instead the algorithm may converge to a local optimum with a resulting reduction in performance in terms of reduction of the unwanted sound.

## 9.6   Acknowledgments

The authors gratefully acknowledge financial support for this work from the Sir Ross and Sir Keith Smith Fund and the Australian Research Council.

# Part III

# Applications

*Chapter 10*

# Active noise control around a human's head using an adaptive prediction method

## S. Honda and H. Hamada

*Department of Information and Communication Engineering, Tokio Denki University, Tokio, Japan, Email: honda@acl.c.dendai.ac.jp*

*This chapter introduces a new active noise control system for optimal control of the sound field around a human's head. Geometrical arrangements of error microphones are proposed around the head in a diffused noise field. In order to evaluate the controlled sound field, a rigid sphere theoretical model is used as a human's head in the computer simulations.*

*In previous studies it was shown that the theoretical sphere model maintains a good approximation of a head-related transfer function (HRTF) below about 5 kHz. In this chapter the relationship between the noise reduction and the corresponding configuration of the error microphone and secondary source arrangement is evaluated. As a result, we found that the closely located secondary sources to the sphere head give significant noise reduction in a large control area. A possible arrangement of the error microphones is found to provide an optimal result according to a given control strategy, such as flatness of reduced sound pressure of the quiet zone.*

## 10.1 Introduction

In this research the aim is to achieve an effective active noise control system at both the listener's ears as well as at a relatively large area around his or her head, so that significant control can be expected even when the listener slightly moves his or her head. For this purpose, we have recently introduced a new strategy for an ANC system, in which the sound field around a human's head has been investigated under noise exposure visualisation techniques.

This was based on the perfect rigid sphere model which was applied as a theoretical model of a human's head. According to the previous work [211, 312, 313], this model maintained a good approximation of the head-related transfer function up to 5 kHz, so that we were now able to define the theoretically calculated HRTF, referred to as the sphere head-related transfer function (SHRTF), to examine acoustic properties around the head.

In an ANC system the effectiveness of noise reduction and its area strongly depend on the primary and secondary sound field as well as on the performance of the controller. If an adequate arrangement between the secondary source and the controlled point can be chosen, it is possible to manipulate the shape of the equalised area. Increasing the number of secondary sources and control points may allow the expansion of the quiet zone, although there is always a restriction on the number of them for practical appliactions [105, 191].

In this Chapter the discussion is focussed on producing a large quiet zone using only a small number of error microphones and secondary sources. The number of secondary sources is chosen to be two for the typical case of the ANC system to control two points at both the listener's ears. Fundamental investigation is carried out for two different sound fields, these are: the free sound field in which the incoming direction of the primary source is known, and the diffused sound field. In addition, we also examine the ANC system for multiple control points to clarify the relationship between controller effectiveness and the number of control points.

## 10.2    Outline of the system

Figure 10.1 shows a block diagram of the single-channel adaptive prediction type active noise controller referred to as the AP-ANC system. $\mathbf{C}$ is the secondary transfer function between the secondary sources and error microphones, and the control filter is $\mathbf{W}$. The AP-ANC system can be divided into two processes: a control process and a learning process. In the learning process shown in the lower part of Figure 10.1 an instantaneous value of the noise signal at the error microphone can be predicted by the time history of the past noise signal. This process is sufficient to update the filter coefficients of the controller, and therefore only information from the error microphone is used to control the noise in the AP-ANC system [89, 120, 233, 279].

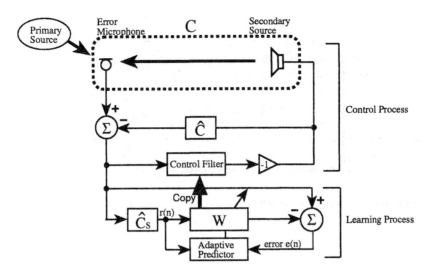

*Figure 10.1    A block diagram of AP-ANC*

We can expand a single-channel AP-ANC system to a multiple-channel system to discuss the area of a controlled sound field in a more general sense. In the following discussions only the control process will be taken into consideration. The control process as in Figure 10.1 can be reformulated to Figure 10.2.

The AP-ANC needs only the primary noise signal without antinoise signals. The AP-ANC has a cancellation filter $\hat{C}$. If the system is capable of performing perfect feedback cancelling, we can rewrite Figure 10.2 as Figure 10.3. Figure 10.3 shows a block diagram of a multiple-channel AP-ANC system.

In this Chapter the primary sound field is assumed to be stationary. This primary noise is actively cancelled by $L$ secondary sources, which are controlled by the adaptive filters to minimise the sum of squares of $M$ error microphone outputs. It is noted that, under low-frequency conditions, surprisingly few error sensors are needed to achieve substantial reductions in total noise energy.

For controlling a noise signal consisting of a pure sinusoid of known frequency, which can be considered a constant value in the frequency domain, we have:

$$D = 1 \qquad (10.1)$$

Thus, by expressing the quantities of Figure 10.3 and omitting the frequency

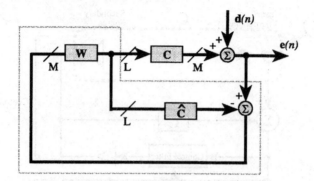

Figure 10.2    *A block diagram of multiple-channel AP-ANC (a part of the control process)*

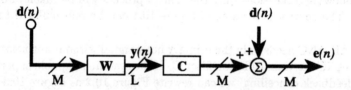

Figure 10.3    *A block diagram of multiple-channel AP-ANC (perfect identification at the control Process)*

index for simplicity, we have:

$$\mathbf{Y} = [\, Y_1 \, Y_2 \, \ldots \, Y_L \,] = \mathbf{W}D = \mathbf{W} \qquad (10.2)$$

where

$$\mathbf{W} \equiv [\, \mathbf{W_1} \, \mathbf{W_2} \, \ldots \mathbf{W_L} \,] \qquad (10.3)$$

is the vector of frequency-domain adaptive weights. Therefore, the error signal can be expressed as:

$$\mathbf{E} \equiv [E_1 \; E_2 \; \ldots \; E_M]^T \; = \; \mathbf{D} + \mathbf{CWD} \tag{10.4}$$

where

$$\mathbf{D} \equiv [D_1 \; D_2 \; \ldots D_M]^T \tag{10.5}$$

and

$$\mathbf{C} = \begin{bmatrix} C_{11} & C_{12} & \cdots & C_{1L} \\ C_{21} & D_{22} & \cdots & C_{2L} \\ \vdots & \vdots & \ddots & \vdots \\ C_{M1} & C_{M2} & \cdots & C_{ML} \end{bmatrix} \tag{10.6}$$

The unconstrained frequency-domain cost function is given in eqn. (10.6) where $\mathbf{E}$ is defined as in eqn. (10.4) and the superscript $H$ (Hermitian) denotes conjugate transpose. By taking the complex gradient of eqn. (10.6), we obtain the multiple-channel frequency-domain FXLMS algorithm.

The filtered reference signal matrix becomes $\mathbf{C}^H$. The reference signal $D$ of eqn. (10.1) is a constant in the frequency domain. It can be easily shown that the steady-state optimal weight solution is given by:

$$\mathbf{W}_{opt} = (\mathbf{C}^H \mathbf{C})^{-1} \mathbf{C}^H \tag{10.7}$$

A theoretical model of $\mathbf{C}$ can be derived by the rigid sphere model (SHRTF). Figure 10.4 shows a relationship between the sound source and the rigid sphere.

We define a transfer function using a rigid sphere as:

$$H = \frac{sound\ pressure\ at\ a\ spherical\ surface}{sound\ pressure\ at\ a\ point\ to\ a\ monopole\ sound\ surface} \tag{10.8}$$

Therefore, the SHRTF used in our simulation is defined as:

$$C_{sf|r=a} = -\frac{1}{ka^2} \sum_{n=0}^{\infty} (2n+1) \frac{h_n^{(2)}(kR)}{h_n^{(2)'}(ka)} P_n(\cos\theta) \tag{10.9}$$

where $k$ is the wave number. The angle $\theta$, distance $r$ and $R$ are defined as shown in Figure 10.4. $a$ is the radius of the rigid sphere; $h_n^{(2)}(z) = j_n(z) - j n_n(z)$ is the spherical Hankel function of the second-order $n$.

If we assume spherical sound waves from the secondary source, the transfer function $C_{sf}$, defined between the secondary sound source and a point located

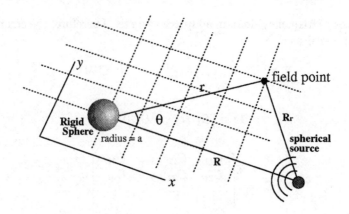

Figure 10.4    *Relationship between spherical source and rigid sphere*

around a rigid sphere, can be written as:

$$C_{sf} = \frac{P_t}{V_s} = \frac{\exp(-ikR)}{R} + ik\sum_{n=0}^{\infty}(2n+1)j_n'(ka)h_n^{(2)}(kr)\frac{h_n^{(2)}(kR)}{h_n^{(2)'}(ka)}P_n(\cos\theta)$$

$$(10.10)$$

Also, in the case of plane wave propagation, the transfer function $C_{pf}$ becomes simpler and is given by:

$$C_{pf} = \frac{P_t}{V_p} = \frac{\exp(-ikR)}{R} - \sum_{n=0}^{\infty}i^n(2n+1)h_n^{(2)}(kr)\frac{j_n'(ka)}{h_n^{(2)'}(ka)}P_n(\cos\theta) \quad (10.11)$$

## 10.3    Simulation

### 10.3.1    *Purpose of simulation*

In this Section we investigate our ANC system using plots of the visualised sound field. First, we deal with the $2L - 2M$ (two secondary sources and two error microphones) ANC system, and examine the difference of the noise reduction due to the primary sound field condition as well as the secondary source locations. We also look at the relationship between the number of error microphones and the controlled area in the diffused sound field.

## 10.3.2   ANC for a single primary source propagation

Here, we deal with the $2L - 2M$ system in a simple sound field, where only a single plane wave propagates as a primary source, and examine how the position of the secondary source can affect the control area of the ANC system. Geometrical arrangements and coordinates used in this simulation are illustrated in Figure 10.5. The incoming direction of the primary source is defined as $0°$ in the right-hand direction.

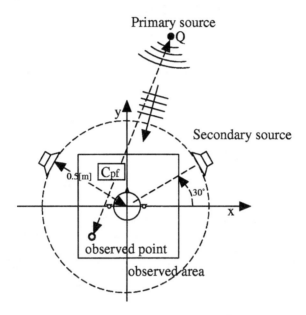

Figure 10.5    *Geometrical arrangements used in the simulation*

The distance between the secondary source and the centre of the rigid sphere is 0.5 m, and the secondary sources in each pair are located symmetrically to $90°$, these are: $0°$ and $180°$, $30°$ and $150°$, $60°$ and $120°$, $85°$ and $95°$. The control points are set to $0°$ and $180°$ on the surface of the rigid sphere with a radius of 0.08 m. The sound pressure level is calculated at points in the observed area of 0.6 m by 0.6 m within the interval of 0.02 m. $Cpf$ is the transfer function between the primary source and the observed point, and it is calculated theoretically.

Figure 10.6 illustrates the original sound field under plane wave propagation with a frequency of 1 kHz, and Figure 10.3.2 shows noise reduction by the ANC system with four different arrangements of the secondary sources. To compare these, we take the ratio of the sound pressure level in the original sound field (i.e. ANC off) and that in the controlled sound field (ANC on) shown in Figure 10.3.2.

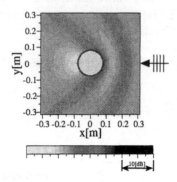

*Figure 10.6     Primary field (0°)*

According to the figures, when the incoming direction of the primary source agrees with the angle of one of the secondary sources, we can see a large controlled area (see Figure 10.3.2*d*), but Figure 10.3.2*a* indicates that a closely placed secondary sources gives a more significant performance of the noise reduction than wide a arrangement. This can also be seen in Figure 10.3.2.

### 10.3.3     Diffuse sound field

In this Section the diffused sound field is taken into account. Geometrical coordinates for the diffused field analysis are illustrated in Figure 10.9, and all the positions of the secondary sources and error microphones are the same as those for the above discussion (see previous Section). To consider the diffused sound field, we assume that plane waves come from every direction around the sphere with 5° interval. Thus 72 plane waves with random phase are considered as the primary sources. The normalised spatial correlation is computed so as to determine the averaging number of the primary sources (see Figures 10.10 and 10.11) [163]. We take the average of 100 simulations to calculate the sound pressure level in the diffused condition.

Figure 10.10 shows the normalised spatial correlation in the primary sound field without a rigid sphere and Figure 10.11 shows the normalised spatial cor-

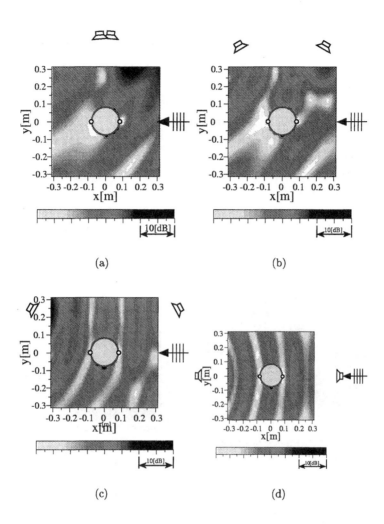

*Figure 10.7    Controlled area (primary area* 0°*)*

relation with a rigid sphere. Figures 10.12, 10.3.3 and 10.3.3 show the results for the $2L-2M$ ANC system. Figures 10.15, 10.3.3 and 10.3.3 present the results for the system with multiple control points on the rigid sphere. Based on the figures it is obvious that the original sound field is more complicated than that due to the single primary source (Figure 10.12). However, we can still ob-

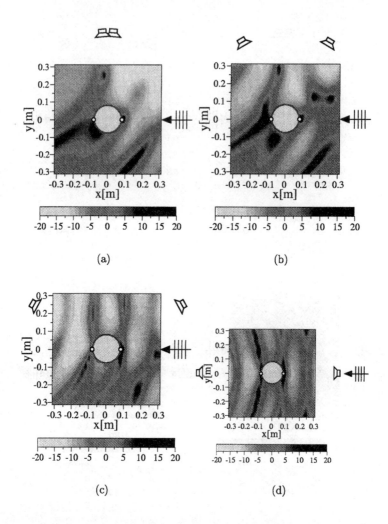

Figure 10.8    *Noise attenuation level (primary area 0°)*

serve that an arrangement of closely placed secondary sources (Figure 10.3.3*a*) gives a large noise reduced area compared to that with wide distances (Figure 10.3.3*d*). Similar results can also be seen in Figure 10.3.3. Concerning the relationship between the effective area and the number of error microphones, from Figures 10.3.3*a* and Figure 10.3.3*b*, we can see the very small area with

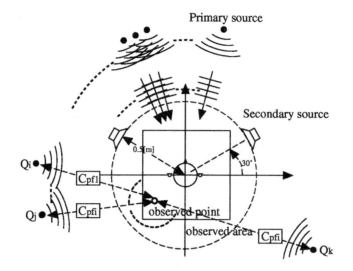

*Figure 10.9    Geometrical arrangements used in the simulation*

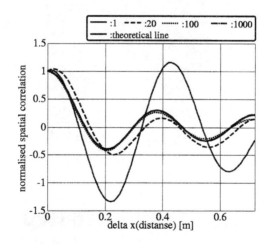

*Figure 10.10    The normalised spatial correlation in the primary sound field
(without the rigid sphere)*

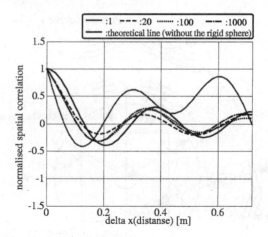

Figure 10.11    *The normalised spatial correlation in the primary sound field(with the rigid sphere)*

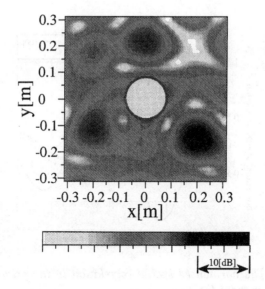

Figure 10.12    *Primary field*

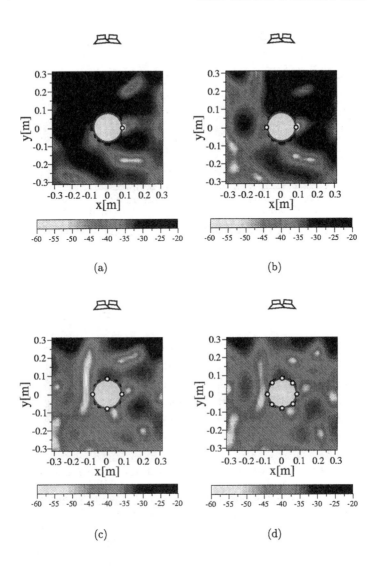

*Figure 10.13    Controlled area (secondary sources 85/95°)*

large noise reduction level, while some noise attenuation is also observed. On the other hand, the multiple-points control system can give a performance

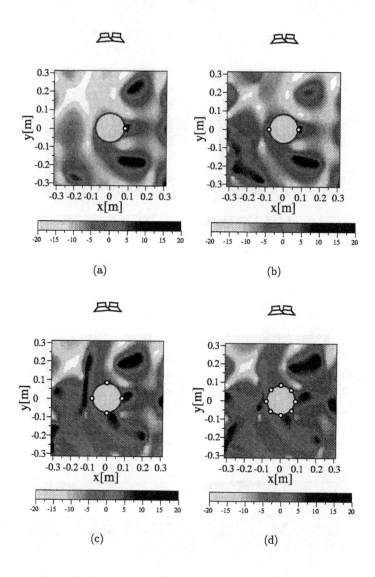

Figure 10.14    *Noise attenuation level (secondary sources 85/95°)*

with very low noise reduction, although it does not give much noise boost.

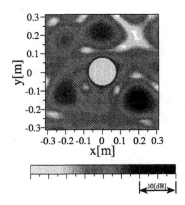

*Figure 10.15     Primary field*

## 10.4  Conclusions

In this Chapter the relationship between noise reduction and the corresponding configuration of error microphones and secondary source arrangements has been evaluated using a rigid sphere theoretical model as a human's head. In the case of the free sound field, in which the incoming direction of the primary source is known, we found that the arrangement with closely placed secondary sources has significant performance both for noise reduction as well as providing a wide control area compared with the distant secondary source arrangements. We have shown the results for a primary source of 1 kHz, but it is obvious that more conspicuous results can also be obtained at higher frequencies. As for the diffused sound field, we confirmed that introducing multiple control points magnifies the control area of noise reduction but the noise attenuation level decreases in general. Using our visualisation technique of sound pressure around a rigid sphere (as a human head) under control, we also indicated the possibility of finding the optimal arrangement of the secondary sources and error microphones for a given specification of performance such as minimum attenuation level of noise and the shape of the control area. It is by no means easy to settle the matter to the satisfaction of both sides, i.e. wide area and noise reduction level. Genetic algorithms may be employed for searching the error microphone positions for the purpose of satisfying the practical specifications which may include the limitation of possible positions and weighted cost functions for each evaluated position.

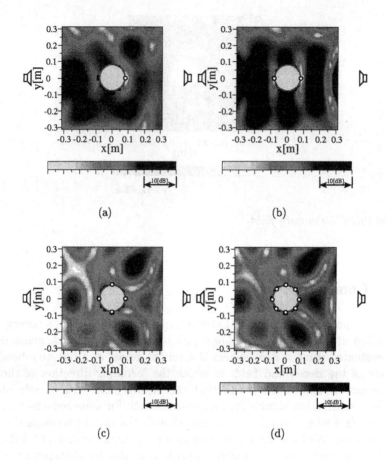

Figure 10.16    *Controlled area (secondary sources: 0/180°)*

...

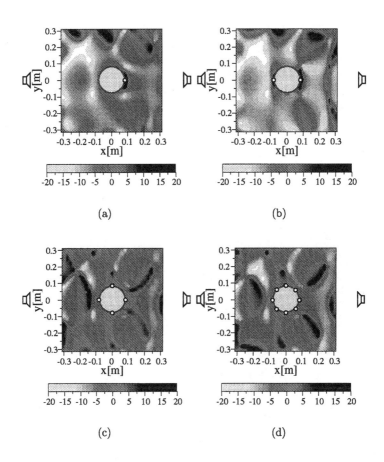

*Figure 10.17    Noise attenuation level (secondary sources: 0/180°)*

*Chapter 11*

# Modelling and feedback control of microvibrations

## G. S. Aglietti[†], J. Stoustrup[‡], R. S. Langley[*], E. Rogers[**], S. B. Gabriel[†]

[†] *School of Engineering Sciences, University of Southampton, UK*
[‡] *Department of Control Engineering, Aalborg University, Denmark*
[*] *Department of Engineering, University of Cambridge, UK*
[**] *Department of Electronics and Computer Science, University of Southampton, UK, Email: etar@ecs.soton.ac.uk*

*Microvibrations, generally defined as low-amplitude vibrations at frequencies up to 1 kHz, are now of critical importance in a number of areas. One such area is onboard spacecraft carrying sensitive payloads, such as accurately targeted optical instruments or microgravity experiments, where the microvibrations are caused by the operation of other equipment, e.g. reaction wheels, necessary for its correct functioning. It is now well known that the suppression of such microvibrations to acceptable levels requires the use of active control techniques which, in turn, require sufficiently accurate and tractable models of the underlying dynamics on which to base controller design and initial performance evaluation. This chapter describes the development of a modelling technique for either mass or equipment loaded panels and the subsequent use of such models in controller design and basic performance prediction of the resulting feedback control schemes.*

## 11.1    Introduction

Microvibrations is the term used to describe low-amplitude vibrations which occur at frequencies up to 1 kHz and which have often been neglected in the past owing to the low levels of disturbances they induce. In recent years, however, the need to suppress the effects of microvibrations has become much greater. This is especially true for spacecraft structures where, due to the ever increasing requirements to protect sensitive payloads, such as optical instruments or microgravity experiments, there is a pressing need to obtain a very high level of microvibration-induced vibration suppression (see, for example, Reference [277] for further background information).

In effect, such vibrations onboard spacecraft are produced by the functioning of onboard equipment such as reaction wheels, gyroscopes, thrusters, electric motors etc. which propagate through the satellite structure towards sensitive equipment (receivers) thereby jeopardising their correct functioning. Figure 11.1 shows a schematic illustration of this process.

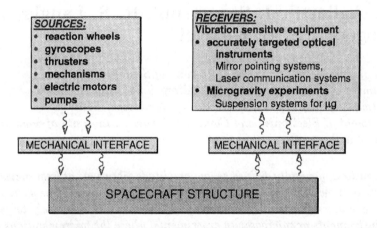

*Figure 11.1    Propagation path of the microvibrations*

In practice, the reduction of the vibration level in a structure can be attempted by action at the source(s), receiver(s) and along the vibration path(s). At the source(s), this action consists in attempting to minimise the amplitude(s) of the vibration(s) by, for example, placing equipment on appropriate mountings. The same approach is commonly attempted at the receiver(s) but with the basic objective of sensitivity reduction. Finally, along the vibration path(s), modifications of structural elements or relocation of equipment is attempted with the aim of reducing the mechanical coupling(s) between source(s) and receiver(s).

All of the approaches described above are based on so-called passive damping technology and, for routine applications, an appropriate combination of them is often capable of producing the desired levels of dynamic disturbance rejection. The use of active control techniques in such cases would only be as a last resort to achieve desired performance.

The requirements of the new generation of satellite-based instruments are such that only active control can be expected to provide the required levels of microvibration suppression. To investigate the use of active control to suppress such vibrations in a structure, computationally feasible models which retain the core features of the underlying dynamics are clearly required. The most obvious approach to the development of such models is to use finite-element methods (FEM) (see, for example, Reference [348]) due to the accuracy available with a sufficiently fine mesh. The only difficulties with this approach are the computational intensity of the models and the fact that they are not in a form directly compatible with feedback control systems design. They can, however, be used, as here, to verify that the modelling strategy employed produces realistic models on which to base controller design and evaluation.

Alternatives to FEM, can be classified as elastic wave methods, variational methods and mechanical impedance-based methods, respectively. A detailed study of the advantages and disadvantages of these methods, together with background references on each of them, can be found in References [3, 4]. Based on this study, a Lagrange-Rayleigh-Ritz (LRR) method is used to develop the mathematical models used as a modelling basis in the research programme on which this chapter is based.

Based on References [3, 4], the first part of this chapter describes how this LRR method, together with supporting software, can be developed to the stage where state-space models in the standard form for controller design (state or output feedback based) are automatically generated given the dimensions, material properties and loading pattern of the structure to be studied. Two generic cases are considered here: a mass loaded panel and the more general case of an equipment loaded panel. A systematic procedure for verifying the models obtained, a critical step before any meaningful controller design studies can take place is also described. This verification procedure is against (industry standard) finite-element (FEM) models of the same structure.

In the second part of this chapter, the use of this modelling technique in controller design and initial evaluation is considered. The focus is on the design of controllers based on linear quadratic optimal control theory. Also it is shown that this modelling approach enables Monte Carlo studies of the effects of uncertainties in the systems properties to be undertaken. This yields estimates of the probability of an unstable plant and statistical measures of expected performance.

## 11.2    System description and modelling

### 11.2.1    Mass loaded panel

We consider first the case of a mass loaded panel, which is an acceptable compromise between problem complexity and the need to gain useful insights into the benefits (and limitations) of active control schemes in this general area. A schematic diagram of the arrangement considered is shown in Figure 11.2, where the equipment mounted on the panel is modelled as lumped masses and the disturbances as point forces.

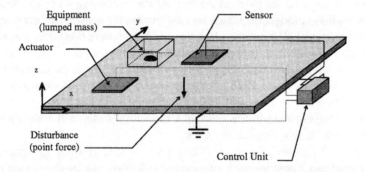

*Figure 11.2    Model layout*

The sensors and actuators employed are twin patches of piezoelectric material bonded onto opposite faces of the panel. The bending vibrations of the panel produce stretching and shrinking of the patches depending on whether they are on the top or the bottom of it, see Figure 11.3.

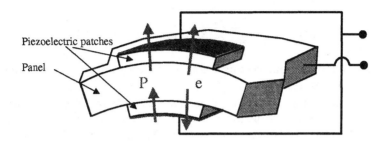

*Figure 11.3    Crosssection of a piezoelectric patch bonded on the panel during deformation. P is poling direction of the piezoelectric material, e is induced electric field*

Owing to the piezoelectric effect, these deformations induce an electric field perpendicular to the panel which is detected by the electrodes of the patches. The outer electrodes of the patches are electrically connected together and the panel, which is grounded, is used as the other electrode for both patches of the pair when acting as a sensor. The same configuration is used for an actuator, but in this case the electric field is applied externally to produce contraction or expansion of the patch, which then induces a curvature of the panel.

The LRR-based procedure used to model this system (Figure 11.2) is based on Lagrange's equations of motion which in the general case take the form:

$$\frac{d}{dt}\left(\frac{\partial T}{\partial \dot{q}_i}\right) - \frac{\partial T}{\partial q_i} + \frac{\partial U}{\partial q_i} = Q_i \qquad (11.1)$$

Here $T$ and $U$ are the kinetic and potential energies of the system, and $q_i$ and $Q_i$ are the $i$th generalised coordinate and force, respectively. For the particular case considered here, the kinetic and potential energies (elastic and electric) can be expressed as:

$$T = T_{pl} + T_{lm} + T_{pz}, \; U = U_{pl} + U_{pz} \qquad (11.2)$$

where $T_{pl}$, $T_{lm}$ and $T_{pz}$ denote the kinetic energies of the panel, lumped masses, and piezoelectric patches, respectively. The terms $U_{pl}$ and $U_{pz}$ denote the potential energies stored in the panel and the piezoelectric patches, respectively.

The displacement field (out-of-plane displacement $w$) is described by a superposition of shape functions $S_{m,n}$ (consisting of the first $N = N_m \times N_n$ modes of the bare panel) multiplied by the time-dependent modal coordinates $\psi_{m,n}$, i.e.:

$$w(x, y, t) = \sum_{m=1}^{N_m} \sum_{n=1}^{N_n} S_{m,n}(x, y)\psi_{m,n}(t) = s^T \Psi \tag{11.3}$$

where the $N \times 1$ column vectors $s$ and $\Psi$ contain the shape functions and modal coordinates, respectively.

As explained below, the full set of generalised coordinates $q_i$ which appears in eqn. (11.1) consists of $\Psi$ together with the voltages at the piezoelectric patches. The external excitation consists of $N_f$ point forces $F_j$ acting on the plate at arbitrary locations. Hence the generalised forces are of the form:

$$Q_i = \sum_{j=1}^{N_f} F_j \frac{\partial w_j}{\partial \psi_i} \tag{11.4}$$

or

$$Q = S_f f \tag{11.5}$$

where $f$ is the $N_f \times 1$ column vector of forces, and $S_f$ is a compatibly dimensioned matrix the columns of which are given by the model shape vector $s$ evaluated at the corresponding force locations.

It is now necessary to compute each of the terms in eqn. (11.2), starting with the kinetic energies. Each of these terms can be calculated using standard formula:

$$T = \frac{1}{2} \int \int \int_{Vol} \rho \dot{w} \, dx \, dy \, dz \tag{11.6}$$

where $\rho$ denotes the material density.

In the case of the transversely vibrating panel, application of eqn. (11.6) yields:

$$T_{pl} = \frac{1}{2}\dot{\Psi}^T M_{pl} \dot{\Psi} \tag{11.7}$$

where $M_{pl}$ is the (diagonal) inertia matrix of the bare panel and is given by:

$$M_{pl} = \int \int \int_{pl} \rho s s^T \, dx \, dy \, dz \tag{11.8}$$

Using the same notation, the kinetic energy of the piezoelectric patches can be written in the form:

$$T_{pz} = \frac{1}{2} \dot{\Psi}^T M_{pz} \dot{\Psi} \tag{11.9}$$

where

$$M_{pz} = \sum_{i=1}^{N_p} \int \int \int_{pz_i} \rho_{pz_i} s s^T \, dx \, dy \, dz \tag{11.10}$$

where the index $i$ denotes one of the $N_p$ patches employed and $\rho_{pz_i}$ its density. Here the inertia matrix ($M_{pz}$) is fully populated, where the off-diagonal entries denote the couplings between the modal coordinates.

Suppose now that there are $N_l$ lumped masses on the panel and let $s_{lm_i}$, $1 \leq i \leq N_l$, denote the shape vector function at lumped mass $i$. Then the total kinetic energy associated with the lumped masses is given by:

$$T_{lm} = \frac{1}{2} \dot{\Psi}^T M_{lm} \dot{\Psi} \tag{11.11}$$

where

$$M_{lm} = \sum_{i=1}^{N_l} M_{lm_i} s_{lm_i} s_{lm_i}^T \tag{11.12}$$

The potential energy of the system is stored as the elastic energy of the panel and the elastic/electric energy of the piezoelectric patches. The elastic energies are directly calculated from the expression:

$$U = \frac{1}{2} \int \int \int_{Vol} \epsilon^T \sigma \, dx \, dy \, dz \tag{11.13}$$

where $\sigma$ and $\epsilon$ are the stress and strain vectors, respectively. Also, by assuming a plane stress condition (see Reference [3] for details of this standard property) for the panel, we can write:

$$U_{pl} = \frac{1}{2} \Psi^T K_{pl} \Psi \tag{11.14}$$

where $K_{pl}$ is the panel stiffness matrix, which is given by:

$$\begin{aligned} K_{pl} = & \int \int \int_{pl} \frac{E z^2}{(1 - \nu^2)} \left( \frac{\partial^2 s}{\partial x^2} \frac{\partial^2 s^T}{\partial x^2} + \frac{\partial^2 s}{\partial y^2} \frac{\partial^2 s^T}{\partial y^2} \right. \\ & + \left. 2\nu \frac{\partial^2 s}{\partial x^2} \frac{\partial s^T}{\partial y^2} + 2(1 - \nu) \frac{\partial^2 s}{\partial x \, \partial y} \frac{\partial^2 s^T}{\partial x \, \partial y} \right) dx \, dy \, dz \end{aligned} \tag{11.15}$$

where $\nu$ denotes the Poisson's ratio for the panel material.

For the piezoelectric patches, the potential energy can be written as the sum of three energy components, i.e.:

$$U_{pz} = U_{pz_{elast}} + U_{pz_{elastelect}} + U_{pz_{elect}} \tag{11.16}$$

where $U_{pz_{elast}}$ is the energy stored due to the elasticity of the material, $U_{pz_{elastelect}}$ represents the additional energy due to voltage-driven piezoelectric effect, and $U_{pz_{elect}}$ is the electric energy stored due to the dielectric characteristics of the piezoelectric material employed. To compute the elastic energy in this case, an appropriate model for the stress-strain pattern in the piezoelectric patches must be selected, and here we make the following assumptions:

(i) The electrodes attached to the piezoelectric patches have negligible stiffness.

(ii) The thickness of the layer of adhesive which connects each of the patches employed to the panel is negligible compared to that of the patches and is able to transfer all of the shear strain.

(iii) The natural boundary conditions at the edges of each patch (i.e. $\sigma = 0$) are not enforced and a strain distribution such as that illustrated in Figure 11.4 is assumed through the whole patch.

This last assumption is particularly appropriate if, as here, the patches employed are very thin and relatively wide.

Given these assumptions, the same procedure as employed for the panel can be used to write:

$$U_{pz_{elast}} = \frac{1}{2}\Psi^T K_{pz_{elast}}\Psi \tag{11.17}$$

where, using $E_{pz_i}$ to denote the Young's modulus for the $i$th patch and $\nu_i$ to denote its Poisson's ratio:

$$
\begin{aligned}
K_{pz_{elast}} = &\sum_{i=1}^{N_p} \int\int\int_{pz_i} \frac{E_{pz_i}z^2}{(1-\nu_i^2)}\left(\frac{\partial^2 s}{\partial x^2}\frac{\partial^2 s^T}{\partial x^2} + \frac{\partial^2 s}{\partial y^2}\frac{\partial^2 s^T}{\partial y^2}\right. \\
&+ \left. 2\nu_i\frac{\partial^2 s}{\partial x^2}\frac{\partial^2 s^T}{\partial y^2} + 2(1-\nu_i)\frac{\partial^2 s}{\partial x\partial y}\frac{\partial^2 s^T}{\partial x\partial y}\right)dx\,dy\,dz
\end{aligned} \tag{11.18}
$$

is the stiffness matrix, which is fully populated.

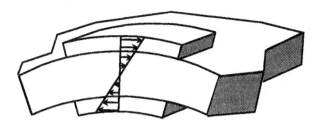

Figure 11.4    *Strain distribution through a piezoelectric patch/panel section*

Suppose now that a patch to be used has a constant thickness $h_{pz}$ which is thin enough to prevent fringe effects and has a voltage $v$ applied at its electrodes. Then a constant electric field $e = \frac{v}{h_{pz}}$ can be assumed across the patch and the further stress due to the applied voltages is given by:

$$\sigma_{elect} = \begin{pmatrix} \sigma_{x_{elect}} \\ \sigma_{y_{elect}} \end{pmatrix} = \frac{E_{pz}}{1 - \nu^2} \begin{pmatrix} d_{xz} + \nu d_{yz} \\ d_{yz} + \nu d_{xz} \end{pmatrix} e \qquad (11.19)$$

Here, $d_{xz}$ and $d_{yz}$ are the piezoelectric constants of the material, which is assumed to have polling direction $z$ perpendicular to the plate. Hence $U_{pz_{elastelect}}$ can be calculated as:

$$U_{pz_{elastelect}} = \int \int \int_{pz} \sigma_{elect}^T \epsilon \, dx \, dy \, dz \qquad (11.20)$$

In the case of $N_p$ patches, the electric field in patch $i$ can be written in the form $e_i(t) = V(t)^T p_i$, where $V(t)$ is the $N_p \times 1$ vector the entries of which are the patch voltages ($v_i$) and the $N_p \times 1$ vector $p_i$ has zero entries except for entry $i$ which is equal to $\frac{1}{h_{pz_i}}$. Also, by assuming that $d_{xz} = d_{yz} = d_z$ and substituting the assumed form of $\sigma_{elect}$ from eqn. (11.19) into eqn. (11.20), it is possible to write the elastoelectric energy stored in the $N_p$ patches as:

$$U_{pz_{elastelect}} = V^T K_{pz_{elastelect}} \Psi \qquad (11.21)$$

where

$$K_{pz_{elastelect}} = \sum_{i=1}^{N_p} \int \int \int_{pz_i} \frac{E_{pz_i} d_{z_i} p_i}{2(1 - \nu_i)} \left( z \frac{\partial^2 s^T}{\partial x^2} + z \frac{\partial^2 s^T}{\partial y^2} \right) dx \, dy \, dz \qquad (11.22)$$

The electrical energy stored in the piezoelectric material can be expressed as:

$$U_{pz_{elect}} = \frac{1}{2} \int \int \int_{pz} e\, d\, dx\, dy\, dz \qquad (11.23)$$

where $e$ is the electric field and $d$ is the electric displacement (charge/area). For each patch, the electric displacement is:

$$d_i = \epsilon_{pz_i} p_i^T v_i \qquad (11.24)$$

where $\epsilon_{pz_i}$ is the dielectric constant of the piezoelectric material which forms the $i$th patch. Hence, an equivalent expression for the stored electric energy is:

$$U_{pz_{elect}} = \frac{1}{2} V^T K_{pz_{elect}} V, \; K_{pz_{elect}} = \sum_{i=1}^{N_p} \int \int \int_{pz_i} \epsilon_{pz_i} p_i p_i^T \, dx\, dy\, dz \qquad (11.25)$$

where the elements of the matrix $K_{pz_{elect}}$ are the capacitances of the piezoelectric patches.

At this stage, all of the energy terms are available as functions of the generalised coordinates, i.e. the modal coordinates and the voltages at the patches written in column vector form as $\Psi$ and $V$ respectively. Hence, straightforward application of Lagrange's equations of motion from eqn. (11.1) yields the second-order matrix differential equation model:

$$(M_{pl} + M_{pz} + M_{lm})\ddot{\Psi} + (K_{pl} + K_{pz_{elast}})\Psi + K_{pz_{elastelect}}^T V = Q$$
$$K_{pz_{elastelect}}\Psi + K_{pz_{elect}}V = 0 \quad (11.26)$$

The first equation in eqns (11.26) results from first differentiating the energy terms with respect to the modal coordinates and writing the results in terms of the column vectors $\Psi$ and $V$ and the second from an identical set of operations but with differentiation with respect to the patch voltages. These operations assume that all modal coordinates and voltages are degrees of freedom (dofs) of the system.

In the case when all patches act as actuators, their voltages are externally driven and hence the second equation in eqns (11.26) is redundant. If all patches are to be used as sensors, the second equation in eqns (11.26) can be used to obtain an expression for the voltages as a function of the modal coordinates. This expression can then be substituted into the first equation in eqns (11.26) to give a complete set of equations in the unknown modal coordinates.

The most general case arises when some of the patches act as actuators and others as sensors, in which case it is necessary to partition the matrix $K_{pz_{elastelect}}$ to separate out actuator and sensor contributions. To do this, let $v_a$ and $v_s$ be the subvectors of the voltages at the actuators and sensors, respectively, and partition $K_{pz_{elastelect}}$ conformally as $K_{pz_{elastelect}} = [K_{pza_{elastelect}}, K_{pzs_{elastelect}}]$. Then the first equation in eqns (11.26) can be rewritten as:

$$M\ddot{\Psi} + C_s\dot{\Psi} + (K_{elas} + K_{pzs})\Psi = -K_{pza_{elastelect}}^T v_a + s_f^T f \qquad (11.27)$$

where all inertia elements are included in the matrix $M$ and all stiffness elements in the matrix $K_{elas}$. Also:

$$K_{pzs} = -K_{pzs_{elastelect}}^T K_{pzs_{elect}}^{-1} K_{pzs_{elastelect}} \qquad (11.28)$$

represents the contribution to the stiffness from the piezoelectric energy stored in the patches acting as sensors, where $K_{pzs_{elect}}$ is the submatrix of $K_{pz_{elect}}$ corresponding to the sensors. In addition, structural damping has been added to the system by including the term $C_s\dot{\Psi}$.

### 11.2.2  Modelling of equipment loaded panels

Modelling equipment mounted on a panel is most easily undertaken by employing the lumped mass assumption as in the previous subsection. If, however, the equipment itself has internal dynamics, or the wavelength of the deformations is less than or equal to the distance between the mounting points (see Figure 11.5), then the lumped mass approximation is no longer valid and a more detailed representation of the mounting geometry needs to be considered. In which case, first note that each particular piece of equipment could have a different mechanical interface securing it to the structure underneath.

By far the most common mounting geometry is four feet positioned at its corners as illustrated in Figure 11.6 where it is assumed that all the feet have identical physical properties. These feet are modelled by a parallel combination of a stiffness, a dash-pot and a piezoelectric prism, where the latter form the actuators for control action at source(s) or receiver(s). Piezoelectric patches bonded onto the panel are again used as the sensors and actuators for control along the structure.

The model of the complete plant in this case can be constructed by assembling together the model of the actively controlled panel, as described in the previous subsection, and the model of the equipment on their suspension systems. In what follows we first summarise the modelling of the equipment on

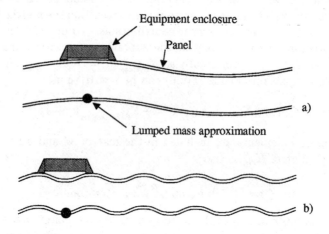

Figure 11.5    *Lumped mass approximation, (a)deformation wavelength longer than equipment mounting feet distance; (b) deformation wavelength shorter than equipment mounting feet distance*

their suspension systems and then the assembly of the model for the complete plant. Complete details can be found in Reference [5].

The enclosure shown in Figure 11.6 is assumed to have three degrees of freedom (dofs), see Figure 11.7, i.e. out-of-plane displacement, pitch angle and roll angle respectively, which are written in column vector form as $\Psi_{\text{eqp}}$. This particular choice of dofs allows us to express the kinetic energy associated with any piece of equipment as:

$$T_{\text{eqp}} = \frac{1}{2}\dot{\Psi}_{\text{eqp}}^T M_{\text{eqp}} \dot{\Psi}_{\text{eqp}} \qquad (11.29)$$

where $M_{\text{eqp}}$ is the associated inertia matrix.

The potential energy associated with the equipment is, in effect, the sum of that stored in the flexible supporting elements (i.e. the stiffness of the piezoelectric prism of each mounting foot). This energy can be evaluated as the sum of the elastic energy, the elastoelectric energy and the electric energy stored in each mounting foot. Next we outline the development of a

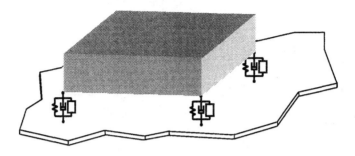

Figure 11.6   *Model of a typical satellite equipment enclosure with active suspensions mounted on a panel*

representation for each of these energies.

Consider first the elastic energy stored in each suspension. Then this energy is proportional to the square of the linear deformation $\delta z_i$, $1 \leq i \leq 4$, of the $i$th mounting foot. This deformation is given by the difference between the out-of-plane displacement of the panel surface $(w(x, y, t))$ evaluated at the mounting foot location $(x_i, y_i)$ and the vertical displacement $(z_i(t))$ of the $i$th corner of the box. Suppose also that $\Psi_c = \left[\Psi^T, \Psi_{\text{eqp}}^T\right]^T$ (where, from the previous subsection, $\Psi$ is the column vector of modal coordinates for the panel.) Then the total (i.e. for all four mounting feet) elastic energy can be written as:

$$U_{\text{eqp}_{\text{elast}}} = \frac{1}{2}\Psi_c^T K_{\text{elast}_{\text{eqp}}}\Psi_c \tag{11.30}$$

Suppose now that an element of the prisms used in the mounting feet has height $h_{pz_i}$ and has a voltage $v_i$ applied across the electrodes on the top and bottom faces. Then a constant electric field $e_i = \frac{v_i}{h_{pz_i}}$ acting in an axial direction can be assumed in the material which has Young's modulus $E_{pz_i}$ and piezoelectric constant $d_{zz_i}$. The stress $(\sigma_{\text{elect}_i})$ produced along the same axial direction will therefore be constant and related to the applied voltage by the equation:

$$\sigma_{\text{elect}_i} = E_{pz_i}d_{zz_i}\frac{v_i}{h_{pz_i}} \tag{11.31}$$

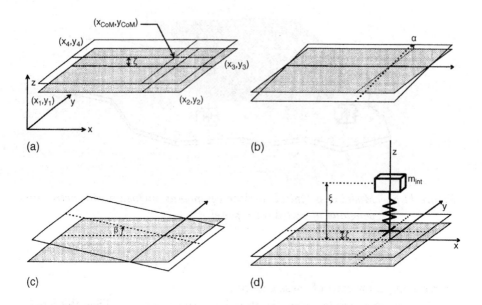

Figure 11.7    *Degrees of Freedom of the equipment enclosures $x_i, y_i$:  co-ordinates of the mounting feet x-CoM, y-CoM: co-ordinates of the box Centre of Mass*

Also, the strain in the material can be assumed to be constant and given by:

$$\epsilon_{z_i} = \frac{\Delta z_i}{h_{pz_i}} \tag{11.32}$$

and the elastoelectric energy stored in each piezoelectric patch used can be computed by application of a known formula.

Suppose now that the column vector $v_{\text{eqp}}$ is formed from the voltages $v_i$ across the electrodes in the mounting feet used. Then the total elastoelectric energy stored in the equipment suspension system can be written as:

$$U_{\text{elastelect}_{\text{eqp}}} = \frac{1}{2} v_{\text{eqp}}^T K_{\text{elastelect}_{\text{eqp}}} \psi_{\text{eqp}} \tag{11.33}$$

Assuming that a uniform electric field exists across the piezoelectric prisms, the electric energy stored in each of them is given by $\frac{1}{2}Cv_i^2$ where $C$ denotes the capacitance of the prisms. Hence, the total stored electric energy can be expressed as:

$$U_{\text{elect}_{\text{eqp}}} = \frac{1}{2}v_{\text{eqp}}^T K_{\text{elect}_{\text{eqp}}} v_{\text{eqp}} \qquad (11.34)$$

In this case, the $4 \times 4$ matrix involved ($K_{\text{elect}_{\text{eqp}}}$) is diagonal with each element equal to $C$.

The presence of dissipative force produced by the dash-pots in the mounting feet means that an extra term must be added to the generalised forces in the application of eqn. (11.1) to this case. This extra term, denoted by $Q_{dp}$ is proportional to $\delta \dot{z}_i$ and can be written as:

$$Q_{dp} = C_{dp}\dot{\Psi}_{\text{C}} \qquad (11.35)$$

where $C_{dp}$ includes the contribution of each foot. It is also necessary to take account of the internal dynamics of the pieces of equipment. This is achieved by adding an extra dof to $\Psi_{\text{eqp}}$ with associated mass and stiffness added to $M_{\text{eqp}}$ and $K_{\text{eqp}}^{\text{elast}}$.

To apply eqn. (11.1) to obtain the final model of the overall system composed of the actively controlled panel and, say, $N_e$ pieces of equipment mounted on it, the generalised coordinate is taken (with obvious notation on the right-hand side) to be:

$$r = \left[\Psi^T, \Psi_{\text{eqp}_1}^T, \cdots, \Psi_{\text{eqp}_{N_e}}^T\right]^T \qquad (11.36)$$

Also introduce (again with obvious notation on the right-hand side):

$$v_e = \left[v_{\text{eqp}_1}^T, \cdots, v_{\text{eqp}_{N_e}}^T\right]^T \qquad (11.37)$$

Suppose also that all stiffness and mass matrices associated with (i) the actively controlled panel and the pieces of equipment, and (ii) those associated with the actuators of the suspensions, have been augmented with rows and columns of zeros as appropriate to be compatible with the dimension of $r$. Then on application of eqn (11.1), the motion of the actively controlled structure can be written as:

$$M_{\text{acs}}\ddot{r} + C_{\text{acs}}\dot{r} + K_{\text{acs}}r = H_1 \qquad (11.38)$$

where

$$H_1 = V_e v_e + V_a v_a + Sf \qquad (11.39)$$

These last two equations govern the motion of the actively controlled structure excited by external sources ($f$), voltage inputs at the active suspensions ($v_e$)

and the piezoelectric patches acting as actuators $(v_a)$. Once the solution $r$ is available, the displacement at any point on the panel can be obtained from:

$$w = \begin{bmatrix} s^T, 0 \end{bmatrix} r \qquad (11.40)$$

and the displacements of the equipment enclosures are simply the corresponding elements in $r$. In the case when the vibration levels of the actively controlled equipment have to be monitored, these are easily inferred from the relevant dofs.

## 11.3    Model verification

In the next Section, it will be shown that the final forms of the models developed in the previous Section for both the mass and equipment loaded panels can be immediately used for controller design. As an essential step before accepting a model derived by this procedure as a realistic basis for controller design/evaluation studies, appropriate model validation studies must be undertaken. This key task is the subject of this section.

Model verification for the mass loaded panel has been covered in considerable detail by Reference [4] where the method devised is based on comparing the results produced by the LRR modelling procedure, via MATLAB-based simulations, for a range of structural and input/output configurations against those produced using standard FE models constructed (in this work) using the commercially available software ANSYS. Note also that this procedure generalises in a natural manner to the case of an equipment loaded panel [5] and hence no details are given here.

Verification of the model is problem specific and here we illustrate the method by comparing the frequency response at the centre of the panel produced by the LRR model against that obtained from an FE model. The test problem consists of a simply supported aluminum panel with a lumped mass mounted on it and two pairs of piezoelectric patches acting as actuators and sensors, see see Figure 11.8 and Table 11.1.

The disturbance is a harmonic point force of 1 N amplitude, i.e. $f = Fe^{\iota\omega t}$, $F = 1$, acting perpendicular to the panel at $x = 254$ mm, $y = 50.8$ mm and, for model testing purposes, a harmonic input voltage of 1 V amplitude applied to the actuator, i.e. $v_a = Ve^{\iota\omega t}$, $V = 1$, is used.

The mode shapes of the bare panel:

$$S_{m,n}(x,y) = \sin(\frac{m\pi x}{a})\sin(\frac{n\pi y}{b}) \qquad (11.41)$$

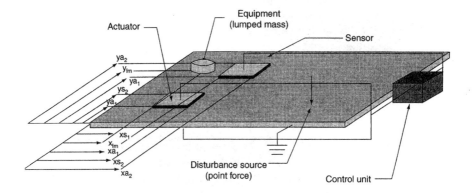

*Figure 11.8    Actively Controlled Panel*

are taken as Ritz functions to model the displacement field. Since the frequency range of interest is up to 1 KHz, and the piezoelectric patches are taken to be $\frac{1}{6}$th of the panel length, the first 36 model shapes were used as Ritz functions, i.e.:

$$w = (s_1, \cdots, s_{36})(\psi_1, \cdots, \psi_{36})^T \qquad (11.42)$$

Note that other shape functions, e.g. the static formed shape generated by a constant input at the actuator, could be used as an alternative to the above reduced modal base in the model validation procedure described here.

Verification in this case is performed by comparing the frequency response at the centre of the panel in the model obtained from the LRR model against that obtained with the FE model. This latter model, see Figure 11.9, was made up of 384 eight-noded layered shell elements (ANSYS-Shell91), where in the areas of the piezoelectric patches there were three layers (piezoelectric material, upper patch/aluminium/piezoelectric material, lower patch) and the remainder of the panel is composed of a single layer of aluminum. A total of 1233 nodes were employed: 49 in the $x$ direction and 33 in the $y$ direction.

Two different test cases were used for model validation:

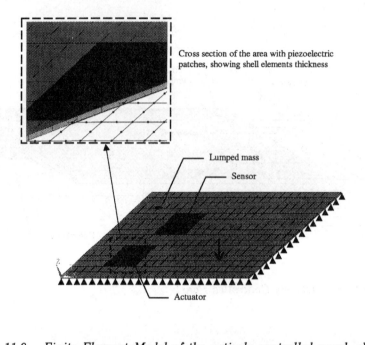

Cross section of the area with piezoelectric
patches, showing shell elements thickness

Lumped mass

Sensor

Actuator

Figure 11.9    *Finite Element Model of the actively controlled panel, element
plot showing point force and constraints*

Test case 1: voltage-driven panel: Here the driving force is produced by the
contractions of the piezoelectric actuator in response to a sinusoidal voltage of
1 V, where, by linearity of the model, the results can be scaled up (or down)
to any input voltage. In the FEM model, the effect of the applied voltage is
produced by applying unit moments along the line edges of the piezoelectric
patch. The value of the applied moment which corresponds to the input voltage is calculated (as in Reference [36]) as an extrapolation of the beam case.
Figure 11.10 shows the frequency response of the panel in the range 50 to 500
Hz. Note that the discrete frequencies employed in the computation of the
FEM results have resulted in the truncation of the resonance peaks - a smaller
(and computationally more expensive) frequency step would fully resolve these
peaks.

Test case 2: point-force-driven panel: In this case, a 50 g mass was added

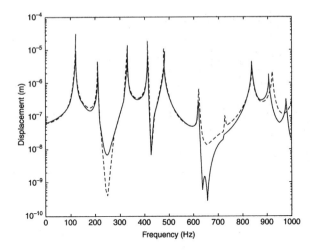

Figure 11.10    *Actively controlled panel displacement at the centre of the panel produced by actuator input, dashed line: FEM; continuous line: LRRM*

to the panel at the location specified in Table 11.1 and the forcing input is a point force acting perpendicular to the panel at the arbitrarily chosen location $x = 254$ mm, $y = 50.8$ mm. The results of the simulations are shown in Figure 11.11.

Comparing the results of these (representative of a large number actually undertaken) tests, it is clear that the LRR-based model gives a good representation of the dynamics of the system. The key point is that good agreement with the results from the FEM has been achieved but with a much smaller model.

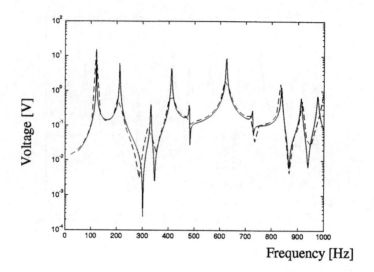

Figure 11.11    *Actively controlled panel voltage at the sensor produced by 1 N*
             *point force dashed line: FEM; continuous line: LRRM*

## 11.4 Control systems design

Consider first the mass loaded panel case. Then the model of eqn. (11.27) can easily be written in state-space form as follows:

$$
\begin{aligned}
\dot{x} &= Ax + B_v v_a + B_f f \\
v_s &= C_v x \\
w_{out} &= C_w x = \begin{bmatrix} s_{out}^T & 0 \end{bmatrix} x
\end{aligned}
\tag{11.43}
$$

where

$$
x = \begin{pmatrix} \Psi \\ \dot{\Psi} \end{pmatrix} \quad A = \begin{bmatrix} 0 & I \\ -M^{-1}K & -M^{-1}C_s \end{bmatrix},
\tag{11.44}
$$

$$
B_v = \begin{bmatrix} 0 \\ M^{-1}K_{pza_{elastelect}}^T \end{bmatrix}, B_f = \begin{bmatrix} 0 \\ M^{-1}s_f^T \end{bmatrix}
\tag{11.45}
$$

$$C_v = \begin{bmatrix} -K^{-1}_{pzs_{elect}} K_{pzs_{elastelect}} & 0 \end{bmatrix} \tag{11.46}$$

Here $w_{out}$ is the output displacement at the particular locations specified by the mode shape vectors $s_{out}$ and $K$ denotes the total stiffness matrix defined from eqn. (11.28) ($K = K_{elas} + K_{pzs}$). A particular feature of the LRR approach is that the computations required for a given example immediately yield the state space description of eqns (41)-(43) which is in the standard form for (state and output feedback based) control systems design. Using this state-space description, it is possible to begin in depth investigations of the potential (or otherwise) of active control schemes in this general area. Here, by way of illustration, the particular control objective considered for the mass loaded panel is to minimise the displacement at a specified point on the panel in the presence of point force disturbances acting at other location(s) on the panel.

There are many possible controller design strategies which could be used at this stage and here we consider, as a representative, a linear quadratic optimal control approach. The cost function has the form:

$$J = \frac{1}{2} \int_0^\infty \left( w_{out}^T Q w_{out} + v_a^T R v_a \right) dt \tag{11.47}$$

where $Q$ and $R$ are symmetric weighting matrices to be selected subject to the requirements that $Q$ is positive semi-definite and $R$ is positive definite. The solution of this problem is the stabilising state feedback law $v_a = G_{fs}x$ where:

$$G_{fs} = R^{-1} B_v P_c \tag{11.48}$$

and $P_c$ satisfies the algebraic Riccati equation:

$$A^T P_c + P_c A - P_c B_v R^{-1} B_v^T P_c + C_w Q C_w^T = 0 \tag{11.49}$$

In practice, an observer will have to be used to implement this control law since not all the states can be directly measured. The observer implementation is:

$$v_a = G_{fs}\hat{x} \tag{11.50}$$

where the the estimated state vector $\hat{x}$ is generated as:

$$\dot{\hat{x}} = A\hat{x} + Bv_a + L(v_s - C\hat{x}) \tag{11.51}$$

where $L$ is the observer gain matrix and the theory of the full-order state observer guarantees that $\lim_{t\to\infty}(x(t) - \hat{x}(t)) = 0$.

To illustrate this design process, consider again the panel defined by Figure 11.8 and Table 11.1 with the disturbance ($f$) again consisting of a harmonic point force of amplitude 1 N acting perpendicular to the panel at $x = 254$ mm and $y = 50.8$ mm, and the control objective is to minimise the displacement at the centre of the panel. In this case, the weighting matrices $Q$ and $R$ in the cost function of eqn. (11.47) are scalars and Figure 11.12 shows a comparison of the displacement response at the centre of the panel with and without control action in the frequency range up to 1 KHz using the weightings $Q = 10^{13}$ and $R = 1$, where these values have been obtained by a small number of design iterations. The continuous line represents the response of the system with no control action and the dotted line that with the active control applied.

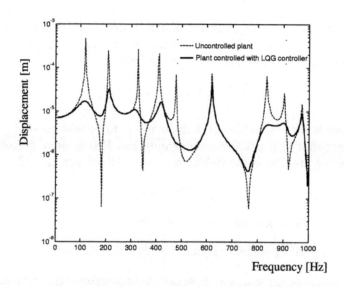

Figure 11.12    *Frequency response to 1 N input disturbance force for plant with and without active control*

The performance potential of the control system can be further highlighted by comparing the deformed shape of the panel with and without the control scheme applied, see Figure 11.13 *a*) and *b*), respectively for an input frequency of 400 Hz.

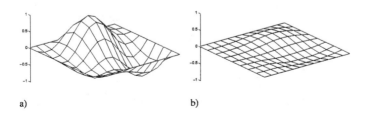

a)                                              b)

Figure 11.13    *Deformed shapes of the panel subject to a point force of 1 New-*
                *ton and frequency 400 Hz (a) response of the system without*
                *control; (b) response with applied active control*

Closer inspection of the the frequency response of Figure 11.13 shows that
in the region of the resonance at 620 Hz, no attenuation is produced on the
panel. The physical explanation of this phenomena can be obtained by studying the contour plot of the deformed shape given in Figure 11.14 which shows
that the actuator patch is cut through the middle by a nodal line, where
$w(x,y) = 0$, $\frac{\partial^2 w}{\partial x^2} = 0$, $\frac{\partial^2 w}{\partial y^2} = 0$ and therefore no control can be applied to this
mode.

The reason why no control action can be applied in this last case is due to
the fact that the effectiveness of the piezoelectric elements (both as actuators
and sensors) decreases when the wavelength of the deformation becomes equal
to the patch length. This is because the signal then produced is partially or
completely cancelled by the opposite field generated by the other part of the
patch which undergoes the opposite kind of deformation. This limiting factor
needs to be considered when attempting to control high-frequency vibrations
which, of course, have very short wavelengths. One possible way of increasing
this limit is to decrease the patch dimensions but this, in turn, would diminish
the control authority at low frequencies.

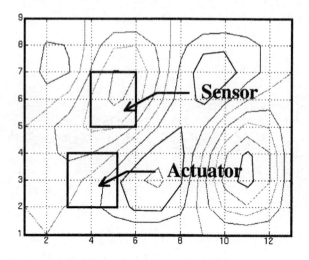

*Figure 11.14    Deformed shape contour plot at 620 Hz*

In the case of the equipment loaded panel, the final model of eqns (11.38)-(11.40) can be written in state-space form as:

$$\dot{x} = Ax + B_1 v + B_2 f$$
$$v_s = C_v x$$
$$w_{out} = C_w x \tag{11.52}$$

where $x = [r^T, \dot{r}^T]^T$, $v = [v_c^T, v_a^T]^T$ and

$$A = \begin{bmatrix} 0 & I \\ -M_{acs}^{-1}K_{acs} & -M_{acs}^{-1}C_{acs} \end{bmatrix} \tag{11.53}$$

$$B_1 = \begin{bmatrix} 0 & 0 \\ M_{acs}^{-1}V_s & M_{acs}^{-1}V_a \end{bmatrix}, \quad B_2 = \begin{bmatrix} 0 \\ M_{acs}^{-1} \end{bmatrix} \tag{11.54}$$

and the observed and controlled output matrices $C_v$ and $C_w$ are defined by:

$$C_v = \begin{bmatrix} -(K_{pzs}^{\text{elect}})^{-1}K_{pzs}^{\text{elastelect}} & 0 \end{bmatrix} \tag{11.55}$$

and

$$C_w = \begin{bmatrix} s^T & 0 \\ s_{rc}^T & 0 \\ s_{sr}^T & 0 \end{bmatrix} \qquad (11.56)$$

Again, controller design by any appropriate method can proceed from this model. Here, for illustrative purposes, we consider the case when the panel, see Figure 11.15 for the layout details, is a simply supported aluminum plate of dimensions (length, breadth, and thickness) 304.8 mm ×203.2 mm ×1.52 mm with four boxes mounted on it. Box 1 is a passive equipment box (e.g. a box of electronic components), box 2 is a source of vibrations, and boxes 3 and 4 are receivers. Boxes 2, 3 and 4 are mounted on active suspensions and box 1 is mounted on springs. The mass of each box is taken to be 0.5 kg equally divided between the mass of the enclosure and the mass of the resonator, and the rotational inertia of the boxes is $10^{-4}$ kg/m². The four suspension springs are of stiffness $k = 10^6$ kg/m and the inertial resonators, which are positioned at the centre of each box to avoid coupling between linear (axial) and rotational (rocking) modes of the boxes, all have the same stiffness value.

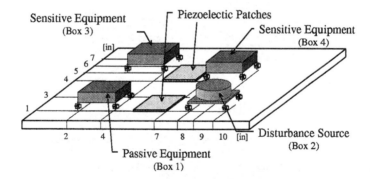

*Figure 11.15    Equipment loaded panel layout*

Note that in this case, the FE model used in the verification studies has (as a typical example) approximately 3500 dofs compared to 480 for the LRR model.

The LRR-based model is also very flexible in terms of investigating the effects of different input and output connections for vibration transmission reduction (in addition to comparing the effectiveness of different control designs for the same control strategy), e.g. minimisation at source(s), or receiver(s) or along transmitting structure(s). Here we again consider observer implemented state feedback design to minimise a standard linear quadratic cost function of the form of eqn. (11.47) (designs undertaken using the MATLAB routine lqg).

In the case of the structure used here (the simply supported aluminum panel with the four boxes detailed above), control along the transmission path, attempted using piezoelectric patches bonded onto the panel acting as sensors and actuators (one each), did not give acceptable results. This is mainly due to the low level of force which can be produced by the patch used as the actuator. In particular, the controller was only able to slightly reduce the amplitudes of the peaks in the frequency response corresponding to the first and third modes and it was not effective in reducing the response at other resonances. The reason for this is due to the position of the actuator which lies along the modal line of most of the nodes and, in particular, modes 2 and 4, see Figure 11.16.

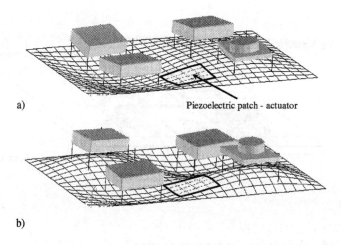

a)                                          Piezoelectric patch - actuator

b)

Figure 11.16   *Modes of vibration of the equipment loaded panel (a) Mode 2,1 (71 Hz), (b) Mode 2,2 (139 Hz)*

This situation does not improve if the controller has other signals available, e.g. the displacements of the receivers, which confirms that the problem with this configuration lies in the position of the actuators. Better results can be obtained when control action is applied at the mechanical interface of the source or receiver.

Control at source: in this case the control system drives (independently) the four suspensions of the equipment enclosure which generates the vibrations (source). The objective is to minimise the displacements and rotations of the two receivers. The signals (potentially) available to the controller are numerous, and the option considered here is to use the displacement and rotations of the source as sensor signals. The controller design is based on a plant model which uses the first $4 \times 4$ modal shape of the bare panel, and then the performance of the controlled structure is simulated by using a more accurate model built using a $6 \times 6$ modal base. Figure 11.17 shows the vertical displacements of the receiver (box 3) with and without control.

Control at receiver: the vibrations at the sensitive equipment can be reduced by control action at the equipment suspensions. In this case the signals available to the control system are taken as the displacements and rotations of the receiver and the controller drives the active suspensions here in order to minimise their displacements and rotations. The results are reported in Figure 11.18 and note that, compared to active control at the source (Figure 11.17), this strategy produces a stronger level of vibration at the receiver location. Also, this type of controller is relatively easy to implement because the sensors and actuators are located very close to each other.

Note that, even if both of the control strategies detailed above are local, the panel plays an important role in the overall dynamics of the plant. The importance of a correct model of the plant is highlighted by the fact that the plant may become unstable for small changes in the panel characteristics. As an example, the controller designed in the first control strategy here, which produces the stable closed-loop response of Figure 11.17, becomes unstable for a 5 per cent change in the panel thickness, see Figure 11.19.

## 11.5   Robustness analysis

As with all physical examples of plants to be controlled, the physical properties of the structure to be controlled here are subject to varying degrees of uncertainty. For example, in the case of the mass loaded panel with similar

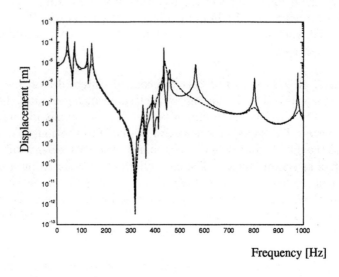

*Figure 11.17    Vertical displacements of receiver 1 without (continuous line)
and with (dashed line) - control at source*

comments holding for the equipment loaded case, the material properties of
the piezoelectric patches and the panel could differ from the nominal design
values. Also, the panel boundary conditions may differ from the ideal simple
supports which were assumed in the development of the models used for controller design here. Similarly, the panel dimensions and the locations of the
piezoelectric patches are subject to a certain degree of uncertainty.

In the work reported here, the variation in the system's dimensions is inten-
tionally used to account for uncertainties in the boundary conditions together
with uncertainties in the actual dimensions, since both of these factors produce changes in the natural frequencies and mode shapes. It is therefore of
interest to study the effects of these uncertainties on the performance of an
active controller designed (of necessity) on the basis of a nominal plant model.

One method of undertaking such a study is by using Monte Carlo simulations. In this approach, statistical properties of the parameters of interest are
specified and used in conjunction with a random number generator to produce

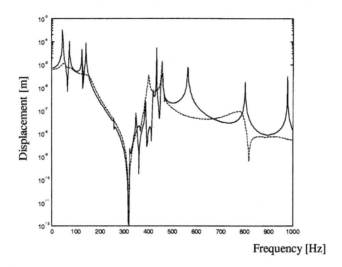

*Figure 11.18*    *Displacement of receiver 1 without (continuous line) and with (dashed line) control - control at receiver*

an ensemble of plant models. The active controller designed on the basis of the nominal model is then applied to each member of the ensemble and hence the probability of an unstable closed-loop system can be obtained.

In practice, the above Monte Carlo-based technique could well be computationally unacceptable since a large ensemble of random plants needs to be analysed to ensure a statistically accurate result, particularly when the probability of an unstable plant is relatively low. An alternative approach is to employ the FORM (first-order reliability method) approach in which uncertain parameters (e.g. the system's dimensions or damping) are considered as components of a vector $X$ and the safety of the system is described by the so-called safety margin, $G(X)$ (see Reference [182]) and, in particular, the system is said to have failed if $g(X) \leq 0$.

In the present application, $g(X)$ can be taken to be the negative of the least stable pole and the exact probability of failure is given by the integral of the joint probability density function of $X$ over the failure region. In gen-

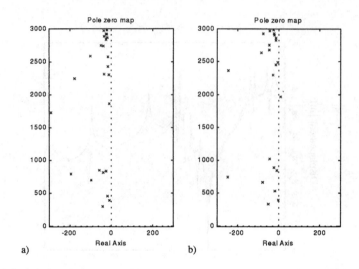

a)                                        b)

Figure 11.19    *Poles for the controlled plant (a) nominal plant (b) plant with thicker panel*

eral, the resulting integral can be a very difficult quantity to compute since $X$ can have a very high dimension and $g(X)$ can have a complex geometry. The FORM approach sets out to obtain an approximate probability of failure by transforming the set of variables in $X$ to a set of uncorrelated Gaussian variables $Z$, each member of which has zero mean and unit standard deviation. The probability of failure is then estimated using:

$$P_f = \Phi(-\beta) \tag{11.57}$$

where

$$\beta^2 = \min\{Z^T Z : g(X) = 0\} \tag{11.58}$$

and $\Phi$ is the cumulative normal distribution function. Geometrically, $\beta$ is the shortest distance between $g(Z)$ and the origin. Eqn. (11.57) is exact if the safety margin $g(Z)$ is a linear function; otherwise the result is an approximation which is based on linearising the safety margin about the point of closest approach to the origin. Eqn. (11.58) is a constrained optimisation problem which can be solved numerically provided $g(X)$ can be evaluated for a specified value of $X$ – the LRR modelling approach used here provides an efficient way of completing this task.

As an illustrative example, consider a simply supported mass loaded panel of dimensions (length, breadth, and thickness in mm) $350 \times 203 \times 1.52$ with piezoelectric patches acting as sensors and actuators. The panel dimensions are taken to be Gaussian random variables with standard deviations of 0.33 mm, 0.33 mm, and 0.066 mm respectively. The damping is also assumed to be random with a standard deviation of 3 per cent of the nominal value. Note also that the relatively large variation in the panel dimensions is intended to make allowance for uncertainties in both the boundary conditions and the actual dimensions themselves.

In a representative result from the Monte Carlo simulations, an ensemble of 1700 plants gave 20 which were unstable and hence a probability of failure of 0.0018. The open and closed-loop poles obtained from a sample of 550 of ensemble are shown in Figure 11.20 and this shows that the poles of the closed-loop system have a very high degree of variability. For example, pole 18 becomes unstable for some of the perturbed plants.

The FORM analysis in this case yielded $\beta = 2.27$ which corresponds to a probability failure of approximately 0.012. Also, the failure point, i.e. the point satisfying eqn. (11.58), lies on the portion of $g(X)$ that is associated with pole 18 and hence both the probability of a failure and the mode of failure are in good agreement with the Monte Carlo results. This analysis also required a small fraction of the computer time needed for the Monte Carlo simulations (typically 30 to 60 calls to the function $g(X)$). Hence, the combination of FORM and the LRR modelling technique is a very efficient method of studying the robustness properties of the actively controlled system.

## 11.6    Conclusions

This chapter has described a Lagrange-Rayleigh-Ritz (LRR) approach to the development of a state-space model of mass and equipment loaded panels on which to undertake the design of feedback control schemes for the suppression of microvibrations. In the case of the equipment loaded panel, the associated enclosures have been modelled as rigid rectangular boxes mounted on the (flexible) panel. The enclosures have internal resonators to simulate internal dynamics and there is provision for rigid or flexible mounting elements to allow for active/passive suspensions. Piezoelectric patches are used as sensors and actuators on the panel (in both cases) and piezoelectric stacks are used for the box suspensions.

One clear advantage of the LRR approach in terms of active feedback control system design is that the resulting model is, essentially, already in state space form for the application of modern control systems design tools (in a MATLAB compatible environment). Also, a technique for model verification has been described and illustrated by an example. A key point here is that the reduced size of the LRR models in comparison to FE models makes them particularly suitable for in depth studies on the active control of microvibrations since, in particular, the essential plant dynamics can be captured in a model of manageable size. Here, by way of illustration, linear quadratic regulator-based designs have been used but others are equally valid. For example, the use of loop transfer recovery methods is reported in Reference [6] and $H_\infty$ methods in Reference [115].

| Panel | a=304.8 mm | E=71e9 Pa |
|---|---|---|
| | b=203.2 mm | $\rho = 2800$ kg/m$^3$ |
| | h=1.52 mm | $\nu = 0.33$ |
| | | $\eta = 0.001^a$ |
| Sensor | $x_{s1} = 0.8$ mm | E=63e9 Pa |
| | $x_{s2} = 101.6$ mm | $\rho = 7650$ kg/m$^3$ |
| | $y_{s1} = 25.4$ mm | $\nu = 0.3$ |
| | $y_{s2} = 76.2$ mm | d=1.66e-10 m/V |
| | $hpz_s = 0.19$ mm | $\epsilon = 1700\epsilon^0$ |
| Actuator | $x_{a1} = 76.2$ mm | E=63e9 Pa |
| | $x_{a2} = 127$ mm | $\rho = 7650$ kg/m$^3$ |
| | $y_{a1} = 101.6$ mm | $\nu = .3$ |
| | $y_{a2} = 152.4$ mm | d=1.66e-10 m/V |
| | $hpz_s = 0.19$ mm | $\epsilon = 1700\epsilon^0$ |
| Lump.mass | $x_{lm} = 50.8$ mm | $W_{lm} = 50$ g |
| | $y_{lm} = 152.4$ mm | |

*Table 11.1*    *Dimensions and Material Properties*

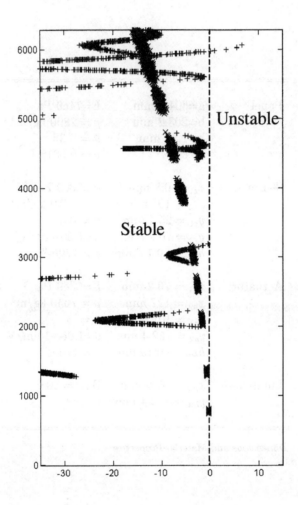

Figure 11.20    *Poles of the plant (x open loop, + closed loop)*

*Chapter 12*

# Vibration control of flexible manipulators

## M. O. Tokhi[†], Wen-Jun Cao[‡], Jian-Xin Xu[‡] and H. Poerwanto[*]

† *Department of Automatic Control and Systems Engineering*
*The University of Sheffield, UK, Email: o.tokhi@sheffield.ac.uk*
‡ *Electrical Engineering Department*
*The National University of Singapore, Singapore*
* *PT PAL Indonesia, Indonesia*

*Control of vibration in flexible robot manipulators is considered in this chapter.*
*A constrained planar single-link flexible manipulator is considered. Open-loop*
*control strategies based on filtered and Gaussian-shaped command inputs and*
*closed-loop control techniques based on partitioning of the rigid and flexible*
*motion dynamics of the system are developed. These incorporate lowpass and*
*bandstop filtered inputs, Gaussian-shaped inputs, switching surface and adap-*
*tive variable structure control, adaptive joint-based collocated and adaptive hy-*
*brid collocated and noncollocated control. The control strategies thus developed*
*are tested within simulations and using a laboratory-scale flexible manipulator*
*test rig.*

## 12.1  Introduction

The demand for the employment of robots in various applications has increased
in line with the increasing demand for system automation. Conventional rigid-

link robots have been successfully used in industrial automation applications, and many control algorithms have been developed to control rigid-link robotic manipulators. However, there are significant limitations associated with rigid-link robots. The dominant factor that contributes largely to performance limitations of a robot is the limited capability of its control system especially in applications requiring high speed and/or large payloads. To obtain high accuracy in the end point position control, the weight to payload ratio must be very high or the operation speed must be very low to prevent link oscillation. A large amount of energy is needed to operate these bulky robots. This drawback greatly limits the applications of rigid robots in the field where high speed, high accuracy and low energy consumption are required. The need for lightweight elastic robot manipulators has increased, as they are capable of improving the speed of operation and handling larger payloads in comparison to rigid manipulators with the same actuator capabilities. Other potential advantages arising from the use of flexible manipulators include lower energy consumption, smaller actuators, safer operation due to reduced inertia, compliant structure, possible elimination of gearing, less bulky design, low mounting strength and rigidity requirements. However, the structural flexibility of such manipulators results in oscillatory behaviour in the system. To overcome this drawback, numerous control schemes have been proposed for flexible manipulators.

Two main approaches can be distinguished when considering the control of flexible manipulator systems. The first approach involves the development of a mathematical model through computation of the necessary geometric, kinematic or kinetic quantities on the basis of assuming rigid body structure. In adopting such an approach, an investigation to reveal the accuracy of the identified parameters is required, so that a satisfactory model is obtained. Alternatively, necessary measurements to yield information on the deflections have to be carried out in addition to the movements of the joints. The second approach accounts, in addition to the factors in the first approach, for deviations caused by the elastic properties of the manipulator and thus requires additional measurements, for example by strain gauges, optical sensors, accelerometers etc. These measurements are to compensate for deviations caused by elasticities and thus are used to improve the control performance.

The vibration control strategies developed for flexible manipulators can broadly be classified as open-loop and closed-loop control. Open-loop control methods for vibration suppression in flexible manipulator systems consist in developing the control input through a consideration of the physical and vibrational properties of the system. The method involves development of suitable forcing functions so as to reduce the vibrations at resonance modes. The methods commonly developed include shaped command methods, computed torque techniques and bang-bang control. The shaped command methods attempt to develop forcing functions which minimise residual motion (vibration) and the effect of parameters that affect the resonance modes [8, 16, 25, 65, 197, 198, 199, 200, 201, 213, 232, 266, 267, 268, 284, 327]. Common problems of concern encountered in these methods include long move (response) time, instability due to unreduced modes and controller robustness in the case of large changes in manipulator dynamics. Frequency shaping is a well known approach to reduce oscillation. Young and Ozguner have introduced a variable structure controller for the frequency-shaped optimal sliding mode [344]. The controller consists of joint position and velocity feedback. A frequency-shaping compensator penalises the inherent resonance modes of a flexible manipulator.

A significant amount of previous work has considered end point regulation in two parts [245, 263, 294]: (1) tracking of the joint angle, (2) suppression of the oscillation of the flexible links. Obviously, there is a trade-off between the two requirements. Siciliano and Book have considered the application of a singular perturbation method to the control of flexible manipulators [263]. In this method, the dynamics of the system are divided into two parts, i.e. a slow subsystem (corresponding to the joint motion) and a fast subsystem (related to the flexible oscillation), and two subcontrollers are designed accordingly. A control law partitioning scheme, which uses an end point sensing device, has been reported by Rattan *et al.* [245]. The scheme uses the end point position signal in an outer loop controller to control the flexible modes, whereas the inner loop controls the rigid body motion independent of the flexible dynamics of the manipulator. Tokhi and Azad have reported a hybrid collocated and noncollocated control strategy incorporating joint angle and velocity feedback for rigid body motion control and end point acceleration feedback for flexible motion control of a single-link flexible manipulator [294]. Ge et al. have investigated the end point regulation performance of a joint-PD controlled single-link flexible manipulator by introducing nonlinear strain feedback [109]. The controller designed consists of a simple joint-PD controller where introducing strain feedback reduces the link oscillation.

Several open-loop and closed-loop control strategies are considered in this chapter and tested with a single-link flexible manipulator under simulated and laboratory-scale experimental environments.

## 12.2   The flexible manipulator system

In this work a constrained planar single-link flexible manipulator is utilised. Accordingly, a dynamic formulation of the characteristics of the system and description of a laboratory-scale test rig are provided in this section.

### 12.2.1   *Dynamic formulation*

A schematic diagram of the flexible manipulator is shown in Figure 12.1, where $\theta$ represents the hub angular displacement, $\tau$ is the input torque, $J_h$ is the hub inertia, $l$ is the length of the flexible link, $m_t$ is the mass of payload, $m$ is the total mass of the flexible link, $EI$ is the flexural rigidity of the flexible link and $\mu$ represents the elastic damping coefficient of the flexible link. It is assumed that the flexible manipulator is only operated in the horizontal plane and consequently the gravity is not considered.

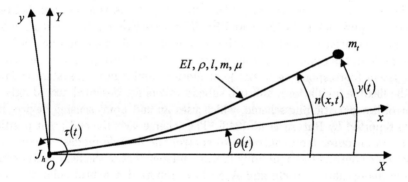

Figure 12.1   *A single-link flexible robot manipulator*

Using the assumed mode method associated with Lagrangian formulation a mathematical model of the flexible manipulator can be derived. A clamped-loaded Euler-Bernoulli beam is selected as the approximate model for the flexible beam. The first $f$ flexible modes are considered. The deflection of the elastic members is represented as:

$$n(x,t) = \varphi^T(x)\mathbf{q}(t), \qquad 0 \le x \le l \tag{12.1}$$

where

$$\varphi(x) = [\phi_1(x)\phi_2(x)\cdots\phi_f(x)]^T, \qquad \mathbf{q}(t) = [q_1(t)q_2(t)\cdots q_f(t)]^T$$

$x$ is any point along the undeformed link. $q_j(t)$ is the $j$th generalised displacement (or modal displacement) for the flexible beam, $\phi_j(x)$ is the $j$th assumed mode shape for the flexible beam, $j = 1, 2, \cdots, f$. The derivation of the dynamic equations for a single-link flexible manipulator accordingly follows [34, 345]:

$$\begin{cases} m_{RR}\ddot{\theta} + \mathbf{m}_{RF}^T\ddot{\mathbf{q}} = \tau - 2\mathbf{q}^T M_{FF}\dot{\mathbf{q}}\dot{\theta} + d(t) \\ \mathbf{m}_{RF}^T\ddot{\theta} + M_{FF}\ddot{\mathbf{q}} = M_{FF}\mathbf{q}\dot{\theta}^2 - C_F\mathbf{q} - D_F\dot{\mathbf{q}} \end{cases} \qquad (12.2)$$

where

$$m_{RR} = J_h + \frac{1}{3}ml^2 + m_t l^2 + \mathbf{q}^T M_{FF}\mathbf{q}$$

$$\mathbf{m}_{RF} = \frac{m}{l}\int_0^l x\varphi(x)dx + m_t l\varphi_e$$

$$M_{FF} = m_t \varphi_e \varphi_e^T + \frac{m}{l}\int_0^l \varphi(x)\varphi^T(x)dx$$

$$C_F = diag\left[ EI\int_0^l \left[\frac{\partial^2\phi_1(x)}{\partial x^2}\right]^2 dx \cdots EI\int_0^l \left[\frac{\partial^2\phi_f(x)}{\partial x^2}\right]^2 dx \right]$$

$$D_F = \int_0^l \mu\varphi(x)\varphi^T(x)dx$$

The subscript $e$ is used to denote the value of the variable occurring at the end point, i.e., the value at $x = l$, and $d(t)$ represents the bounded exogenous disturbance. Introducing the orthogonality properties [51] about mode shapes yields:

$$M_{FF} = diag\left( \frac{EI}{\omega_1^2}\int_0^l \left[\frac{\partial^2\phi_1(x)}{\partial x^2}\right]^2 dx \quad \cdots \quad \frac{EI}{\omega_f^2}\int_0^l \left[\frac{\partial^2\phi_f(x)}{\partial x^2}\right]^2 dx \right)$$

where $\omega_j$ is the natural frequency of the $j$th flexible mode given by:

$$\omega_j = \beta_j^2\sqrt{\frac{EI}{m/l}}$$

and $\beta_j$ is the $j$th eigenvalue of the following characteristic equation:

$$m(1 + \cos\beta_j l \cosh\beta_j l) = \beta_j m_t l(\sin\beta_j l \cosh\beta_j l - \cos\beta_j l \sinh\beta_j l)$$

The dynamic model in eqn. (12.2) can be rewritten as:

$$M(\xi)\ddot{\xi} + D(\xi, \dot{\xi})\dot{\xi} + C\xi = \mathbf{b}_1\tau + \mathbf{d} \qquad (12.3)$$

where

$$\xi = \left[\begin{array}{cc} \theta & \mathbf{q}^T \end{array}\right]^T, \qquad \mathbf{b}_1 = \left[\begin{array}{cc} 1 & \mathbf{0}_{1\times f} \end{array}\right]^T, \qquad M = \left[\begin{array}{cc} m_{RR} & \mathbf{m}_{RF}^T \\ \mathbf{m}_{RF} & M_{FF} \end{array}\right]$$

$$C = diag\left[\begin{array}{cc} 0 & C_F \end{array}\right], \qquad D = \left[\begin{array}{cc} \mathbf{q}^T M_{FF}\dot{\mathbf{q}} & \mathbf{q}^T M_{FF}\dot{\theta} \\ M_{FF}\mathbf{q}\dot{\theta} & D_F \end{array}\right], \qquad \mathbf{d} = \left[\begin{array}{c} d \\ \mathbf{0}_{(2r+1)\times 1} \end{array}\right]$$

Defining the state vector:

$$\mathbf{x} \equiv \left[\begin{array}{cc} \xi^T & \dot{\xi}^T \end{array}\right]^T$$

yields the state-space representation:

$$\dot{\mathbf{x}} \equiv A\mathbf{x} + \mathbf{b}\tau + \mathbf{v} \tag{12.4}$$

where $\mathbf{x}$ is measurable and:

$$A = \left[\begin{array}{cc} \mathbf{0}_{(f+1)\times(f+1)} & \mathbf{I}_{(f+1)\times(f+1)} \\ -M^{-1}C & -M^{-1}D \end{array}\right], \qquad \mathbf{b} = \left[\begin{array}{c} \mathbf{0}_{(f+1)\times 1} \\ M^{-1}\mathbf{b}_1 \end{array}\right], \qquad \mathbf{v} = \left[\begin{array}{c} \mathbf{0}_{(f+1)\times 1} \\ M^{-1}\mathbf{d} \end{array}\right]$$

$A$ and $\mathbf{b}$ can be decomposed into the linear part $A_l$ and $\mathbf{b}_l$ which are constant matrices, and the nonlinear part $\Delta A$ and $\Delta \mathbf{b}$:

$$A = A_l + \Delta A, \qquad \mathbf{b} = \mathbf{b}_l + \Delta \mathbf{b}$$

where

$$A = \left[\begin{array}{cc} \mathbf{0}_{(f+1)\times(f+1)} & \mathbf{I}_{(f+1)\times(f+1)} \\ -M_l^{-1}C & -M_l^{-1}D_l \end{array}\right], \qquad \mathbf{b} = \left[\begin{array}{c} \mathbf{0}_{(f+1)\times 1} \\ M_l^{-1}\mathbf{b}_1 \end{array}\right], \qquad M_l = \left[\begin{array}{cc} m_{RRl} & \mathbf{m}_{RF}^T \\ \mathbf{m}_{RF} & M_{FF} \end{array}\right]$$

$$m_{RRl} = J_h + \tfrac{1}{3}ml^2 + m_t l^2, \qquad D_l = diag\left[\begin{array}{cc} 0 & D_F \end{array}\right]$$

The end point angular displacement can be obtained by the assumed mode method as:

$$y = \theta + \frac{\varphi_e^T \mathbf{q}}{l} = \mathbf{c}^T \mathbf{x} \tag{12.5}$$

where

$$\mathbf{c} = \left[\begin{array}{cccc} 1 & \varphi_e^T/l & 0 & \mathbf{0}_{1\times f} \end{array}\right]^T$$

The angular displacement $y$ uniquely determines the end point position on the XOY plane (see Figure 12.1). The general form of the end point transfer function, considering the linear plant, is:

$$g(s) = \mathbf{c}^T(s\mathbf{I} - A)\mathbf{b} = k_g \frac{1 + b_1 s + b_2 s^2 + \cdots + b_{2f-1}s^{2f-1} + b_{2f}s^{2f}}{s^{2f+2} + a_{2f-1}s^{2f+1} + \cdots + a_1 s^3 + a_0 s^2} \tag{12.6}$$

which possesses $f$ zeros outside the stability region, hence representing a nonminimum-phase system [42]. The conjugate poles result in sinusoidal oscillatory behaviour of end point trajectory.

## 12.2.2 The flexible manipulator test rig

A schematic diagram of the experimental flexible manipulator test rig is shown in Figure 12.2. This consists of two main parts: a flexible arm and measuring devices. The flexible arm contains a flexible link driven by a printed armature motor at the hub. The measuring devices are a shaft encoder, a tachometer and an accelerometer. The shaft encoder is used for measuring the hub angle of the manipulator, the tachometer is used for measurement of the hub velocity and the accelerometer is located at the end point of the flexible arm measuring the end point acceleration. The flexible arm is constructed using a piece of thin aluminium alloy, and its parameters are given in Table 12.1.

An IBM compatible PC/AT microcomputer based on a 486DX2 50 MHz CPU, with 20 Mbytes of dynamic and 540 Mbytes of static memory, is utilised in conjunction with an RTI-815 I/O board. The RTI-815 I/O board is used to provide a direct interface between the microcomputer and the actuator and transducers, through an AM9513A counter/timer chip for a variety of data acquisition, analogue output, digital I/O and time-related digital I/O applications.

The experimental setup requires one analogue output to the motor drive

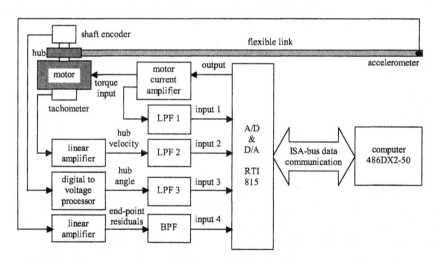

Figure 12.2   Schematic diagram of the flexible manipulator test rig

amplifier, four analogue inputs from the hub angle and velocity transducer,

Table 12.1    *Physical parameters and characteristics of the flexible manipulator system*

| Parameter | Value |
|---|---|
| Length | 960.0 mm |
| Width | 19.008 mm |
| Thickness | 3.2004 mm |
| Mass density per unit volume | 2710 $kgm^{-3}$ |
| Second moment of inertia, $I$ | $5.1924 \times 10^{-11}$ $m^4$ |
| Young's modulus, $E$ | $71 \times 10^9$ $Nm^{-2}$ |
| Moment of inertia, $I_b$ | 0.04862 $kgm^2$ |
| Hub inertia, $I_h$ | $5.86 \times 10^{-4}$ $kgm^2$ |

accelerometer and motor current sensor. This is provided through the RTI-815 I/O board. The interface board is used with a conversion time of 25 $\mu$s for A/D conversion and settling time of 20 $\mu$s for D/A conversion, which are satisfactory for the system under consideration.

## 12.3    Open-loop control

In this Section, shaped command inputs based on filtering techniques are developed and applied in an open-loop configuration to suppress system response oscillations. The aim of this investigation is to develop control methods to reduce motion-induced vibrations in flexible manipulator systems during fast movements. The assumption is that the motion itself is the main source of system vibrations. Thus, torque profiles which do not contain energy at system natural frequencies do not excite structural vibration and hence require no additional settling time. The procedure for determining shaped inputs which generate fast motions with minimum residual vibration has previously been addressed [25, 200, 293]. The torque input needed to move a flexible manipulator from one point to another without vibration must have several properties: (i) it should have an acceleration and deceleration phase, (ii) it should be able to be scaled for different step motions and (iii) it should have as sharp a cut-off frequency as required. These three properties of the required torque input will allow the manipulator system to be driven as quickly as possible without exciting its resonance modes. Two types of open-loop shaped input torque are developed on the basis of extracting the energies around the natural frequencies so that the vibration in the flexible manipulator system during and after the movement is reduced. In the first approach, the extraction of energy at the system resonances is based on filter theory. The filters are used for preprocessing the input to the plant, so that no energy is ever put into the system near its resonances. In the second approach a Gaussian-shaped input torque is developed.

### 12.3.1 *Filtered torque input*

The filtered torque input strategy adopted here is to use a single cycle of a square wave, which is known to give optimal response, and filter out any spectral energy near the natural frequencies. The simplest method to remove energy at system natural frequencies is to pass the square wave through a lowpass filter. This will attenuate all frequencies above the filter cut-off frequency. The most important consideration is to achieve a steep roll-off rate at the cut-off frequency so that energy can be passed for frequencies nearly up to the lowest natural frequency of the flexible manipulator. An alternative method of removing energy at system natural frequencies will be to use (narrowband) bandstop filters with centre frequencies at selected (dominant) resonance modes of the system.

There are various types of filter, namely, Butterworth, Chebyshev and elliptic, which can be employed. Here Butterworth filters are used. These filters have the desired frequency response in magnitude, allow for any desired cut-off rate and are physically realisable. The magnitude of the frequency response of a lowpass Butterworth filter is given by [147]:

$$|H(j\omega)|^2 = \frac{1}{1 + \left[\frac{\omega}{\omega_C}\right]^{2n}} = \frac{1}{1 + \varepsilon \left[\frac{\omega}{\omega_P}\right]^{2n}} \qquad (12.7)$$

where $n$ is a positive integer signifying the order of the filter, $\omega_C$ is the filter cut-off frequency $-3$ dB frequency), $\omega_P$ is the passband edge frequency and $(1+\varepsilon^2)^{-1}$ is the band edge value of $|H(j\omega)|^2$. Note that $|H(j\omega)|^2$ is monotomic in both the passband and stopband. The order of the filter required to yield an attenuation $\delta_2$ at a specified frequency $\omega_S$ (stopband edge frequency) is easily determined from eqn. (12.7) as:

$$n = \frac{\log\left\{\frac{1}{\delta_2^2} - 1\right\}}{2\log\left\{\frac{\omega_S}{\omega_P}\right\}} = \frac{\log\left\{\frac{\delta_1}{\varepsilon}\right\}}{\log\left\{\frac{\omega_S}{\omega_P}\right\}} \qquad (12.8)$$

where, by definition, $\delta_2 = (1 + \delta_1^2)^{-0.5}$. Thus, the Butterworth filter is completely characterised by the parameters $n$, $\delta_2$, $\varepsilon$ and the ratio $\frac{\omega_S}{\omega_P}$.

Eqn. (12.8) can be employed with arbitrary $\delta_1$, $\delta_2$, $\omega_C$ and $\omega_S$ to yield the required filter order $n$ from which the filter design is readily obtained. The Butterworth approximation results from the requirement that the magnitude response be maximally flat in both the passband and the stopband. That is, the first $(2n - 1)$ derivatives of $|H(j\omega)|^2$ are specified to be equal to zero at $\omega = 0$ and at $\omega = \infty$.

To provide a comparative assessment of the performance of the system with filtered torque inputs, the system was excited with a bang-bang torque. This is shown in Figure 12.3. The corresponding system responses as hub angle, hub velocity, end point acceleration and motor torque are shown in Figures 12.4, 12.5, 12.6 and 12.7, respectively. It is noted that the first three resonance modes of the system occur at 12.07 Hz, 34.48 Hz and 62.07 Hz.

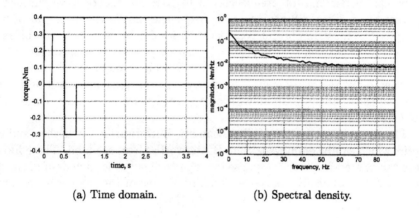

(a) Time domain.                    (b) Spectral density.

*Figure 12.3    The bang-bang torque input*

To study the performance of the system with a lowpass Butterworth filter design, a lowpass filtered bang-bang torque input is used and the system response is measured at the hub and end point. A third-order lowpass Butterworth filter with a cut-off frequency at 5.0 Hz was designed and used for preprocessing the bang-bang torque input. To study the performance of the system with bandstop filters, three third-order bandstop filters of 5.0 Hz bandwidth and centre frequencies at the first, second and third resonance frequencies of the system were designed and used to process the bang-bang input torque. The lowpass and bandstop torque profiles with the corresponding system responses are shown in Figures 12.8–12.12 and 12.13– 12.17, respectively.

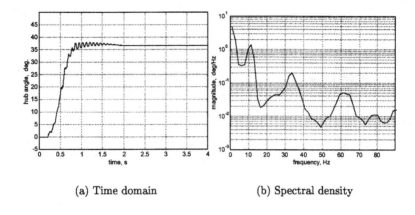

(a) Time domain    (b) Spectral density

*Figure 12.4    The hub angle with bang-bang torque input*

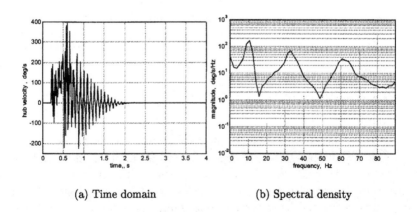

(a) Time domain    (b) Spectral density

*Figure 12.5    The hub velocity with bang-bang torque input*

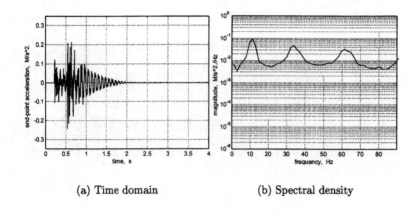

(a) Time domain                    (b) Spectral density

Figure 12.6    *The end point acceleration with bang-bang torque input*

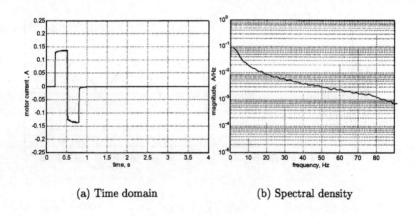

(a) Time domain                    (b) Spectral density

Figure 12.7    *The motor current with bang-bang torque input*

Comparing these results accordingly with those obtained with the bang-bang torque, it is noted that the vibration at the first, second and third modes of the system was attenuated by 26.21, 24.81 and 30.37 dB, respectively, with the lowpass filtered command input. The attenuation at these modes with the bandstop filtered input was 22.44, 21.40 and 18.33 dB, respectively.

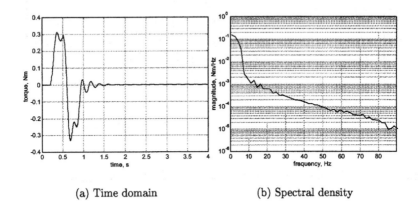

(a) Time domain                    (b) Spectral density

*Figure 12.8    Butterworth lowpass filtered torque input*

Comparing the results achieved with the bandstop and lowpass filtered torque inputs reveals that better performance at a reduced level of vibration of the system is achieved with lowpass filtered torque inputs. This is due to the indiscriminate spectral attenuation in the lowpass filtered torque input at all resonance modes of the system. Utilisation of bandstop filters, however, is advantageous in that spectral attenuation in the input at selected resonance modes of the system can be achieved. Thus, the open-loop control strategy based on bandstop filters is optimum in this sense. Note that the strategy can also be viewed as being equivalent to designing a controller with zeros which cancel out the system poles, which represent the resonance modes.

### 12.3.2    Gaussian-shaped torque input

A Gaussian-shaped input torque, i.e. the first derivative of the Gaussian distribution function, is examined in this section. The application of this function in the form of an acceleration profile, to develop an input torque profile through inverse dynamics of the system, has previously been shown [25]. Here, the behaviour of the function as an input torque profile for the system is investigated

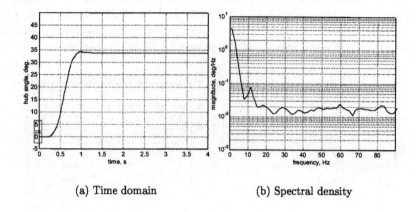

(a) Time domain                (b) Spectral density

Figure 12.9    *The hub angle with Butterworth lowpass filtered torque input*

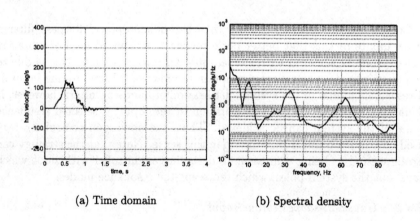

(a) Time domain                (b) Spectral density

Figure 12.10    *The hub velocity with Butterworth lowpass filtered torque input*

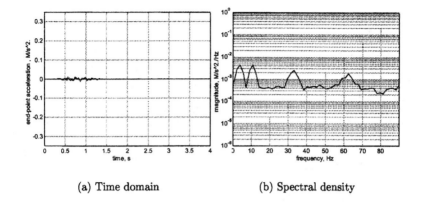

(a) Time domain  (b) Spectral density

Figure 12.11   *The end point acceleration with Butterworth lowpass filtered torque input*

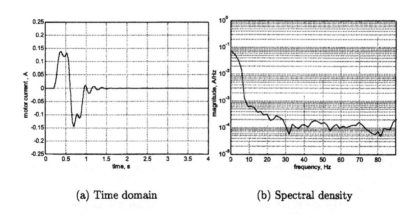

(a) Time domain  (b) Spectral density

Figure 12.12   *The motor current with Butterworth lowpass filtered torque input*

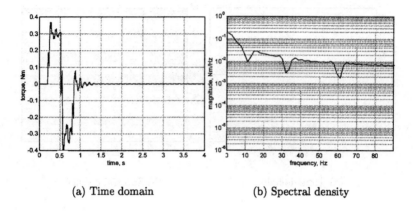

(a) Time domain    (b) Spectral density

Figure 12.13    *Butterworth bandstop filtered torque input*

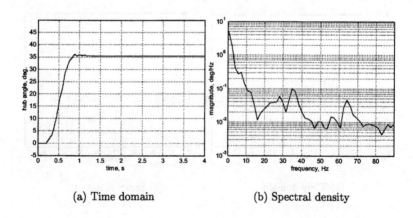

(a) Time domain    (b) Spectral density

Figure 12.14    *The hub angle with Butterworth bandstop filtered torque input*

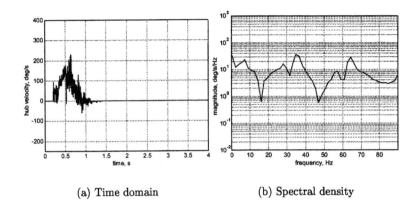

(a) Time domain                    (b) Spectral density

Figure 12.15    *The hub velocity with Butterworth bandstop filtered torque input*

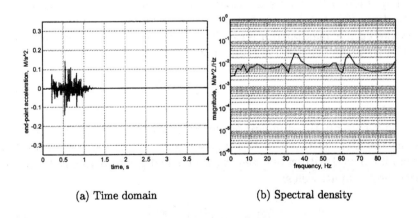

(a) Time domain                    (b) Spectral density

Figure 12.16    *The end point acceleration with Butterworth bandstop filtered torque input.*

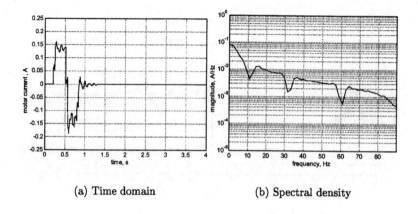

(a) Time domain                    (b) Spectral density

Figure 12.17    *The motor current with Butterworth bandstop filtered torque input*

by adopting a much simpler method of developing an input torque profile for a flexible manipulator system. Variations of frequency distribution, duty cycle and amplitude of the Gaussian-shaped torque input with various parameters are studied. This enables the generation of an appropriate input trajectory to move the flexible manipulator for a given position with negligible vibration.

The Gaussian distribution function is given as:

$$P(x) = \frac{1}{\sqrt{2\pi}\sigma}e^{\left[\frac{-(x-\mu)^2}{2\sigma^2}\right]} \tag{12.9}$$

where $\sigma$ represents the standard deviation and $\mu$ the mean of the variable $x$. Taking the first derivative of this function and considering this as a system torque input $\tau(t)$ with $x$ representing time $t$ and $\mu$ and $\sigma$ as constants for a given torque input, yields:

$$\tau(t) = \frac{(t-\mu)}{\sqrt{2\pi}\sigma^3}e^{\left[\frac{-(t-\mu)^2}{2\sigma^2}\right]} \tag{12.10}$$

The essential properties of the torque profile, such as amplitude, cut-off frequency and duty cycle can be affected by the parameters $\mu$ and $\sigma$ [238]. Accordingly, a suitable set of $\mu$ and $\sigma$ can be used for a desired system response.

To investigate the effectiveness of the Gaussian-shaped input torque on the performance of the flexible manipulator system, a Gaussian-shaped input torque was developed with a cut-off frequency at 10.0 Hz, $\sigma = 0.15$ and $\mu = 10\sigma$. The performance of the manipulator was studied experimentally with this Gaussian-shaped torque input in comparison with a bang-bang input torque for a similar angular displacement, keeping the peak torque at a similar level in each case. The Gaussian-shaped input torque profile and the corresponding system responses are shown in Figures 12.18 and 12.19–12.22, respectively. It is noted that a smooth transition from acceleration to deceleration is achieved with the Gaussian-shaped input, and consequently, a significant amount of reduction in the level of vibration is achieved at the resonance modes of the system. As noted, the reduction in vibration at the first, second and third resonance modes of the system achieved in comparison with the bang-bang torque input was 29.92, 25.14 and 23.52 dB, respectively.

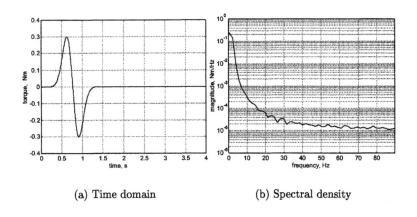

(a) Time domain        (b) Spectral density

*Figure 12.18   The Gaussian-shaped torque input*

## 12.4  Switching surface and variable structure control

The difficulties associated with the controller design for flexible manipulators include: the nonminimum-phase property of the end point transfer function making a conventional robust controller with end point position feedback unstable, the oscillatory end point motion, the presence of nonlinearities $\Delta A$

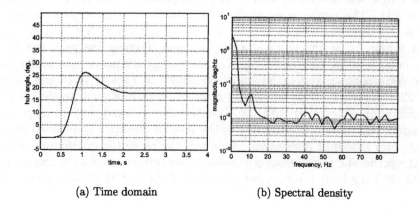

(a) Time domain    (b) Spectral density

Figure 12.19    *The hub angle with Gaussian-shaped torque input*

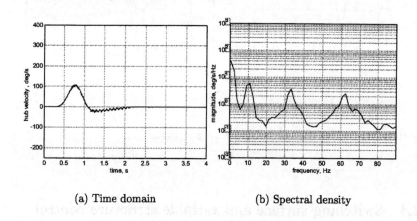

(a) Time domain    (b) Spectral density

Figure 12.20    *The hub velocity with Gaussian-shaped torque input*

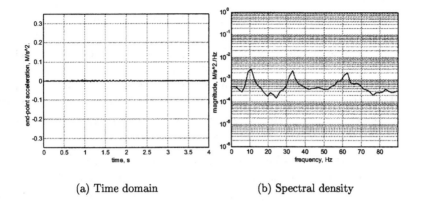

(a) Time domain         (b) Spectral density

*Figure 12.21    The end point acceleration with Gaussian-shaped torque input*

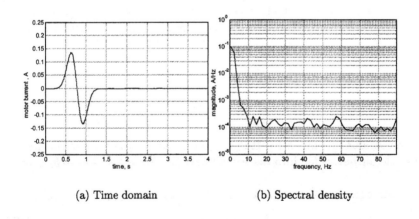

(a) Time domain         (b) Spectral density

*Figure 12.22    The motor current with Gaussian-shaped torque input*

and $\Delta \mathbf{b}$ and uncertainty $\mathbf{d}$. An attempt is made in this section to solve these difficulties as follows:

(i) For the linear plant, the corresponding transfer function is factorised into two parts. The first part contains all plant poles and the second part retains all plant zeros. For the first part, a reference model can be selected which leads to a switching surface. The reference model contains stable real poles and no finite zeros to eliminate link oscillation. The output of the first part can be fed into the second part and the zeros of the second part that are outside of the stability region have no effect on the tracking error in the steady state.

(ii) For the residual motion control a variable structure controller (VSC) is utilised to suppress the perturbation, which includes $\Delta A$, $\Delta \mathbf{b}$ and $\mathbf{d}$. An adaptive method can be used to reduce the estimated upper bound of the perturbation.

### 12.4.1   Switching surface design

In the design of a switching surface, the linear plant is considered under nominal condition:

$$\dot{\mathbf{x}} = A_{l,n}\mathbf{x} + \mathbf{b}_{l,n}\tau \qquad (12.11)$$

where $n$ denotes the nominal values. To eliminate link oscillation, a straightforward approach would be to relocate plant poles onto the negative real axis. The end point transfer function $g(s)$ can accordingly be factorised into:

$$g(s) = h_1(s)(1 + b_1 s + b_2 s^2 + \cdots + b_{2f-1} s^{2f-1} + b_{2f} s^{2f}) \qquad (12.12)$$

where the first part of $g(s)$ is:

$$h_1(s) = \frac{k_g}{s^{2f+2} + a_{2f-1} s^{2f+1} + \cdots + a_1 s^3 + a_0 s^2}$$

which contains no finite zero. It is supposed that the output $\eta$ of the first part, a pseudo output, is fed into the second part $(1 + b_1 s + b_2 s^2 + \cdots + b_{2f-1} s^{2f-1} + b_{2f} s^{2f})$.

Given eqns (12.6) and (12.12) and the output of $h_1(s)$ as $\eta = \mathbf{m}^T \mathbf{x}$:

$$h_1(s) = \mathbf{m}^T(s\mathbf{I} - A_{l,n})\mathbf{b}_{l,n} \qquad (12.13)$$

The zero polynomial of eqn. (12.13) is [153]:

$$\begin{aligned}
\mathbf{m}^T \, adj(s\mathbf{I} - A_{l,n})\mathbf{b}_{l,n} &= \mathbf{m}^T\mathbf{b}_{l,n}s^{f-1} + (\mathbf{m}^T A_{l,n}\mathbf{b}_{l,n} + a_1\mathbf{m}^T\mathbf{b}_{l,n})s^{f-2} + \cdots \\
&+ (\mathbf{m}^T A_{l,n}^{2f+1}\mathbf{b}_{l,n} + a_1\mathbf{m}^T A_{l,n}^{2f}\mathbf{b}_{l,n} + \cdots + a_{f-1}\mathbf{m}^T\mathbf{b}_{l,n})
\end{aligned}$$

Since $h_1(s)$ does not contain any finite zero, the following holds:

$$\mathbf{m}^T A_{l,n}^{2f+1} \mathbf{b}_{l,n} = k_g, \quad \mathbf{m}^T A_{l,n}^{i-1} \mathbf{b}_{l,n} = 0, \quad i = 1, 2, \ldots, 2f+1 \quad (12.14)$$

As the pair $(A_l, \mathbf{b}_l)$ is controllable, eqn. (12.14) can be rewritten as:

$$\mathbf{m}^T = \begin{bmatrix} 0 & 0 & \cdots & 0 & k_g \end{bmatrix} \begin{bmatrix} \mathbf{b}_{l,n} & A_{l,n}\mathbf{b}_{l,n} & \cdots & A_{l,n}^{2f} & A_{l,n}^{2f+1}\mathbf{b}_{l,n} \end{bmatrix}^{-1} \quad (12.15)$$

For the linear plant, eqn. (12.11), the above yields:

$$\dot{\eta} = \mathbf{m}^T \dot{\mathbf{x}} = \mathbf{m}^T A_l \mathbf{x}, \quad \ddot{\eta} = \mathbf{m}^T A_l \dot{\mathbf{x}} = \mathbf{m}^T A_l^2 \mathbf{x}, \cdots, \eta^{(2f+1)} = \mathbf{m}^T A_l^{2f+1} \mathbf{x} \quad (12.16)$$

A reference model is accordingly chosen for $h_1(s)$ as:

$$\hat{h}_1(s) = \frac{\eta(s)}{r(s)} = \frac{k_g}{s^{2f+2} + \alpha_1 s^{2f+1} + \cdots + \alpha_{2f+1} s + \alpha_{2f+2}} \quad (12.17)$$

with negative real poles and no finite zeros, where $r$ is a constant reference input. From eqn. (12.17) the following can be obtained:

$$\eta^{(2f+2)} + \alpha_1 \eta^{(2f+1)} + \cdots + \alpha_{2f+1}\dot{\eta} + \alpha_{2f+2}(\eta - r) = 0$$

which can be rewritten as:

$$\frac{d}{dt}\left\{ \eta^{(2f+1)} + \alpha_1 \eta^{(2f)} + \cdots + \alpha_{2f+1}\eta + \alpha_{2f+2}\int_0^t [\eta(v) - r]\, dv \right\} = 0. \quad (12.18)$$

Substituting for $\eta^{(i)}$; $i = 1, \ldots, 2f+1$, from eqn. (12.16) into eqn. (12.18) yields:

$$\frac{d}{dt}\left\{ \mathbf{m}^T \left[ A_{l,n}^{2f+1} + \alpha_1 A_{l,n}^{2f} + \cdots + \alpha_{2f} A_{l,n} + \alpha_{2f+1}\mathbf{I}_{(2f+1)\times(2f+1)} \right] \mathbf{x} \right.$$
$$\left. + \alpha_{2f+2}\int_0^t [\eta(v) - r]\, dv \right\} = 0 \quad (12.19)$$

The switching surface can be chosen according to eqn. (12.19) as:

$$\mathbf{m}^T \left[ A_{l,n}^{2f+1} + \alpha_1 A_{l,n}^{2f} + \cdots + \alpha_{2f} A_{l,n} + \alpha_{2f+1}\mathbf{I}_{(2f+1)\times(2f+1)} \right] \mathbf{x} + \alpha_{2f+2}\int_0^t [\eta(v) - r]\, dv =$$

To reduce the system nonlinearities, a nonlinear term is introduced and the following switching surface is adopted:

$$\sigma = \mathbf{s}^T \mathbf{x} + \alpha_{2f+2}\int_0^t \left[ \mathbf{m}^T \mathbf{x} - r \right] dv + \int_0^t \mathbf{c}_4^T \mathbf{q}\, \dot{\theta}^2 \mathbf{dv}$$

$$\quad (12.20)$$

$$= c_1 \theta + \mathbf{c}_2^T \mathbf{q} + c_3 \dot{\theta} + \mathbf{c}_4^T \dot{\mathbf{q}} + \alpha_{2f+2}\int_0^t \left[ \mathbf{m}^T \mathbf{x} - r \right] dv + \int_0^t \mathbf{c}_4^T \mathbf{q}\, \dot{\theta}^2 dv = 0$$

where

$$\mathbf{s}^T = \mathbf{m}^T \left[ A_{l,n}^{2f+1} + \alpha_1 A_{l,n}^{2f} + \cdots + \alpha_f A_{l,n} + \alpha_{f+1} \mathbf{I}_{(2f+1)\times(2f+1)} \right]$$

$$\mathbf{m}^T = \begin{bmatrix} 0 & 0 & \cdots & 0 & k_g \end{bmatrix} \begin{bmatrix} \mathbf{b}_{l,n} & A_{l,n}\mathbf{b}_{l,n} & \cdots & A_{l,n}^{2f} & A_{l,n}^{2f+1}\mathbf{b}_{l,n} \end{bmatrix}^{-1}$$

$$c_1 = s_1, \qquad \mathbf{c}_2 = \begin{bmatrix} s_2 & s_3 & \cdots & s_{f+1} \end{bmatrix}^T$$

$$c_3 = s_{f+1}, \qquad \mathbf{c}_4 = \begin{bmatrix} s_{f+3} & s_{f+4} & \cdots & s_{2f+2} \end{bmatrix}^T$$

and $s_j$ denotes the $j$th element of vector $\mathbf{s}$.

### 12.4.2    Stability analysis of the switching surface

It can be shown from eqn. (12.1) that:

$$\Delta\ddot{\theta} = \tau - \mathbf{m}_{RF}^T \mathbf{q}\,\dot{\theta}^2 + \mathbf{m}_{RF}^T M_{FF}^{-1} C_F \mathbf{q} + \mathbf{m}_{RF}^T M_{FF}^{-1} D_F \dot{\mathbf{q}} - 2\mathbf{q}^T M_{FF}\dot{\mathbf{q}} + d \quad (12.21)$$

$$\begin{aligned} \Delta\ddot{\mathbf{q}} &= -M_{FF}^{-1}\mathbf{m}_{RF}\left( \tau - \mathbf{m}_{RF}^T \mathbf{q}\,\dot{\theta}^2 + \mathbf{m}_{RF}^T M_{FF}^{-1} C_F \mathbf{q} + \mathbf{m}_{RF}^T M_{FF}^{-1} D_F \dot{\mathbf{q}} - 2\mathbf{q}^T M_{FF}\dot{\mathbf{q}} + d \right) \\ &+ \Delta\mathbf{q}\,\dot{\theta}^2 - \Delta M_{FF}^{-1} C_F \mathbf{q} - \Delta M_{FF}^{-1} D_F \dot{\mathbf{q}} \end{aligned} \quad (12.22)$$

where

$$\Delta = m_{RR} - \mathbf{m}_{RF}^T M_{FF}^{-1} \mathbf{m}_{RF}$$

Differentiating eqn. (12.20) with respect to time yields:

$$\dot{\sigma} = c_1\dot{\theta} + \mathbf{c}_2^T \dot{\mathbf{q}} + c_3\ddot{\theta} + \mathbf{c}_4^T \ddot{\mathbf{q}} + \alpha_{2f+2}\left[ \mathbf{m}^T \mathbf{x} - r \right] + \mathbf{c}_2^T \mathbf{q}\,\dot{\theta}^2 \quad (12.23)$$

Combining eqns (12.21), (12.22) and (12.23) yields:

$$\lambda^{-1}\Delta\dot{\sigma} = \tau + \mathbf{k}^T \chi \quad (12.24)$$

where $\lambda = c_3 - \mathbf{c}_4^T M_{FF}^{-1} \mathbf{m}_{RF}$ and:

$$\mathbf{k} = \begin{bmatrix} \lambda^{-1}\Delta c_1 \\ -\mathbf{m}_{RF} \\ -\lambda^{-1}\Delta M_{FF}^{-1} C_F \mathbf{c}_4 + M_{FF}^{-1} C_F \mathbf{m}_{RF} \\ \lambda^{-1}\Delta\mathbf{c}_2 - \lambda^{-1}\Delta M_{FF}^{-1} D_F \mathbf{c}_4 + M_{FF}^{-1} D_F \mathbf{m}_{RF} \\ -2\frac{EI}{\omega_1^2} \int_0^l \left[ \frac{\partial^2 \phi_1(x)}{\partial x^2} \right]^2 dx \\ \vdots \\ -2\frac{EI}{\omega_f^2} \int_0^l \left[ \frac{\partial^2 \phi_f(x)}{\partial x^2} \right]^2 dx \\ \lambda^{-1}\Delta\alpha_{2f+2} \\ d \end{bmatrix}, \qquad \chi = \begin{bmatrix} \dot{\theta} \\ \mathbf{q}\,\dot{\theta}^2 \\ \mathbf{q} \\ \dot{\mathbf{q}} \\ q_1\dot{q}_1 \\ \vdots \\ q_f\dot{q}_f \\ \eta - r \\ 1 \end{bmatrix}$$

Since $\lambda^{-1}\Delta\dot{\sigma} = \tau + \mathbf{k}^T\chi$, substitution of $\tau = -\mathbf{k}^T\mathbf{X} + \lambda^{-1}\Delta\dot{\sigma}$ into eqns (12.21) and (12.22) yields:

$$\ddot{\theta} = \xi_1^T\mathbf{x} + \lambda^{-1}\alpha_{2n+2}(\eta - r) + \lambda^{-1}\dot{\sigma} \tag{12.25}$$

$$\ddot{\mathbf{q}} = \xi_2^T\mathbf{x} + \lambda M_{FF}^{-1}\mathbf{m}_{RF}\alpha_{2n+2}(\eta - r) + \mathbf{q}\,\dot{\theta}^2 + \lambda M_{FF}^{-1}\mathbf{m}_{RF}\dot{\sigma} \tag{12.26}$$

where

$$\xi_1^T = \begin{bmatrix} 0, & \lambda^{-1}\mathbf{c}_4^T M_{FF}^{-1}C_F, & -\lambda^{-1}c_1, & \lambda^{-1}\mathbf{c}_4^T M_{FF}^{-1}D_F - \lambda^{-1}\mathbf{c}_2^T \end{bmatrix}$$

$$\xi_2^T = \begin{bmatrix} 0, & \lambda^{-1}M_{FF}^{-1}\mathbf{m}_{RF}\mathbf{c}_4^T M_{FF}^{-1}C_F - M_{FF}^{-1}C_F, \\ \lambda^{-1}M_{FL}^{-1}\mathbf{m}_{RF}c_1, & \lambda^{-1}M_{FF}^{-1}\mathbf{m}_{RF}(\mathbf{c}_2^T - \mathbf{c}_4^T M_{FF}^{-1}D_F) - M_{FF}^{-1}D_F \end{bmatrix}$$

Define $\bar{\mathbf{x}} := \mathbf{x} - \begin{bmatrix} r & \mathbf{0}_{1\times n} & 0 & \mathbf{0}_{1\times n} \end{bmatrix}^T$. Making $\dot{\sigma} = 0$ and substituting for $\eta$ from eqn. (12.21) into eqns (12.25) and (12.26) yields the following system equation for the closed-loop system dynamics during the ideal sliding phase:

$$\dot{\bar{\mathbf{x}}}_d = A_{cl}\bar{\mathbf{x}}_d + \mathbf{f}_{h.o.t.}(\bar{\mathbf{x}}_d) \tag{12.27}$$

where

$$A_{cl} = \begin{bmatrix} \mathbf{0}_{(f+1)\times(f+1)} & \mathbf{I}_{(f+1)\times(f+1)} \\ \xi_1^T + \lambda^{-1}\alpha_{2f+2}\mathbf{m}^T \\ \xi_2^T + \lambda^{-1}\alpha_{2f+2}M_{FF}^{-1}\mathbf{m}_{RF}\mathbf{m}^T \end{bmatrix}, \quad \mathbf{f}_{h.o.t} = \begin{bmatrix} 0 \\ \mathbf{0}_{f\times 1} \\ 0 \\ \mathbf{q}_d\,\dot{\theta}_d^2 \end{bmatrix}$$

The subscript $d$ is used in the above to denote the value of the variable under ideal sliding mode.

From the above form of $\mathbf{f}_{h.o.t}(\bar{\mathbf{x}}_d)$, it can be obtained that:

$$\lim_{\|\mathbf{x}_d\|\to 0} \frac{\|\mathbf{f}_{h.o.t}(\bar{\mathbf{x}}_d)\|}{\|\mathbf{x}_d\|} = 0 \tag{12.28}$$

Hence, the controlled system in the sliding mode is uniformly asymptotically stable around the equilibrium point:

$$\mathbf{x}_d = \begin{bmatrix} r^* \\ \mathbf{0}_{(2r+1)\times 1} \end{bmatrix}$$

provided that $A_{cl}$ is Hurwitz (i.e. has all its eigenvalues strictly in the left half plane).

According to eqn. (12.5), as $\bar{\mathbf{x}}_d \to \mathbf{0}_{(2f+2)\times 1}$, the end point angular displacement under ideal sliding mode $y_d \to r$, and this produces zero steady-state output error.

### 12.4.3    Adaptive variable structure control scheme

To implement a stabilising VSC for the derived switching surface, estimates of system uncertainties must be made. Overestimation will result in unnecessarily high gains and large chattering, which consequently will degrade the system's performance, and underestimation is not permitted as it may incur instability. To solve this problem, an adaptive technique is incorporated into the VSC to estimate suitable upper bounds of system uncertainties, similar to that reported by Yoo and Chung [343]. A lower switching gain at the beginning of adaptation reduces the impact to the system and a higher gain in the end lowers steady-state error. To further improve the control system's robustness, a dead zone modification is introduced.

Under nominal conditions, where $m_t$, $J_h$, $\mu$, $m$ and $l$ take their nominal values at the equilibrium in the absence of the exogenous disturbance $d$, the $\mathbf{k}$ in eqn. (12.24) can be decomposed into:

$$\mathbf{k} = \mathbf{k}_n + \delta\mathbf{k} \qquad (12.29)$$

where $\mathbf{k}_n$ can be calculated by setting $d = 0$, $\mathbf{q} = \mathbf{0}$, $\dot{\mathbf{q}} = \mathbf{0}$ and using nominal values of $m_t$, $J_h$, $\mu$, $m$ and $l$, and $\delta\mathbf{k}$ is the residual. Denoting $|\mathbf{w}|_1^T = \begin{bmatrix} |w_1| & |w_2| & \cdots & |w_m| \end{bmatrix}$ where $w_j$ is the $j$th element of vector $\mathbf{w}$, the detailed adaptive variable structure control scheme is described as:

$$\tau = -\mathbf{k}_n^T \chi - (\hat{\mathbf{k}}^T |\chi|_1 + \rho)sign(\lambda)sat(\sigma, \varepsilon) \qquad (12.30)$$

$$\dot{\hat{\mathbf{k}}} = \begin{cases} \gamma |\sigma| \, |\chi|_1 & |\sigma| > \varepsilon \\ 0 & |\sigma| \leq \varepsilon \end{cases} \qquad (12.31)$$

where $\gamma > 0$ is the adaptation rate. $\varepsilon$ is a small positive value defining the dead zone's size.

*Theorem 12.1:* For the system in eqn. (12.1) and the proposed switching surface in eqn. (12.20), the control law in eqn. (12.29) and the adaptive law in eqn. (12.30) ensure that the dead zone defined by $|\sigma| \leq \varepsilon$ can be reached in a finite time. The system states may leave this region due to disturbance and system uncertainties, but stay inside the dead zone forever after a finite accumulation of the time spent outside the region.

*Proof of Theorem 12.1:* It has been shown by Chen that $\Delta$ is positive definite, hence $\Delta$ is bounded [51]. Define:

$$\Delta_{\max} := \sup \left( m_{RR} - \mathbf{m}_{RF}^T M_{FF}^{-1} \mathbf{m}_{RF} \right)$$

It follows from eqns (12.24) and (12.29) that $\delta\mathbf{k}$ is related to the system parameters $m_t$, $J_h$, $\mu$, $m$, $l$ and the disturbance $d(t)$. Since perturbation for system parameters and disturbance are all bounded, therefore $\frac{\Delta_{\max}}{\Delta}\delta\mathbf{k}$ is bounded. Define:

$$\mathbf{k}^* := \sup \frac{\Delta_{\max}}{\Delta} |\delta\mathbf{k}|_1 \qquad (12.32)$$

where $\|\mathbf{k}^*\| < \infty$. Let $\hat{\mathbf{k}}$ be the estimate of $\hat{\mathbf{k}}^*$ and consider the Lyapunov function candidate:

$$V(\sigma, \mathbf{k} - \mathbf{k}^*) = \begin{cases} \frac{1}{2}\Delta_{\max}\varepsilon^2 + \frac{1}{2}\gamma^{-1}|\lambda|\left(\hat{\mathbf{k}} - \mathbf{k}^*\right)^T & |\sigma| \le \varepsilon \quad (t \in \Omega_1) \\ \frac{1}{2}\Delta_{\max}\sigma^2 + \frac{1}{2}\gamma^{-1}|\lambda|\left(\hat{\mathbf{k}} - \mathbf{k}^*\right)^T & |\sigma| > \varepsilon \quad (t \in \Omega_2) \end{cases} \qquad (12.33)$$

where

$$\Omega_1 \equiv \{t \mid |\sigma| \le \varepsilon\}, \qquad \Omega_2 \equiv \{t \mid |\sigma| > \varepsilon\}$$

When $t \in \Omega_2$, differentiating eqn. (12.33) with respect to time and using eqn. (12.29), the control law in eqn. (12.30), the adaptation law in eqn. (12.31) and eqn. (12.32) yields:

$$
\begin{aligned}
\dot{V} &= \frac{\Delta_{\max}}{\Delta}\lambda\sigma\left(\tau + \mathbf{k}^T\chi\right) + \gamma^{-1}|\lambda|\left(\hat{\mathbf{k}} - \mathbf{k}^*\right)^T\dot{\hat{\mathbf{k}}} \\
&= \frac{\Delta_{\max}}{\Delta}\lambda\sigma\left[-\left(\hat{\mathbf{k}}^T|\chi|_1 + \rho\right)sign\,(\lambda)\,sign\,(\sigma) + \delta\mathbf{k}^T\chi\right] + \gamma^{-1}|\lambda|\left(\hat{\mathbf{k}} - \mathbf{k}^*\right)^T\dot{\hat{\mathbf{k}}} \\
&= \frac{\Delta_{\max}}{\Delta}|\lambda||\sigma|\hat{\mathbf{k}}^T|\chi|_1 - \frac{\Delta_{\max}}{\Delta}|\lambda||\sigma|\rho + \frac{\Delta_{\max}}{\Delta}\lambda\delta\mathbf{k}^T\chi + |\lambda||\sigma|\hat{\mathbf{k}}^T|\chi|_1 - |\lambda||\sigma|\hat{\mathbf{k}}^{*T} \\
&\le -|\lambda||\sigma|\hat{\mathbf{k}}^T|\chi|_1 - |\lambda||\sigma|\rho + \frac{\Delta_{\max}}{\Delta}|\lambda||\sigma||\delta\mathbf{k}|_1^T|\chi|_1 + |\lambda||\sigma|\hat{\mathbf{k}}^T|\chi|_1 - |\lambda||\sigma|\hat{\mathbf{k}}^{*T} \\
&\le -|\lambda||\sigma|\rho - |\lambda||\sigma|\left(\mathbf{k}^* - \frac{\Delta_{\max}}{\Delta}|\delta\mathbf{k}|_1\right)^T|\chi|_1 \\
&\le -|\lambda||\sigma|\rho
\end{aligned}
$$

The negative definiteness of $\dot{V}$ results in a finite time to reach the dead zone and the total time during which adaptation takes place is finite.

Owing to system perturbations, it is possible that the system states leave $\Omega_1$ at some time $t_1$. As a result of eqn. (12.34), the control law in eqn. (12.30) and adaptation law of eqn. (12.30) will drive the states to reenter the dead zone at a finite time $t_2 > t_1$. Integration of eqn. (12.34) from $t_1$ to $t_2$ gives:

$$V(t_2) \le V(t_1) - |\lambda|_1\,\varepsilon\rho\delta t_1, \qquad \delta t_1 = t_2 - t_1$$

Assume that the system states move out of $\Omega_1$ at time $t_3$ again and reenter it at $t_4$. In general, the system states leave the dead zone at $t_{2i-1}$ and reenter at $t_{2i}$. It can be shown that such motion can only repeat for finite times ($i < \infty$) and the system states will stay within the dead zone $\Omega_1$ forever. For an arbitrary number $i$, the following holds:

$$V(t_{2i}) \leq V(t_{2i-1}) - |\lambda_1|\varepsilon\rho\delta t_i \leq V(t_{2i-2}) - \sum_{j=0}^{1} |\lambda_1|\varepsilon\rho\delta t_{i-j} \leq \cdots$$

$$\leq V(t_1) - \sum_{j=0}^{i-1} |\lambda_1|\varepsilon\rho\delta t_{i-j}, \qquad \delta t_{i-j} = t_{2i-2j} - t_{2i-2j-1}$$

Since $V(t_{2i}) \geq 0, \forall i \in Z_+$, then:

$$\sum_{j=0}^{i-1} \delta t_{i-j} \leq \frac{V(t_1)}{|\lambda_1|\varepsilon\rho}$$

holds. When $i$ approaches infinity, the following stands:

$$\lim_{i \to \infty} \sum_{j=0}^{i-1} \delta t_{i-j} \leq \frac{V(t_1)}{|\lambda_1|\varepsilon\rho} < \infty$$

The above concludes: (1) $\lim_{i \to \infty} \delta t_i = 0$, i.e. the system states will stay inside the dead zone forever, and (2) the accumulation of the time spent outside the dead zone is finite.

### 12.4.4  Simulations

In order to demonstrate the effectiveness of the proposed scheme, numerical simulation is performed on the model in eqn. (12.2) with the numerical data of the beam in nominal condition as in Table 12.2. The first flexible mode of the system is considered, i.e. $f = 1$. The control target is to move the end point from $(X, Y) = (l, 0)$ to $(X, Y) = (\frac{l}{\sqrt{2}}, \frac{l}{\sqrt{2}})$ quickly without exciting the flexible modes.

The first simulation was performed using a conventional sliding mode $\sigma_1 = \dot{\theta} + c_e(\theta - \theta_d)$ where $c_e = 1.5$, $\theta_d = \pi/4$, and $m_t = 0.5kg$. An adaptive type VSC is adopted which is similar to the one proposed in Section 12.4.3. The result of this simulation is shown in Figure 12.23, which indicates that conventional sliding mode fails to achieve a smooth end point trajectory since it does not penalise the inherent resonance mode.

The second simulation was carried out for a sliding mode constituted with end point position, i.e., $\sigma_2 = \dot{y} + c_y(y - y_d)$ where $c_y = 1.5$, $y_d = \frac{\pi}{4}$, $y =$

Table 12.2    The simulated flexible link parameters

| Property | Nominal value |
|---|---|
| beam length ($l$) | 1 m |
| flexural rigidity ($EI$) | 45.36 Nm$^2$ |
| mass ($m$) | 0.5859 kg |
| actuator's moment of inertia ($J_h$) | 1.0 kgm$^2$ |
| viscous damping coefficient ($\mu$) | 0.02 |
| payload ($m_t$) | 0.5 kg |

$\theta + \frac{1}{l}\varphi_e^T\mathbf{q}$; the aim was to directly control the end point position. It was found from the simulation that the system went to instability immediately from the beginning. The conventional output tracking made this nonminimum-phase system unstable.

The third simulation was performed using the proposed method. The state-space representation capturing the first flexible mode is:

$$\dot{\mathbf{x}} = \begin{bmatrix} 0 & 0 & 1 & 0 \\ 0 & 0 & 0 & 1 \\ 0 & 68.5190 & 0 & 0.0052 \\ 0 & -164.4062 & 0 & -0.0125 \end{bmatrix} \mathbf{x} + \begin{bmatrix} 0 \\ 0 \\ 0.9904 \\ -0.9609 \end{bmatrix} \tau$$

$$y = \begin{bmatrix} 1 & 1.0668 & 0 & 0 \end{bmatrix} \mathbf{x}$$

and the corresponding transfer function is:

$$g(s) = \frac{96.9772(1 + 0.000076s - 0.00036s^2)}{s^2(s^2 + 0.0125s + 164.4062)}$$

The reference model was selected as:

$$\hat{h}_1(s) = \frac{256}{(s+4)^4} = \frac{256}{s^4 + 16s^3 + 96s^2 + 256s + 256}$$

The resultant switching surface is accordingly calculated as:

$$\sigma(t) = \begin{bmatrix} 256 & 1350.1 & 96.0 & -2.0 \end{bmatrix} \mathbf{x}(t) + 256 \int_0^t [\eta(\tau) - r]dv - 2 \int_0^t q_1\,\dot{\theta}^2 dv$$

where $r = 45°$ and $\eta(t) = [\ 1\quad 1.0306\quad -0.0001\quad -0.0001\ ]\mathbf{x}(t)$. The adaptation rate was chosen as $\gamma = 20$, $\mathbf{k}_n$ was calculated as:

$$\mathbf{k}_n^T = \left[\ 2.6142\quad 0\quad -0.7270\quad 71.1864\quad -13.7807\quad -1.4563\quad 2.6142\quad 0\ \right]^T$$

The dead zone size and boundary layer were determined by $\varepsilon = 0.01$. The sampling interval was chosen as $\tau_S = 1ms$ and $\rho = 0.01$. The initial feedback gain was a zero matrix, i.e. $\hat{\mathbf{k}}\big|_{t=0} = \mathbf{0}_{8\times1}$. Figure 12.24 shows the end point trajectory achieved, which is fairly smooth with no oscillation, and the response is as fast as in the first case. From Figure 12.25 it can be noticed that the oscillation of the first flexible mode was eliminated. Figure 12.26 shows that the control torque is smooth. When the system state reaches a prescribed bound, the dead zone scheme switches the adaptation off to maintain robustness. This feature can be found from Figure 12.27 for $m_t = 1.0$ kg. For simplicity, it is only shown that the eighth element of $\hat{\mathbf{k}}$, $\hat{k}_8$, stops increasing once the system enters the prescribed bound. After entering and leaving the dead zone for two instances, the system state stays inside the dead zone forever. The dead zone scheme further improves the robustness of the system.

To demonstrate the robustness to parameter uncertainties and the disturbance, the variation range of $m_t$ was considered as $[0,\ 1.0]$ kg and the exogenous disturbance defined below was applied:

$$d(t) = \left\{\begin{array}{cc} 0 & t < 1s \\ 0.1Nm & t \geq 1s \end{array}\right\}$$

The eigenvalues of $A_{cl}$ for $m_t = [\ 0,\quad 1.0\ ]$ kg, plotted in Figure 12.28, are all located in the left half plane. The proposed sliding mode remains stable for the specified payload variation. The variation of $\lambda$ for the specified payload variation is plotted in Figure 12.29 from which it can be determined that $\lambda > 0$ and hence the control law in eqn. (12.30) can be implemented. Two extreme cases, i.e., $m_t = 0$ kg and $m_t = 1.0$ kg were individually tested in the fourth simulation: $\mathbf{k}_n$, $\varepsilon$ and $\tau_S$ were chosen to be the same as in the third simulation and the corresponding end point trajectories, shown in Figures 12.30 and 12.31, are much the same as in the nominal case. This implies that the system robustness to parameter uncertainties occurred even in the linear model. The control inputs for $m_t = 0$ kg and $m_t = 1.0$ kg shown in Figure 12.32 and Figure 12.33 are quite smooth.

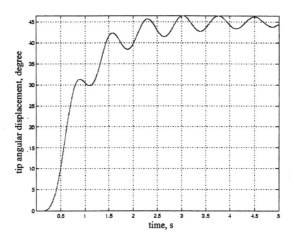

Figure 12.23    *End point trajectory y(t) using conventional sliding mode control,* $m_t = 0.5$ *kg*

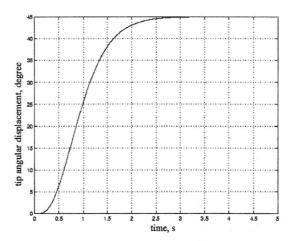

Figure 12.24    *End point trajectory y(t) using the proposed method,* $m_t = 0.5$ *kg*

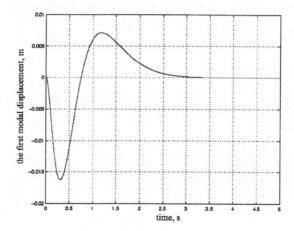

Figure 12.25    *The first modal displacement $q_1(t)$ using the proposed method, $m_t = 0.5$ kg*

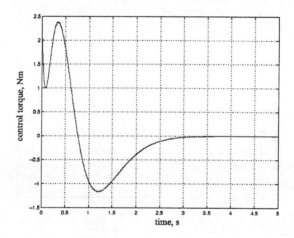

Figure 12.26    *Control torque $\tau(t)$ using the proposed method, $m_t = 0.5$ kg*

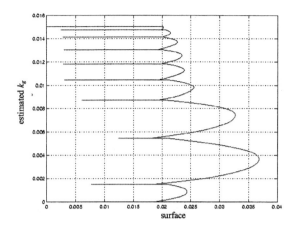

Figure 12.27    $\hat{k}_8$ *versus* $|\sigma|$ *using the proposed method,* $m_t = 0.5$ *kg*

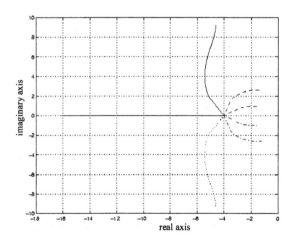

Figure 12.28    *Variations of eigenvalues of $A_{cl}$ with respect to $m_t \in [0, \ 0.5]$ kg (solid line – eigenvalue 1, dotted line – eigenvalue 2, dashed line – eigenvalue 3, dash dot line – eigenvalue 4*

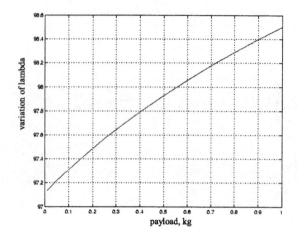

Figure 12.29    *Variations of eigenvalues of λ with respect to* $m_t \in [0, \ 0.5]$ *kg*

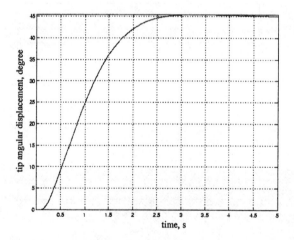

Figure 12.30    *End point trajectory* $y(t)$ *using the proposed method,* $m_t = 0$ *kg with exogenous disturbance* $d(t)$

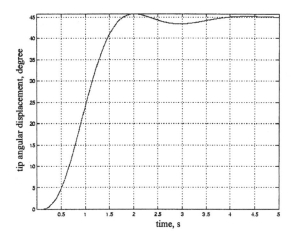

Figure 12.31    *End point trajectory y(t) using the proposed method, $m_t = 1.0$ kg with exogenous disturbance d(t)*

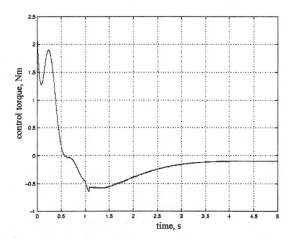

Figure 12.32    *Control torque $\tau(t)$ using the proposed method, $m_t = 0$ kg with exogenous disturbance d(t)*

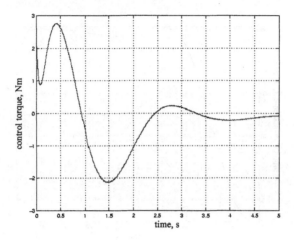

*Figure 12.33    Control torque $\tau(t)$ using the proposed method, $m_t = 1.0$ kg
with exogenous disturbance $d(t)$*

## 12.5    Adaptive joint-based collocated control

The flexible manipulator motion typically consists of three phases: acceleration, constant speed and deceleration. Both acceleration and deceleration require the application of a force of either positive or negative magnitude. Furthermore, a combination of a position controller and a vibration cancellation controller are needed to support the idea of having a suitable controller to position the end point of the flexible manipulator from one point to another with minimum vibration. A fixed joint-based collocated (JBC) controller has previously been proposed by utilising proportional and derivative (PD) feedback of collocated sensor signals [294]. This control strategy has been developed within a simulation environment and reported as resulting in a significant improvement in controlling the rigid body movements of the flexible manipulator with reduced vibrations at the end point as compared with a bang-bang open-loop control strategy. In this Section, the fixed JBC-control strategy based on a pole assignment design is proposed and utilised to formulate and develop a self-tuning JBC control scheme.

A block diagram of the adaptive JBC-controller is shown in Figure 12.34, where $\hat{K}_P = A_C K_P$ and $\hat{K}_V = A_C K_V$ with $K_P$ and $K_V$ representing the proportional and derivative gains, respectively. The hub angle is represented by $\theta(t)$, $R_f$ is the desired hub angle and $A_C$ is the amplifier gain. The input-output signals are sampled at a sampling period $\tau_S$, within 5.0 ms. The control signal, $u(t)$, can be defined as:

$$u(t) = \hat{K}_P R_f - \left[\hat{K}_P + \frac{\hat{K}_V}{\tau_S}(1 - z^{-1})\right]\theta(t) \qquad (12.35)$$

where $z$ is the time-shifting operator.

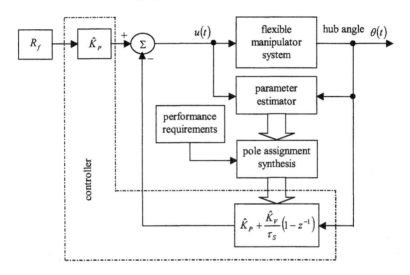

*Figure 12.34   Adaptive joint-based collocated controller*

In a general design setting where the underlying system may have complex dynamics, a number of rules of thumb exist which assist in the selection of the controller coefficients $A_C$, $K_P$ and $K_V$. In a synthesis situation, however, the requirements on the underlying system are quite strict. In particular, in order to synthesise exactly the JBC controller coefficients, the system to be controlled must be assumed to have a special structure of the form:

$$\theta(z) = \frac{b_0 z^{-1}}{1 + a_1 z^{-1} + a_2 z^{-2}} u(z) \qquad (12.36)$$

The restriction on the system model form is to ensure that only one set of JBC controller coefficients result from the design.

Combining the system model in eqn. (12.36) with the controller eqn. (12.35) results in the closed-loop equation relating $R_f$ and $\theta(t)$. Thus:

$$\frac{\theta}{R_f} = \frac{A_C K_P b_0 z^{-1}}{(1 + a_1 z^{-1} + a_2 z^{-2}) + A_C \left[ K_V \tau_S^{-1} (1 - z^{-1}) + K_P \right] b_0 z^{-1}} \qquad (12.37)$$

The controller coefficients can be determined by equating the characteristic equation of the closed-loop system with a desired closed-loop characteristic equation:

$$T = 1 + t_1 z^{-1} + t_2 z^{-2} \qquad (12.38)$$

where

$$t_1 = -2 \exp\left(-\zeta \omega_n \tau_S\right) \cos\left\{ \tau_S \omega_n \left(1 - \zeta^2\right)^{1/2} \right\}, \qquad t_2 = \exp\left(-2\zeta \omega_n \tau_S\right)$$

with $\zeta$, $\tau_S$ and $\omega_n$ representing, respectively, the damping factor, sampling-period and natural frequency of the desired closed-loop second-order system. This yields:

$$\hat{K}_P = \frac{(t_1 + t_2 - a_1 - a_2)}{b_0} \qquad (12.39)$$

$$\hat{K}_V = \frac{(a_2 - t_2)\, \tau_S}{b_0} \qquad (12.40)$$

Eqns (12.39) and (12.40) are the controller design rules, realisation of which results in a pole assignment self-tuning control scheme, as shown in Figure 12.34.

The performance of the adaptive JBC controller for positioning the experimental flexible manipulator was investigated with various payloads. Figure 12.35 shows the performance of the adaptive JBC controlled system with 0, 40, 60 and 100 g payloads. It can be seen that the flexible manipulator comfortably reaches the demanded angular position with the adaptive JBC controller. In these experiments a recursive least-squares (RLS) algorithm was used for estimating the parameters of the plant model.

## 12.6    Adaptive inverse-dynamic active control

In this Section, a noncollocated feedback control strategy for end point vibration suppression of the system is developed. A hybrid collocated and non-collocated controller has previously been proposed for control of the flexible manipulator [294]. The controller design utilises end point displacement feedback through a PID control scheme and a PD configuration for control of

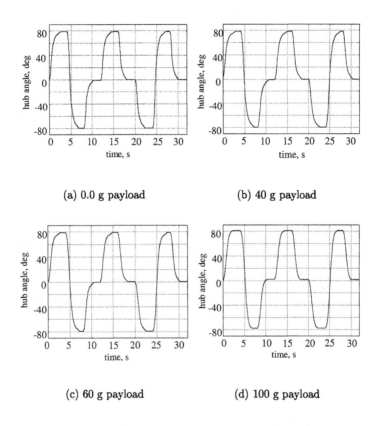

(a) 0.0 g payload

(b) 40 g payload

(c) 60 g payload

(d) 100 g payload

*Figure 12.35* Performance of an adaptive JBC-controlled system with various payloads

the rigid body motion of the flexible manipulator. These two loops are then summed to give a command torque input. A hybrid collocated noncollocated control scheme incorporating the JBC control and an inverse end point model control is proposed and further developed in this Section to provide an adaptive hybrid control scheme for flexible manipulators.

The method adopted is illustrated in Figure 12.36. The purpose of the adaptive inverse control is to drive the flexible manipulator with an additional signal from a controller the transfer function of which is the inverse of that of the plant itself. Note that the adaptive inverse control is active when the flexible manipulator is in motion, so that the computed torque is used to force the end point vibration to a minimum level. Since the plant is generally unknown, it is necessary to adapt or adjust the parameters of the controller in order to create the true plant inverse.

In implementing the adaptive control algorithm, in addition to the practical issues related to the properties of the disturbance signal, robustness of the estimation and control, system stability and processor-related issues such as word length, speed and computational power, a problem commonly encountered is that of instability of the system, especially when nonminimum phase models are involved. Thus, to avoid this problem of instability, either the estimated model can be made minimum phase by reflecting its noninvertible zeros into the stability region and using the resulting minimum phase model to design the controller, or once the controller is designed the zeros that are outside of the stability region can be reflected into the stability region. In this manner, a factor $1 - pz^{-1}$ corresponding to a controller-pole/model-zero at $z = p$, in the complex $z$-plane, that is outside of the stability region, can be reflected into the stability region by replacing the factor with $p - z^{-1}$.

The adaptive inverse controller was implemented and tested with the experimental flexible manipulator. To estimate the end point acceleration up to the first three resonance modes, a sixth-order model would be required. Using a 5.0 ms sampling interval, the computational power of a 486DX2 50 MHz system is not enough to perform the online parametric estimation using the RLS algorithm with this model order. To resolve this computational problem, a fixed JBC controller was used instead of the adaptive JBC controller and a second-order model was chosen for the end point acceleration. This will allow only the first resonance mode to be identified. Thus, the adaptive inverse control was considered to control the end point vibrations up to the first mode. To avoid spillover problems due to higher modes, a lowpass filter with 20 Hz cut-off frequency was used to suppress the higher modes.

The performance of the adaptive inverse controlled system at the end point vibration reduction is shown in Figures 12.37. The dotted line shows the un-

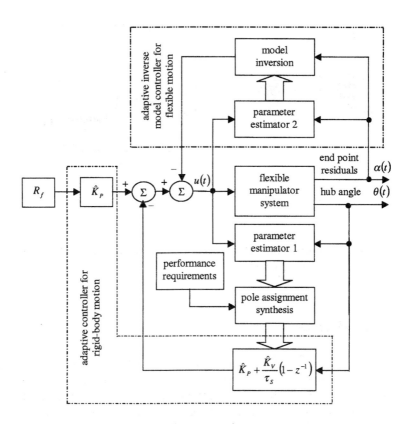

*Figure 12.36    Collocated adaptive JBC and noncollocated adaptive inverse control scheme*

compensated end point acceleration and the solid line shows the compensated end point acceleration. It is noted that when using the adaptive inverse controller the end point vibrations were reduced by 11.48 dB for the first resonance mode. Figure 12.38 shows the corresponding torque input to the system.

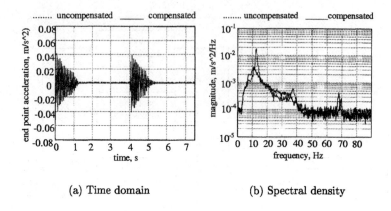

(a) Time domain                    (b) Spectral density

Figure 12.37    *The end point acceleration with the adaptive inverse dynamic control scheme*

## 12.7    Conclusions

Open-loop and closed-loop control strategies have been developed and verified within a flexible manipulator system under simulated and experimental conditions. Open-loop control strategies using lowpass and bandstop filtered torque inputs and Gaussian-shaped torque input have been applied and verified within a constrained planar single-link flexible manipulator experimental rig.

Open-loop control methods involve the development of the control input by considering the physical and vibrational properties of the experimental flexible manipulator system. The shaped torque input is used to minimise the energy input at system resonance modes so that system vibrations are reduced. Lowpass and bandstop filtered torque input functions have been developed and investigated in an open-loop control configuration. Significant improvement in the reduction of system vibrations has been achieved with these control functions as compared with bang-bang torque input.

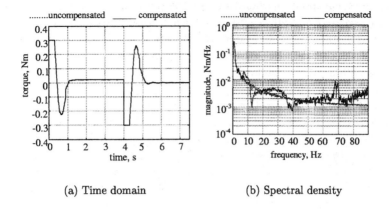

(a) Time domain        (b) Spectral density

*Figure 12.38    Torque input with the adaptive inverse dynamic control scheme*

Comparing the results achieved with the lowpass and bandstop filtered torque inputs reveals that better performance at a reduced level of vibration of the system is achieved with lowpass filtered torque inputs. This is due to the indiscriminate spectral attenuation in the lowpass filtered torque input at all the resonance modes of the system. Utilisation of bandstop filters, however, is advantageous in that spectral attenuation in the input at selected resonance modes of the system can be achieved. Thus, the open-loop control strategy based on bandstop filters is optimal in this sense. Note that this strategy can also be viewed as being equivalent to designing a controller with zeros which cancel out the system poles (resonance modes). However, if the spectral energy around a large number of resonance modes of the system contributes significantly to system vibrations, it will be more desirable to use lowpass filtered command inputs.

In terms of a smooth transition from acceleration to deceleration, the Gaussian-shaped torque input provides a good candidate. A significant reduction in the level of vibration at the first three resonance modes of the system has been achieved with Gaussian-shaped torque input.

A direct approach for end point regulation has been developed and verified within a simulation environment characterising a single-link flexible manipulator. The switching surface introduced according to a selected reference model relocates system poles and the link oscillation is accordingly reduced. The adaptive variable structure controller design avoids the difficulty arising from estimation for system uncertainties and disturbance. Stability analysis of the proposed switching surface and controller has been performed. The dead zone introduced further improves system robustness. It has also been demonstrated that the proposed sliding mode remains stable for a certain variation of system parameters such as payload. Moreover, the system is robust to exogenous disturbances.

An adaptive JBC controller and combined adaptive JBC and inverse control have been developed and tested in the control of the experimental flexible manipulator. Good positioning control and significant reduction in system vibration have been achieved with these schemes. It can be seen that the control torque is very much dependent on the estimated parameters. The problem of system instability due to the nonminimum-phase behaviour of the system has been resolved by reflecting the noninvertible zeros of the system model into the stability region before generating the control signal. It has been demonstrated that significant reduction in the level of vibration of the system is achieved with this control strategy.

## Chapter 13

# An active approach to interior noise reduction for a modern electric locomotive

## M. Viscardi†, M. Fontana∗, A. Vecchio∗, B. Mocerino L. Lecce∗

† Active S.r.l., Napoli, Italy
Email: massimo.viscardi@tin.it University of Naples, Federico II, Italy
∗ Universita di Napoli, Napoli, Italy
‡ Ansaldo Trasporti, Napoli, Italy

*This chapter presents the application of active noise control techniques inside the driver cabins of electric locomotives. A full analysis of the problem is provided. Algorithmic solutions and DSP hardware architectures are included and experimental results are explained.*

## 13.1 Introduction

The problem of noise and vibration control is becoming everyday more and more acute, and this concerns several modern means of transport, including the modern electric locomotive. Among the variety of control approaches, active suppression techniques may represent an innovative and effective solution when compared with a passive approach, since they could have, in theory, good results in the low-frequency range and a short retrofitting time. The main objective of the present study is to point out the feasibility of such an approach and provide its first experimental testing [13.7]. The first part of the work has been focused on an interior noise field characterisation aimed at the identification of the different sources which contribute to interior noise. In the second part of the work, different active noise control strategies have been assessed and experimentally tested, both on laboratory mock ups and on an operative locomotive, to verify their own feasibility as well as the real possibility of application on board an electric train.

## 13.2    Noise sources in electric trains

Train noise has several causes, thus it is difficult to extrapolate a law which completely defines all the parameters of noise. To evaluate outdoor annoyance due to a passing train, measurements of noise are usually done outside of the train. Turbulent boundary-layer air flow acting on the external surface of a train, rail-wheel interaction, electronic devices on board with their respective cooling systems and the engine are the main sources of noise. Secondary sources are exhausts, in diesel locomotives, and friction of the pantograph trolley for electric locomotives; the latter become important at high frequencies (8 kHz) [13.1]. Occasional sources of noise are impulsive, due to rails gap and switches, and tonal, due to small radius turns and brakes.

### 13.2.1    Aerodynamic noise

Turbulent flow in the boundary layer is the aerodynamic excitation which causes noise mainly on the forward surface of the locomotive. The effect is transmitted directly inside the cabin through the windscreen. This noise has generally no noticeable tonal components, but its sound pressure level is nearly constant in the entire audible frequency band (broadband noise). Aerodynamic noise can be considered as background noise and its mean spectral level rises with the speed of the train. Analysis has shown that the effects of aerodynamic noise become of the same magnitude, in sound emission, as other noise sources inside the locomotive at velocities higher than 300 km/h, and it even exceeds them at the highest velocities [13.2]. This is due to the aerodynamic sound spectrum which covers tonal components of other noise sources, when the speed increases, making an active noise control approach more difficult. This is another important reason, apart from the problems related to aerodynamic drag, for studying optimal frontal aerodynamic shapes especially for high-speed trains.

## 13.2.2   Wheel-rail noise

One of the main noise sources in an electric train is the wheel-rail contact area. This kind of source is both in the locomotive and in the carriages but, since the intensity of this noise depends on factors such as wheel radius and the weight on each of them, the more critical situation is inside the locomotive. During the rolling of the wheel on the rail, both are elastically deformed by weight, driving torque and friction. Strain energy is then converted into noise and vibrations which are transferred through the structure inside the train. The phenomena are amplified by micro or macro surface defects of the wheel and rail, and are present even in the ideal case of perfectly smooth surfaces [13.3-13.4]. The characteristic of this noise is that the wheels are the main noise sources at the high-frequency range, and the rail is prevalent at medium frequency: their prevalence depends on their surface roughness and on train velocity.

## 13.2.3   Equipment on the locomotive

Equipment is able to produce high levels of noise inside an electric locomotive. Fans for the engine cooling system, static transformers and their cooling system, electric devices in which magnetostrictive phenomena are present, are sources of typical tonal noise and high-frequency electric noise. Fans, present usually on these types of locomotive, are high-power devices able to produce huge airflow to cool traction engines and other apparatus. For example, inside the French TGV train eleven fans are able to make a 105 m$^3$/h flow. These are sources of high tonal levels of noise at low frequency inside the locomotive, thus becoming one of the main causes of cabin noise.

## 13.2.4   Electric motors

Modern locomotives operate with electric motors and their number and their power vary with the features of the locomotive. Motors are usually three-phase asynchronous, with power varying between 1000 and 2000 kW. For example, each one of the electric motors used on the French TGV train has 1100 kW power, and the locomotive described in this active noise control application has four motors capable of 1500 kW each. The TGV is a high-speed passenger train able to reach 300 km/h velocity, and the application locomotive is a high power machine able to reach 220 km/h both with passenger trains and heavy freight trains. Noise produced in electric motors is due to several factors; among these the magnetic field is the main one. The discrete distribution of coils in the rotor and stator produces a nonuniform magnetic field, which makes a periodic motion that is characterised by a frequency spectrum with pure tones that are proportional to rotor angular velocity. When the tones are close to the natural frequencies of the structure of the engine, the corresponding natural modes can be excited causing resonance. Periodic deformation of the core of the engine, due to magnetostriction and magnetoattraction, are also the cause of tonal noise. Other causes of noise are mechanical vibration due to dynamic unbalance of the rotor. The transmission of vibration and noise are due to the dual behaviour of the stator and of the external case: they can act as an acoustical cavity, and as an acoustical radiators transferring vibration and acoustic energy to the external structures and to the environment.

## 13.3   Locomotive noise characterisation

The described noise sources have been analysed on a new-generation electric locomotive in which an active noise control approach has been investigated. The main noise transmission paths have been analysed onboard this locomotive to evaluate the influence of each single noise source on the overall noise inside its cabin. A locomotive with asynchronous motors (four axial inverters and two motor trucks of 5200 kW continuous power and 6000 kW max power) is used in the test. It is able to speed heavy freight trains up to 160 km/h and light passenger trains up to 220km/h. The machine has two opposite drivers' cabins which are linked by an aisle down the length of the locomotive along which the electric systems and engine cooling compressors are located. All the

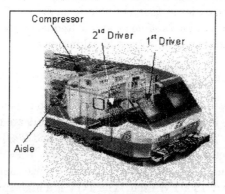

*Figure 13.1   Inside view of the locomotive cabin*

doors of cabins, to the aisle and to outside, are fitted with pneumatic gaskets to avoid unwanted opening during the trip and, moreover, to isolate the cabin from the noise sources. In each one of the two symmetric cabins two big electric racks are present, as well as two seats for drivers and the instrument consoles. The two centrifugal compressors for engine cooling systems are next to the drivers' cabins and are separated by a metallic wall (Figure 13.1).

### 13.3.1   Noise measurements inside the locomotive

First measurements of overall sound pressure levels (SPL) were taken inside the entire locomotive at two different velocities. Noise measured under the normal conditions of operation was used to detect the main sources. At 200 km/h speed and with working fan compressors, the noise inside the cabin is nearly 79.5 dB(A), as shown in Figure 13.2. The measurements in the aisle between the cabins showed that, among the apparatus in the locomotive, the main noise sources are the fan compressors. Their presence close to the cabins allowed them to be identified as one of the main noise sources in the cabin.

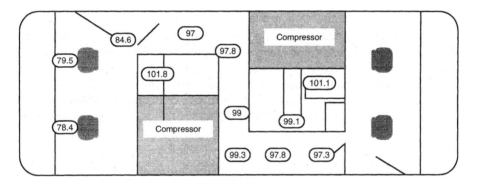

Figure 13.2   *Acoustic map inside the locomotive at 200 km/h [dB(A)]*

### 13.3.2   Noise characterisation

Characterisation of cabin noise and analysis of transmission paths were obtained by investigating the time histories of seven microphones and three accelerometers by a ten-channel digital recorder. The digital recorder was able to measure a frequency band of 0–4 kHz for each channel. Previous analysis showed that the highest noise emissions are below the 2 kHz level. The dis-

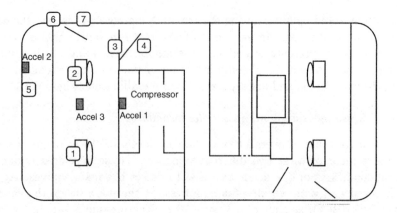

*Figure 13.3    Position of transducers during noise characterisation*

tribution of microphones and accelerometers is shown in Figure 13.3: three microphones are inside the cabin, near to the drivers' heads and on the door to the aisle (microphones 1, 2 and 3). Microphone 4 was placed in the aisle near to the fan compressor of the engine cooling system; number 5 was near the windscreen to measure noise from the outlet of the air conditioner. Microphone 6 was outside to measure airborne noise, and so was microphone 7, which was near the wheels to measure wheel-rail noise. Accelerometer 1 was on the wall between the fan compressor and the cabin; number 2 was on the inner side of the windscreen and number 3 was on the floor of the cabin. The characterisation and identification of sound transmission paths have been made by the analysis of signals from the noise sources and from the two microphones on the drivers' seats. Evaluation of coherence between the signals allowed insight into the sources which have influence on the noise level. The characteristics of the spectra of the investigated noise sources are shown in Figure 13.4.

### 13.3.3   Cabin noise analysis (microphones 1, 2, 3)

The analysis of spectra of signals from the microphones in the cabin clarified how the noise is influenced by train speed. Figure 13.5 shows the frequency spectrum for microphones 1, 2 and 3 at 200 km/h. The difference between the same microphone signals at 120 km/h is nearly 10 dB(A) for all microphones in the cabin. Moreover mean SPL from microphone 2 is always a little bit higher

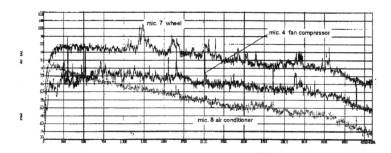

*Figure 13.4   Noise sources spectra at 200 km/h*

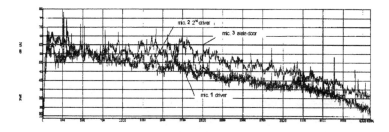

*Figure 13.5   Noise spectra in the cabin at 200 km/h*

than that from microphone 1. Noise characterisation has been performed at the train speed of 200 km/h, at which higher sound pressure levels are present.

### 13.3.4   Fan compressor noise (microphone 4)

The centrifugal compressor, which is located besides the cabin, is able to produce a typical spectrum of rotating machines. The blade passage frequencies (BPF) can be distinguished at 300 Hz, at which an SPL of 95 dB(A) is present. Other tonal components are related to BPF higher-order harmonics. The SPL at 300 Hz is attenuated by the door to the cabin and to a lesser degree by the partition wall with the cabin, in which microphones 1 and 2 sound pressure levels are, respectively, 62 dB(A) and 65 dB(A) at that frequency. The door isolation is not very efficient, since SPL on microphone 3 is at 73 dB(A). The sound emission at 300 Hz causes strong vibration of the separating wall between the compressor and the cabin and, as indicated by the accelerometric signal and its coherence with the signal from microphone 2, it is a good sound transmission path for the compressor noise. The 82.5 dB(A) at 200 Hz doesn't cause important effects inside the cabin, since the door and the wall produce better isolation at higher frequencies.

### 13.3.5   Air conditioner noise (microphone 5)

The fundamental characteristic of the spectrum of the signal of the microphone near the outlet of the conditioner is a high broadband SPL, in particular at low frequency, reaching 78 dB(A) between 50 and 90 Hz. Analysis of the coherence function and of the signal from accelerometer 2 allows us to identify the influence of this noise on the drivers' microphones (1 and 2) at low frequencies. In particular, at 90 Hz sound pressure levels are, respectively, 67.5 and 75 dB(A) for microphones 1 and 2. The signal of the accelerometer reveals the contribution of the windscreen in reflecting sound waves at this frequency towards the drivers. At higher frequencies the influence of this device decreases.

### 13.3.6   Aerodynamic noise (microphone 6)

This noise showed a typical broadband spectrum with no significant tonal components. Moreover, the coherence analysis of this signal with respect to those in the cabin did not show any significant influence of this noise on cabin tonal components. Thus, at the investigated speed (200 km/h), aerodynamic noise is not yet important, even if present as broadband noise. As already stated earlier, it is confirmed that airborne noise becomes much more relevant for speeds greater than 300 km/h. The spectrum of this noise is not plotted in the figures, since its influence can be considered irrelevant to the characterisation of the noise tonal components inside the cabin.

### 13.3.7   Wheel-rail noise (microphone 7)

This contact noise has a flat spectrum with a high mean SPL and sound peaks at nearly 1200 Hz. This is a complex signal since, apart from tonal components due to rotating parts, impulsive components are present. The sound transmission path of this noise is mainly structural, as the accelerometer 3 signal and unitary coherence function with microphones 1 and 2 confirm, especially at 1092 Hz and 1246 Hz.

The analysis above shows that a direct sound transmission path is present starting with the compressor to the cabin through the door and the wall, and noise transmission is due to wheel-rail contact, mainly through the entire structure of the locomotive. Using the described results, noise attenuation inside the driver's cabin should be performed at a low-frequency range, since noise reduces gradually at frequencies greater than 1200 Hz. Thus, for this type of locomotive, it can finally be said that the main causes of cabin noise are due to the fan compressor and the air conditioner. Both of these produce intense noise at low frequency, both directly through air transmission and indirectly through structural vibrations.

## 13.4 Generalities of active control approaches for cabin noise reduction

A first attempt at noise reduction using passive noise control techniques could be performed by: applying damping materials to the wall between the compressor and the cabin; structural modification of the door by inserting absorbing materials; dynamic balancing of rotor parts of the centrifugal compressor and of the air conditioner fan; redesign of the air conditioner outlets; better smoothing of wheel surfaces. All these solutions result in a mean noise reduction of nearly 5–6 dB, especially at high frequency. A much more effective solution for low-frequency noise reduction could be based on active control technologies, especially when tonal components are present in the noise spectrum.

During the development of the present study, an active control approach was tested to verify advantages and drawbacks in comparison with a passive approach [13.7]. From a general point of view, the noise control techniques may be divided into three main categories: at source, on the noise transmission path and at target. For both the passive and active approaches, the different methods imply different operative methodologies and offer different expected results. In particular, the basic physics of active control requires a superposition of the primary and secondary noise fields, the relative shape of which greatly influences the obtainable results. This driving concept and operative consideration will influence, as will be seen, the choice of approach.

## 13.5 Noise control at source

A source control strategy tries to reduce the phenomena before it interacts with the external environment; when applicable such an approach may generally guarantee the best results in terms of reduction and spatial distribution. One of the areas where this concept finds application is air moving devices. These offer probably the most considerable advantages of active control over passive treatments at the moment, especially with regard to low-frequency acoustic fields (for which the cut-off low-frequency upper limit can be assumed to be the frequency at which the transversal acoustic mode starts to be excited). Let

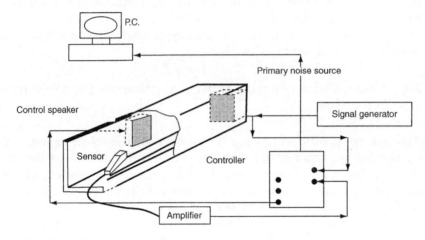

*Figure 13.6    Line duct test setup*

us assume a simple line duct, as shown in Figure 13.6, at the end of which is positioned a loudspeaker, generating pure sine tones, and a sensor microphone which is free to move along the duct itself. At low frequencies the acoustic pressure measured along the duct shows a longitudinal shape with an acoustic wave front parallel to the duct main axis and constant pressure inside the front.

In Figure 13.7, as an example, the acoustic distribution within the duct is shown. For a defined frequency range, depending from the duct dimensions when the frequency is rising, we may observe an attenuation of the pressure distribution along the duct. This behaviour can be observed up to the so-called cut-off frequency, when transversal modes appear.

If the disturbance signal lies below the cut-off frequency, a control architecture using one control actuator and one (or two) error sensors generally performs well. It appears evident that the quality of the expected results (especially in relation to global rather than local attenuation) is a function of relative device positioning. Many factors can be taken into account regarding the device's optimisation process, the discussion of which is beyond the scope of the present chapter. It can only be stated that many tools are available to the researcher based on both deterministic (for example minimum

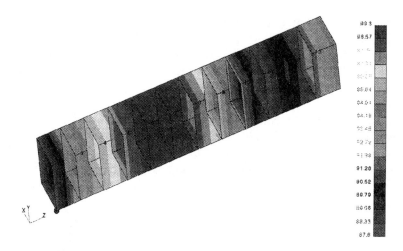

Figure 13.7    *SPL distribution inside the duct, control off*

least squares and/or progressive selection techniques) and stochastic methods (genetic algorithms, neural networks and others). With regard to the current

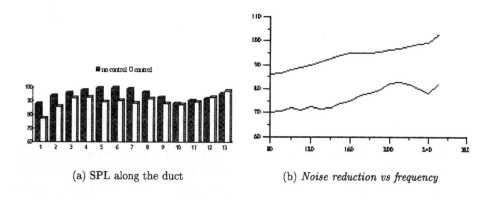

(a) SPL along the duct

(b) *Noise reduction vs frequency*

Figure 13.8

setup, appropriate data sampling/recording and a successive optimisation procedure have been performed aimed at guaranteeing control results as effective and spatially distributed as possible. The main results of the first control test are summarised in the following diagrams reporting the SPL (Figure 13.8*a*) measured along the duct in the presence of (and without) the active control system working (at fixed frequency) and the noise reduction (Figure 13.8*b*) at the duct free end as a function of the frequency variation of the disturbance signal.

It appears that the control action produces a uniform behaviour at frequency variation (inside a well defined range) with enough uniform noise reduction along the whole duct except at the sections most distant to the error sensor's locations. This observation appears more clearly in Figure 13.9, where the SPL distribution is again presented. A similar behaviour is gener-

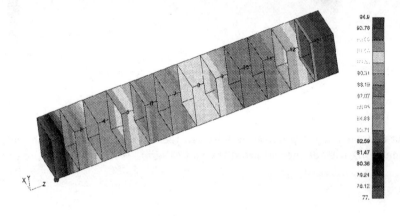

Figure 13.9    *SPL distribution inside the duct, control on*

ally present under harmonic noise excitation, as in air moving devices where harmonic peaks (related to the blade passage frequency and its upper components) emerge from broadband noise (related to system aerodynamics and turbulent flow). In Figure 13.10, noise measurements are presented when only the fan compressor is working. Spectral analysis shows that the noise presents a classical air moving device shape; this is evident from the microphone located close to the source, but appears clearly inside the cabin (measured at the driver and co-driver seats) where the acoustic cavity coupling also appears.

The two drivers' spectra are very similar apart from a small amplitude difference attributable to the different acoustic paths from the noise source. From the previous picture it emerges also that the noise reduction due to the

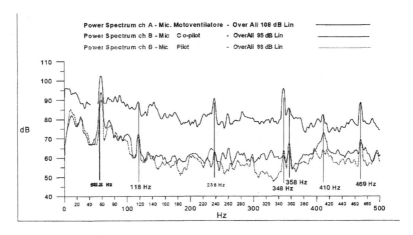

*Figure 13.10    Noise spectra due to motor ventilation fan*

cabin wall does not exceed an overall level (related to the 0–500 Hz range) of 15 dB and that the wall effect is lower with decreasing frequency. This fact gives an alternative explanation to why the active control approaches have to be verified.

On the basis of what has been previously said, direct action on the fan source seems to be the most appropriate approach since this would reduce the noise before it reaches the cabin interior. Because of some operative problems related to locomotive operation and system modification, laboratory experimentation was planned to be held on a line duct structure presenting an axial fan on one of its side (Figure 13.11), as being representative of this approach. This has all the limitations that a mock-up approach carries, but is a first attempt to verify some ideas.

The control architecture [13.9] incorporated one error microphone at the free end of the duct and one loudspeaker as the control source. An optical signal was used as the reference signal, due to the high correlation with the primary noise field. The control action target was in fact decided to be only the noise harmonic peaks without affecting the broadband aerodynamic noise. Figure 13.5 shows the noise field without and in the presence of the control system working as measured, at 50 cm distance from the free end of the duct. It is interesting to note that all around the duct a similar behaviour may be observed.

As a consequence of such experiences [13.8], and on the basis of other

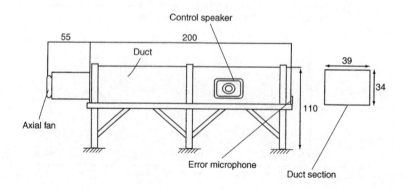

*Figure 13.11    Test setup scheme on line duct mock-up*

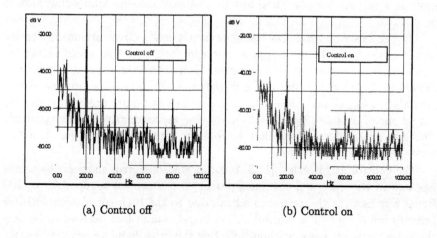

(a) Control off                    (b) Control on

*Figure 13.12    SPL at the free end of the duct*

similar approaches, the idea of a control system working at source on the motor fan duct emerged. The present results made this approach very attractive and suggested its use with other noise sources (such as the air conditioning system). From an operational point of view, system unavailability, in the sense of required nonmodification of the motor fan present on board and the difficulties related to lab tests (due to required power and safety problems), required a different approach to the problem, consisting in noise control at target instead of at source.

## 13.6   A target noise control strategy

Regarding an active control scheme working at target, which is an environment to be preserved, two general classes may be identified: a global approach (focused on noise reduction over all the environment) and a local one (focused on the creation of silent areas around the location points to be preserved). The first approach is obviously to be preferred when applicable because it may guarantee homogeneous attenuation; unfortunately it is very often bypassed because of the application difficulties. In general, active noise control in three-dimensional space, and under a multiple number of control sources and error sensors requires a very deep understanding of both the exciting noise as well as acoustic system prerogatives. As for the mechanical system, a closed acoustic space has its own natural frequencies and related natural modes which may be excited by an external incident source. In particular, the number of acoustic modes below a defined frequency $f$, for an acoustic volume $V$, may in general be approximated as:

$$N \cong \frac{4\pi V}{3c^3} f^3 \tag{13.1}$$

where $c$ and $\pi$, respectively, are the sound speed and the medium density. Every one of these modes has its own frequency, damping coefficient and related mode shape. The acoustic spatial distribution, which follows the incidence of an external acoustic field, is dependent on the coupling of the external field and the acoustic volume. This fact may be illustrated by taking as a reference an acoustic box with a parallelepiped shape.

As pointed out, the acoustic box exhibits natural modes the complexity of which grows with frequency. If the disturbing noise coincides with a natural frequency (and moreover if it occurs at low frequencies) global noise reduction is possible through realisation of destructive interference of the noise fields.[1]

---

[1] As the omeopatic medicine founder said "Similia similibus curantur", that is to say that only similar things may interfere with success

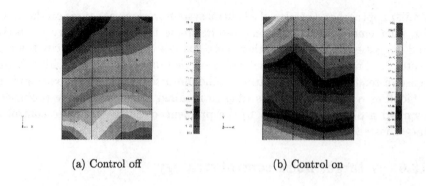

(a) Control off                    (b) Control on

*Figure 13.13    Acoustic box test set up*

As the frequency rises, and if it does not coincide with a natural frequency, the complexity of the noise field increases with a higher number of sensors and actuators (which have to be representative of the noise field) needed by the controller. The approach becomes almost impossible to apply in the presence of broadband or high-frequency narrowband signals occurring in the noise, and under these circumstances a local approach seems to be more appropriate. This is the case for the locomotive cabin where the exciting incidence signal presents a middle frequency harmonic spectrum on a broadband base. The large acoustic volume shows high modal density which has been qualitatively analysed by transfer function acoustic measurements reported in the Figure 13.14.

A global approach would require an enormous number of sensors and actuators as well as a powerful hardware platform and algorithms; it has not been assessed. The creation of two silent regions in the area surrounding the seats of the driver and second driver seems to be the best compromise, especially because of real operational conditions which imply fixed locations for both crew. To better understand the performance of such an approach, a test setup has been built in the laboratory before implementation on the locomotive. Figure 13.15 shows the setup which represents active seats with fitted heads in the middle simulating the people's presence; a set of loudspeakers, distributed all over the laboratory, simulated the primary noise sources.

Two secondary source locations have been tested (at the extremities of the head-set and on the top) as well as three different locations for the error sensors to verify the relative influence with respect to the noise reduction at

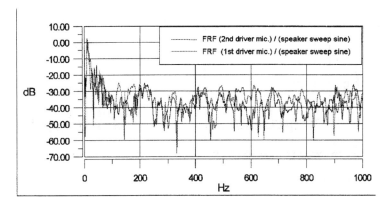

Figure 13.14    Typical FRF measured in the cabin

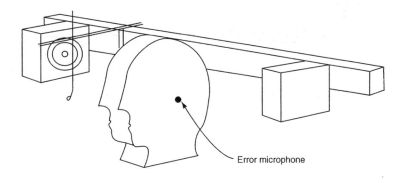

Figure 13.15    Active seats, test setup

the target points (the drivers' ears), because it does not appear possible to put the sensors at the ear locations. Figures 13.16 – 13.18 show some of the main results.

Due to head movement, it is important to guarantee dimensions for the silent region which avoid strong variations of sound pressure with small movements of the head. Through a moving sensor, a spatial characterisation of the

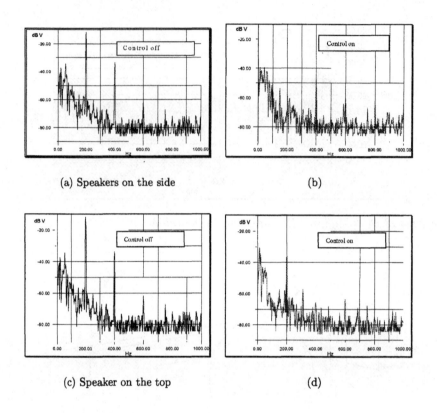

(a) Speakers on the side                              (b)

(c) Speaker on the top                                (d)

*Figure 13.16    Control test main results*

control space has been performed. Figures 13.17 and 13.6, respectively, show the instrumentation setup and some of the main results.

At the end of the simulation stage, on the basis of best implementation possibilities on control hardware, a setup based on two separate systems for the driver and co-driver were applied. Each presented one actuator positioned on the ceiling of the cabin and two microphones linked at the seats. A functionality control has been first planned to verify that the acoustic power of the speakers is comparable to that of the primary field, to measure the crosscorrelation of the two control systems and to position the reference sensor so as to guarantee the appropriate correlation with the noise to be controlled. With regard to this problem it was decided to take an accelerometric signal from a

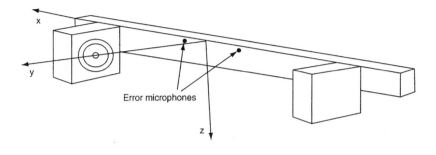

Figure 13.17    *Active seat instrumentation setup*

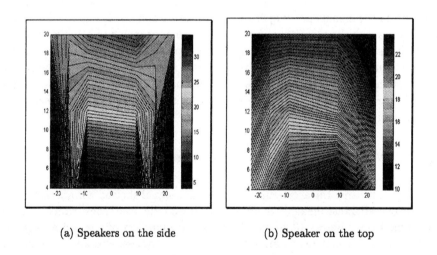

(a) Speakers on the side        (b) Speaker on the top

Figure 13.18    *Active seat noise reduction (dB) inside the control region*

sensor positioned on the motor-fan, the spectral shape of which is reported in Figure 13.19.

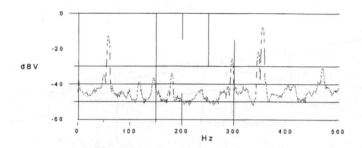

*Figure 13.19    Accelerometer spectral shape*

It can be seen from comparison with Figure 13.10 that this signal presents all the harmonic peaks identified in the noise field and is therefore a good reference signal for the control algorithm; structural limits did not permit the use of an optical signal derived from the blades. As a drawback, the accelerometric signal carries spurious components and background noise which does not favour the algorithm's role. A parallel approach was also tested using a synthesised reference signal, consisting of a pure tone. This approach allows a lower number of filter taps to be used and results in a better reduction of the noise at that frequency [13.9]; as a drawback, the reference signal is not strictly correlated with the noise field and this could lead to inefficiencies in the control algorithm.

## 13.7    Main results of the field experimentation

The first test using the accelerometric reference signal showed a main reduction of the overall noise field of 3 dB in the 0–500 Hz frequency range. As is apparent from Figures 13.20 and 13.21, an attenuation of the noise peaks is present in the reference signal on both the microphones, and an increment of the uncorrelated components appears. This behaviour is in perfect agreement with the working fundamentals of the algorithm.

The results of the experience using the synthesised signal (Figure 13.22) showed that when a pure tone reference signal is present (or a harmonic one), the system is able to reduce the peak to the level of the background noise. This result gives some idea of the results which could be obtainable with an optic reference.

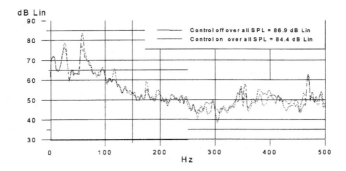

*Figure 13.20    Effects of noise control on driver's microphone*

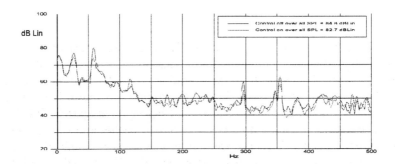

*Figure 13.21    Effects of noise control on co-driver's microphone*

Another interesting investigation may be made regarding the substantial independence of the two control systems, the crosscorrelation of which is close to zero in practice. This is essentially due to the relative low power required at the control sources in comparison to the distances and directivity of the sources themselves.

## 13.8   Conclusions

Many more investigations could have been made at the end of this research mainly focusing on locomotive interior noise characterisation and reduction by

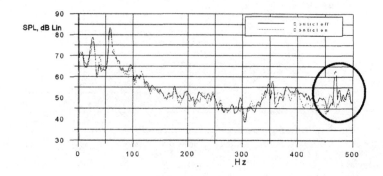

*Figure 13.22     Effects of noise control on driver's microphone (synthesised mono tonal reference signal)*

the use of alternative active approaches. The noise fields measured on the locomotive showed a very complex shape because of the summation of the many noise sources which contribute to the overall noise level. With reference to the target locomotive that has been chosen, the compressor motor fan was identified as one of the main sources of noise and better investigated. This kind of source allowed the design of different control approaches and their assessment in laboratory tests which opened the way to new areas of application. Among the several approaches studied, operational consideration forced an active seats solution and testing, the results of which allowed the implementation of many improvements.

## 13.9    Acknowledgements

The authors would like to thank all the people involved in the development of the present work and in particular: Dr. Antonio Vecchio and Dr. Fabio Ciraolo whose degree thesis at the University of Naples has deeply contributed to the research activities, and Dr. Roberto Setola, Dr. Nicola Marino and Dr. Gianluca Isernia for their substantial contribution in the development and assessing of the controller hardware and software.

Special thanks are due to Dr. Biagio Mocerino and his staff, from Ansaldo Trasporti, for their full availability and participation in the work.

Finally we would like to thank our families for their love and for encouraging us daily throughout the work.

## 13.10 Appendix

### 13.10.1 Introduction

The aforementioned active control experimental tests have been performed by the use of a self-build digital controller.

The hardware of the proposed board, Hermes, for the sake of flexibility, is composed of two parts:

- the main processor board, containing the DSP, the memories, the A/D and D/A converters for a total of eight inputs and eight outputs, able to work as stand-alone or in a multiprocessor system (being configurable as master or slave without any hardware change).

- an analogue board containing the interfaces, I/O conditioning filters and amplifiers. These circuits also have the capability to short inputs with outputs in a testing phase.

All the boards of the system are put in a rack, with at least one processor and an interface board.

### 13.10.2 Digital processing circuit

The heart of the circuit is a last-generation DSP, the ADSP-21060 SHARC produced by Analog Devices. Its ultra large scale of integration allows one to integrate 4 Mbit of RAM in a dual port memory (DPM) with an I/O processor charged with the management of access in the same DPM. Internally, there are three address buses (PM address bus, DM address bus, I/O address bus) and three data buses (PM data bus, DM data bus, I/O data bus). This architecture enables us to obtain a throughput of 40 MIPS, with a peak of 120 MFLOPS and a sustained performance of 80 MFLOPS. The precision is according to the standard IEEE floating-point data format 32-bit single precision and 40-bit extended precision or the 32-bit fixed-point data format.

Around the main processor there are circuits providing full functionality to the system:

*EPROM area:* the EPROM area contains the elaboration algorithm, this software comes loaded, at every power-up of the circuit, in the program area of the SRAM contained inside the DSP. The circuit realisation chosen for the EPROM area allows the use of bank from 512 Kbit to 8 Mbit.

*Control logic area:* the control logic area manages the control signals which the DSP uses in external accesses and performs address decoding and interrupt handling. It is implemented via two complex PLD XC9536. Beyond the functions of control logic management, the two CPLD also carry out the functions of address decoder and interrupt handler.

*Analogue input/output area:* the A/D conversion bench comprises two analog devices AD7874 converters with 12-bit resolution, each able to acquire four incoming channels simultaneously. The tension level applicable to the eight inputs must be contained in the range of 10 V. The maximum sampling speed is equal to 20 kHz and is under the direct control of the operating software. In fact the signal to begin conversion, called start conversion, is driven by the signal of count expired produced by the inner timer of the DSP. This ties the speed of sampling to the demanded application, optimising the several phases comprising the elaborative process. Once the conversion is completed, which happens in the maximum time of 35 $\mu s$, the interrupt signal to the DSP is generated. The D/A conversion bench comprises two analog devices DAC8412 converters, each one of which is a quadruple DAC with a conversion resolution of 12 bits. The maximum obtainable amplitude of tension on each of the eight outputs is in the range 2.5 V. The conversion procedure of the DAC8412 permits the simultaneous updating of all the outputs. Such a procedure expects that every digital value comes written directly from the DSP to a support register inside the DACs. Once the loading of this buffer has been completed the activation of a single signal allows the contemporary D-to-A conversion of the eight analogue outputs.

*Nonvolatile RAM area:* the nonvolatile RAM area is formed by two CMOS nvSRAM STK12C68 produced by SIMTEK. For the DSP it constitutes a buffer, the contents of which are not lost with power off. Such a buffer bit is seen by the DSP to be like a RAM memory bank with a capacity of 8 kword by 16 bits. The maximum access time to this RAM is equal to 25 ns, this allows the DSP to carry out read/write operations with a single wait cycle.

*RS232 interface area:* this area has the scope to supply to the circuit a standard interface useful for connection with external units.

*Multiprocessor interface area:* the multiprocessor interface area allows the connection of up to four Hermes circuits to increase computational power and the number of analogue inputs/outputs. It is based on data exchange between the elaboration units by means of a shared memory bank and the definition of one master and one or more slaves. The master or slave information for every circuit in the system is given by a configuration register the inputs of which are directly linked to the external connector. In this way, initialisation is executed by means of rack cabling, and a board can be a master or slave by simply changing its position in the rack. The shared memory buffer is a dual port memory IDT7024L produced by Integrated Device Technology.

Its capacity is 16 kwords of 16 bit, with an access time of 20 ns. Peculiar to this component is the on-chip capability to manage the last two memory locations as mailboxes, one per port. This control demands the generation of an interrupt flag for every access at one side of the port; this flag is disabled when the same location is read on the other side, allowing syncronisation of the data flux based on the interrupt flag, enabling access in the shared memory.

*BITE area:* this last area is used to flag the user for eventual anomalies. It is formed by two LEDs the state of which is directly driven by the DSP.

### 13.10.3  Analogue interface board

The interface board is composed of a mainboard, which connects the rack with the filtering circuits, and up to eight filtering circuits which condition the amplitude of the input signal and filter the upper harmonics. The output signal is also conditioned and filtered by a Butterworth filter and is provided with a processor-controlled switch to directly connect the inputs to the outputs to test the circuit.

*Mainboard:* the mainboard aim is to keep the rack connected with the conditioning circuits. It provides the power supplies, the input and output lines and all the triggering signals. On it there is place for up to eight interface circuits by means of eight SIMM 72 connectors; the interface circuit could be different from the one presented here to adapt to other end user needs (sensors for piezoelectric or accelerometers).

*Interface circuits:* the circuits presented here are designed for use with microphones and loudspeakers, although it is easy to design circuits which accept other sources; each circuit manages one input and one output of Hermes. There are six function blocks for each circuit.

The two amplitude conditioning blocks scale the signals by the desired factor (from the microphone and from the D/A converter). The circuit gain is to be established during the on-site tests and can be changed by varying the retroaction resistors. The two filter blocks remove the harmonics beyond the Nyquist frequency. Output filters, required for interpolation of the sampled data, are then present.

After the filter, in the input chain, we find a clamp filter, to prevent excessive voltages to the analogue-to-digital converters.

The last block is a test switch which, below microprocessor command, links the output directly with the input. This can be used, during tests, to verify the input and output lines.

### 13.10.4    Software environments

The user may interact with Hermes *via* a user friendly interface realised on a host PC. The interface allows the user to select the desired algorithm, define the parameter of the system, select the variables to monitor and start/stop the DSP. However Hermes, once configured, is able to work in a standalone unattended configuration.

At the moment the authors have developed all the software needed to implement FIR or IIR-based ANC algorithms.

Currently, off-line identification is used to evaluate FIR estimation of each secondary path (i.e. from each actuators to each sensors). The user may choose the number of taps, say $m$, used to estimate the channel,[2] the sample rate, the convergence parameters of the algorithm, the type of signal used to excite the channel (white noise, bandlimited white noise, tone, multitones) where the amplitude of the signal is automatically defined by the algorithm in order to achieve the best SNR (it is possible to disable this option). Moreover, it is possible to use as the exciting signal an external source or a time history stored in the RAM of Hermes. The algorithm also generates the estimation error signal, i.e. the difference between the response of the estimated and of the actual channel, for an immediate check of the convergence, *via* a conventional oscilloscope. The algorithm may be stopped as soon as the convergence is achieved or when a predefined amount of time has elapsed. It is possible to pass to Hermes directly the coefficients of the FIR and to store the estimated coefficients on the host PC for further use. It is also possible to evaluate the frequency response of each secondary path (and of the path from the reference signal to each sensor) at a given set of frequencies.

The online control algorithm section allows the designer to implement an FXLMS control scheme. The user may choose to use an FIR or IIR control algorithm, define the number of taps, say $l$, of the control, select the sample rate, select the error signal (up to seven) and the control signal (up to eight), define the convergence parameters, select the parameter of the feedforward action (if needed).

Work in progress is also devoted to developing a software environment able to automatically translate an algorithm written in some high-level graphical language (e.g. Simulink/MATLAB) into C and assembler format compatible with Hermes.

---

[2]In the present version the number of taps must be the same for all the secondary paths

*Chapter 14*

# Active noise control for road booming noise attenuation

## Y. Park, H. S. Kim and S.H. Oh

*Department of Mechanical Engineering*
*Korea Advanced Institute of Science and Technology*
*Science Town, Taejon, Enail: osh@cais.kaist.ac.kr*

*This chapter presents an application of active noise control techniques for passenger comfort in cars. At the analysis stage nonstationary vibrations of the front and rear wheels are generated by nonuniform road profiles and change of vehicle speed. These propagate through the tyre and the complicated suspension system and finally generate structure-borne noise, impulsive noise, and other low-frequency noise in the interior of the passenger vehicle. These noises produce acoustical resonance in the interior of the passenger car and such resonant noise is called "road booming noise". Cancellation of this noise is the topic of this chapter.*

## 14.1  Introduction

There are several characteristics of road booming noise. Independent vibrations of the four wheels generate it, so it can hardly be reduced by ANC with just one or two reference sensors. Second, the properties of road booming noise change as vehicle speed or road profile varies. And the system from the wheel vibration to road booming noise is not linear because of the complexity of the suspension system, nonlinear interaction between the structural vibration and the interior acoustics etc. We want to compensate for road booming noise, which has a strong correlation with the vibration signals measured at the suspension system. Active noise control of road booming noise is rather difficult to achieve because of its nonstationary characteristics. We developed a multi-input multi-output constraint filtered-$x$ least-mean-square algorithm using an IIR-based filter. The proposed algorithm can track nonstationary processes and concentrate the control efforts on the desired control frequency range which can be selected arbitrarily by the designers.

In a feedforward-type active noise control approach, a number of secondary sources (usually loudspeakers) are controlled to cancel the noise of the desired space. The physical principle of this approach is destructive interference of acoustic waves. For active control of road booming noise, a set of reference signals presumed to be the cause of the booming in the car interior is measured from the car suspension system. This is natural, because the road inputs must pass through the suspension system before they produce noise inside the vehicle. Controller $W(z)$ is inserted between these reference inputs and secondary sources. Denoting the number of reference signals as $N$ and the number of secondary sources as $K$, the controller $W(z)$ is characterised by a $N$-input $K$-output filter. If one locates $M$ microphones in the desired quiet zone to monitor the performance, the auxiliary system between the secondary sources and the measuring microphones can be represented as a $K$-input $M$-output secondary path $H(z)$. If we let $X(k)$ be the $N$ by 1 reference vector, and denoting the system from the reference to the error microphone as $N$-input $M$-output plant $P(z)$, the control problem becomes to find optimal $W(z)$ which minimises the $M$ by 1 error vector $E(k) = [P(z) + H(z)W(z)]X(k)$. If we choose the cost function as $E(k)^T E(k)$ (the squared sum of the error signal from the $M$ microphones) and construct $W(z)$ as an FIR-based adaptive filter, we can get an MFX (multiple filtered-$x$) LMS algorithm [82]. The MFX LMS algorithm has been a basic feedforward control algorithm in the active control of road booming noise applications [93, 128, 208, 281, 282, 283].

There are some problems in applying the MFX LMS algorithm to the road booming noise attenuation application. The delay in the secondary path decreases the upper limit of the convergence speed in this algorithm, and this may degrade the performance considerably for nonstationary reference input cases, such as in road booming noise applications. In order to recover the convergence bound of the MFX LMS algorithm, a CMFX (constraint multiple filtered-$x$) LMS algorithm can be used [281]. The frequency band of road booming is usually fixed and is not greater than a few tens of hertz and is independent of driving speed. However, the FIR-filter-based MFX LMS algorithm tries to optimise $W(z)$ with equivalent weighting for all frequency ranges resulting in a waste of control effort. In this chapter, to overcome the aforementioned drawbacks, a novel active noise control algorithm is proposed. Every SISO channel of $W(z)$ is organised as a linear combination of the IIR base filters which are second-order narrow bandpass filters. The centre frequency of each IIR filter covers the road booming frequency ranges. Also the error signals used for updating weights in $W(z)$ are reconstructed to maximise the convergence speed as originally introduced in the CMFX LMS algorithm. To demonstrate the effectiveness of the proposed algorithm to the road booming noise attenuation application, experiments are performed for a rough asphalt road profile.

## 14.2   Constraint multiple filtered-x LMS algorithm

When the noise to be attenuated is caused by several independent inputs such as with road booming noise inside a car, it is necessary to deal with multiple reference signals to actively control the undesired noise. Each of the four wheel vibrations due to irregular road surface excitation can be considered as an independent noise source. We can place multiple control sources and error microphones to enlarge the quiet zone or increase the noise reduction level. For this purpose the CFX LMS algorithm should be extended to a multi-input multi-output system. In CMFX LMS, the $m$th error $e_m(k)$ and the constraint error $e'_m(k)$ are:

$$e_m(k) = d_m(k) + \sum_{n=1}^{N}\sum_{k=1}^{K}\sum_{j=0}^{L_h}\sum_{i=0}^{L} h_{mkj}w_{kni}(k-j)x_n(k-i-j) \qquad (14.1)$$

$$e'_m(k) = d_m(k) + \sum_{n=1}^{N}\sum_{k=1}^{K}\sum_{j=0}^{L_h}\sum_{i=0}^{L} h_{mkj}w_{kni}(k)x_n(k-i-j) \qquad (14.2)$$

where $d_m(k)$ is the undesired noise signal at the $m$th microphone, $x_n(k)$ is the $n$th reference and $w_{kni}$ is the $i$th coefficient of FIR-type adaptive filter $W_{kn}(z)$, i.e. $W_{kn}(z) = \sum_{i=0}^{i=L} w_{kni}(k)z^{-i}$ the input of which is the $n$th reference and the output of which goes to the $k$th control speaker. $h_{mkj}$ is the $j$th coefficient of the secondary path $H_{mk}(z)$, i.e. $H_{mk}(z) =$ which is located between the $k$th control speaker and the $m$th error microphone. The constraint error $e'_m(k)$ expressed in eqn. (14.2) cannot be measured directly, but can be calculated by modifying $e_m(k)$ by eliminating $d_m(k)$ from eqns (14.1) and (14.2). Weights of CMFX LMS are updated by the steepest descent method. Derivation of the CMFX LMS algorithm which tries to minimise the cost function $e'^2_1(k) + e'^2_2(k) + \cdots + e'^2_M(k)$ is straightforward and the results are shown in Table 14.1. The signal $fx_{mkn}(k)$ is a filtered signal of $x_n(k)$ through

Table 14.1    *Constraint multiple filtered-x LMS algorithm*

| given $x_n(k)$, $e_m(k)$, $n = 1, 2, \cdots, N$, $m = 1, 2, \cdots, M$ at $k$ step |
| --- |

$$fx_{mkn}(k) = \sum_{j=0}^{L_h} h_{mkj}x_n(k-j)$$

$$e'_m(k) = e_m(k) - \sum_{k=1}^{K}\sum_{j=0}^{L_h} h_{mkj}y_k(k-j) + \sum_{n=1}^{N}\sum_{k=1}^{K}\sum_{i=0}^{L} w_{kni}(k)fx_{mkn}(k-i)$$

$$w_{kni}(k+1) = w_{kni}(k) - 2\mu \sum_{m=1}^{M} e'_m(k)fx_{mkn}(k-i)$$

$$y_k(k) = \sum_{n=1}^{N}\sum_{i=0}^{L} w_{kni}(k)x_n(k-i)$$

the error path model $H_{mk}(z)$. The CMFX LMS algorithm is different from the MFX LMS algorithm only in that it uses $e'_m(k)$ instead of $e_m(k)$.

## 14.3    Constrained multiple filtered-x LMS algorithm using an IIR-based filter

We consider an adaptive filter $W(z)$ consisting of a linear combination of stable IIR filter bases the impulse responses of which are exponentially developed sinusoidal functions. Let us define $B_i(z)$ as one of these filter bases. To consider $L$ frequency components in an IIR-based filter, there must be $2L$ IIR filter bases of $B_i(z)$. Each IIR base can be expressed in the discrete time domain as follows:

$$B_i(z) = [B_{i,c}(z), \quad B_{i,s}(z)]^T \tag{14.3}$$

where

$$B_{i,c}(z) = \frac{1 - e^{-\sigma_i T}z^{-1}\cos\omega_i T}{1 - 2e^{-\sigma_i T}z^{-1}\cos\omega_i T + e^{-2\sigma_i T}z^{-2}}$$

$$B_{i,s}(z) = \frac{e^{-\sigma_i T}z^{-1}\sin\omega_i T}{1 - 2e^{-\sigma_i T}z^{-1}\cos\omega_i T + e^{-2\sigma_i T}z^{-2}} \tag{14.4}$$

Denoting $y_{nk}(k)$ as the control output to the $k$th secondary source due to the $n$th reference input, the control output to the $k$th secondary source denoted as $y_k(k)$ can be obtained as the summation of $y_{nk}(k)$ for $n = 1, 2, \cdots, N$ as follows:

Table 14.2    *IIR-based constraint multiple filtered-x LMS algorithm*

---

$$u_{ni}(k) = B_i(z)x_n(k)$$

$$fx_{mkn}(k) = \sum_{j=0}^{L_h} h_{mkj}x_n(k-j)$$

$$fu_{mkni}(k) = B_i(z)fx_{mkn}(k)$$

$$e'_m(k) = e_m(k) - \sum_{k=1}^{K}\sum_{j=0}^{L_h} h_{mkj}y_k(k-j) + \sum_{n=1}^{N}\sum_{k=1}^{K}\sum_{i=0}^{L} w_{kni}^T(k)fu_{mkni}(k)$$

$$w_{kni}(k+1) = w_{kni}(k) - 2\mu \sum_{m=1}^{M} e'_m(k)fu_{mkni}(k)$$

$$y_k(k) = \sum_{n=1}^{N}\sum_{i=0}^{L} w_{kni}^T(k)u_{ni}(k)$$

---

$$y_k(k) = \sum_{n=1}^{N} y_{nk}(k) \tag{14.5}$$

where

$$y_{nk}(k) = \sum_{i=1}^{L} w_{kni}^T(k)u_{ni}(k) \tag{14.6}$$

$$u_{ni}(k) = B_i(z)x_n(k) \tag{14.7}$$

Scalar controller coefficients $w_{kni}(k)$ are adaptive weights of $u_{ni}(k)$ which is the $i$th IIR filter output of $x_n(k)$. The proposed IIR-based filter has an infinite impulse response, but it does not have any nonlinearity or instability problems on updating filter weights unlike the conventional adaptive IIR filters. So, it is easy to model a lightly damped system with an IIR-based filter. The IIR-based CMFX LMS algorithm is summarised in Table 14.2, and its block diagram representation is shown in Figure 14.1.

## 14.4    Experimental results

To carry out an experiment, four acceleration signals ($y$ and $z$ directions of two front wheels) and acoustic pressure signals from two microphones near the headrest of the two front seats were used as the reference signals and the error signals, respectively. We developed an ANC system of four references, two secondary speakers and two error microphones. The test car was a Sonata

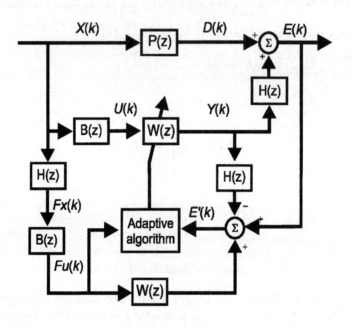

Figure 14.1    *Block diagram of CMFX LMS algorithm using IIR base B(z)*

II from Hyundai with a 2000 cc engine. ANC was performed while driving on a rough asphalt road with 60 km/h speed. Four reference signals were lowpass filtered with a cut-off frequency of 500 Hz and then A/D converted along with two microphone signals which are considered as undesired noise $D(k)$. The sampling frequency was 1000 Hz, and four secondary paths were identified using two secondary speakers located behind the front seats. A DSP board equipped with a TMS320c40 chip was used. After several driving tests on various road profiles, changing the vehicle speed from 40 up to 80 km/h, it was found that road booming frequency in our test vehicle is about 250 Hz with a 15 Hz bandwidth. FIR adaptive filter lengths and the FIR model for the secondary path were 130 and 50, respectively. 18 IIR bases with centre frequencies of 230, 235, 240, 245, 250, 255, 260, 265 and 270 Hz and damping coefficient of $\sigma_i = 25$ were selected. The length of the FIR filter could not be increased to more than 130 because of the calculation power limit of DSP hardware, although the filter length for one secondary source and one error microphone configuration could be increased up to 300. The overall experimental setup is given in Figure 14.2.

The attenuation of the overall level of the A-weighted spectrums is plotted in Figure 14.3. The spectrum of $e(k)$ before ANC, after ANC using conventional CMFX and using an IIR-based one are represented with a dotted line, solid line, and thick solid line, respectively. No remarkable reduction is achieved by using the conventional CMFX LMS algorithm, and even an increment of the noise level is observed. This is because the signal power of the road booming component in the error microphone signal before A-weighting is significantly smaller than that of the noise component under 200 Hz. The reason for the poor performance of the CMFX LMS is thought to be due to the short weight length used; another important reason is that the FIR filter tries to treat all frequency ranges with more weighting for higher power components in the error signal in the least-squares sense. The power under 200 Hz is dominant from the microphone signal in the car interior and adaptive filters will try to attenuate these components first. Since there is low coherence between references and error signals in this frequency range, weights of the adaptive filter fluctuate rapidly consuming a lot of the control effort. The effort to attenuate road booming near 250 Hz is comparatively very small despite higher coherence than in other frequency ranges. These make the performance of CMFX LMS worse in this application. 5–7 dB reduction was achieved at road booming frequency region (around 250 Hz) when the IIR-based filter was used. There is no considerable change of noise level out of the booming frequency range where IIR bases are not located. Because of the road booming noise characteristics, the CMFX LMS algorithm using an FIR filter tries to reduce

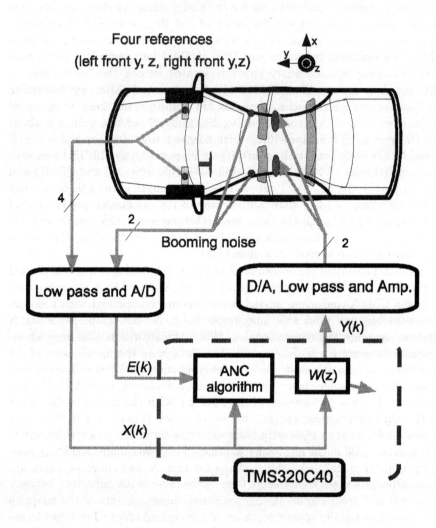

Figure 14.2    *Experimental setup*

the low-frequency range and little sound attenuation is achieved, whereas the IIR-based filter reduced noise effectively in the range of the road booming frequency.

## 14.5   Conclusions

Road booming noise of a vehicle is a naturally time varying signal due to the change of vehicle speed and road profile and it appears in a certain frequency range which is not wideband. The conventional CMFX LMS algorithm because of its characteristics achieves little sound attenuation of nonstationary road booming noise. So, an IIR-based filter incorporating a stable and narrow bandpass IIR filter is proposed to effectively reduce the road booming noise in a passenger vehicle. With the IIR-based filter, it is possible to control undesired noise only in a selected frequency range and enhance the control efficiency in road booming noise control. To compare the performance of the algorithm using the IIR-based and the conventional FIR filter, an experiment was performed. Driving the rough asphalt with constant 60 km/h speed, we attenuated the noise in the headrest areas of the two front seats with two error microphone, two control speakers and four references which are acceleration signals near the suspension system. In the experiment, it is difficult to achieve noise reduction with the FIR adaptive filter, however, 5–7 dB reduction of noise in booming frequency range was achieved with the IIR-based filter. If the desired control frequency range is not wideband such as in the case of road booming noise, the IIR-based filter is more effective than FIR filter and this is verified by the experiment.

## 14.6   Acknowledgments

The authors would like to express appreciation to Hyundai motor company and KATECH for their financial support.

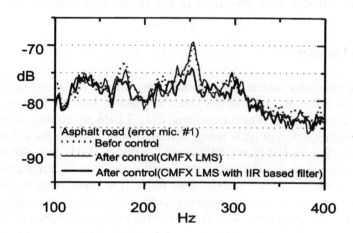

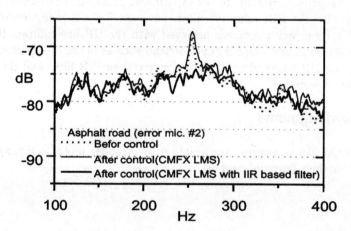

Figure 14.3    *Experimental results for rough asphalt road at (a) right-hand front seat and (b) left-hand front seat with two different adaptive filters*

*Chapter 15*

# Sequential and parallel processing techniques for real-time active control

## M. O. Tokhi [†] and M. A. Hossain [‡]

† *Department of Automatic Control and Systems Engineering*
*The University of Sheffield, Sheffield, UK, Email: o.tokhi@sheffield.ac.uk*
‡ *School of Engineering, Sheffield Hallam University, Sheffield, UK*

*Sequential and parallel processing techniques for real-time adaptive active control are considered in this chapter. Three different algorithms, namely simulation, control and identification, are involved in the adaptive control algorithm. These are implemented on a number of uniprocessor and multiprocessor computing platforms. The interprocessor communication speed and the impact of compiler efficiency on processor performance are investigated. A comparative assessment is provided, on the basis of the real-time communications performance, computation performance and compiler performance, to lead to merits of design of parallel systems incorporating fast processing techniques for real-time active control applications.*

## 15.1  Introduction

The performance demands required in modern control systems have led to complex processes, such as adaptive control, optimisation, failure tolerance, rule-based systems and neural networks, becoming essential control engineering tools. These processes are almost invariably implemented by digital computing systems, and for online control the implementation is required to be in real time. This means that it is required to compute the control algorithm within the loop sample time. Despite the vastly increased computing power which is now available there can still be limitations in the computing capability of digital processors in real-time control applications for two reasons:

(i) sample times have become shorter as greater performance demands are imposed on the system

(ii) algorithms are becoming more complex as the development of control theory leads to an understanding of methods for optimising system performance.

To satisfy these high performance demands, microprocessor technology has developed at a rapid pace in recent years. This is based on:

- processing speed

- processing ability

- communication ability

- control ability

Digital signal processing (DSP) devices are designed in hardware to perform concurrent add and multiply instructions and execute irregular algorithms efficiently, typically finite-impulse response (FIR) and infinite-impulse response (IIR) filter algorithms. Vector processors are designed to efficiently process regular algorithms involving matrix manipulations. However, many demanding complex signal processing and control algorithms cannot be satisfactorily realised with conventional computing methods. Alternative strategies where high-performance computing methods are employed could provide suitable solutions in such applications [296, 297, 298].

In the main, multiprocessor systems based on homogeneous architectures, where the main processing is carried out by processors of the same nature, are considered. To add extra flexibility into the computing system it is useful to consider heterogeneous architectures where the main processing is carried out by processors of different nature. This allows the exploitation of the computing capabilities of the various processors to execute the tasks to which they are most suited. There are many types of processing device available, for example RISC, ASIC and DSP devices. Another processor suited to parallel processing

(PP) using point-to-point communication is the transputer, which has been extensively discussed in the literature. This is a fairly flexible processor, which allows easy coding and execution of complex algorithms. However, it has been found not to be as efficient as DSP devices for performing regular tasks. For general engineering purposes it is normal that the overall computing task can be split into routine and specialised tasks. It is important that parallel architectures incorporating devices suited to both these types of operation should be considered, and this chapter explores a few possibilities. The requirements of control engineering applications normally demand signals to be analysed and suitable control action generated. This leads to the conclusion that a useful heterogeneous architecture to consider for control engineering applications is one incorporating DSP devices to perform the signal analysis and general-purpose PP devices to perform tasks of a specific nature. Such architectures incorporating DSP and PP devices are considered in this chapter.

Programmable DSP chips appear to perform poorly on the more general computational features required in control applications because their architecture is optimised for fast signal processing. The coefficient resolution is limited by a fixed register length, e.g. 8-bit or 16-bit, and by fixed-point arithmetic. Moreover, the performance is sensitive to number representation [247, 301]. Investigations have previously been carried out in devising parallel processing architectures using DSP chips [52, 104]. These have revealed that, although in theory the improvement in data throughput achieved with $N$ processors will be $N$ times that of one processor, due to communication and process management overheads this is reduced as the number of processors increases. Nevertheless, such architectures can successfully be applied to small regular tasks [104].

Research experience in the use of transputers for real-time control has shown that despite this processor's much vaunted potential for real-time embedded systems, the special demands of real-time control (short sampling interval and correspondingly modest computational task sizes) are sometimes illmatched to the granularity of the transputer [106, 144]. Indeed, as the architecture evolves (T2 series to T9 series), this granularity mismatch widens. A system integrating finer-grain architectures (DSP devices) with the transputer (exploiting its facility for handling irregular computational tasks) will create a more versatile system.

For digital processors with widely different architectures, performance measurements such as MIPS (million instructions per second), MOPS (million operations per second) and MFLOPS (million floating-point operations per second) are meaningless. Of more importance is to rate the performance of a processor on the type of program likely to be encountered in a particular application [296]. The different processors and their different clock rates, memory cycle times etc. bring in confusion to the issue of attempting to rate the processors. In particular, there is an inherent difficulty in selecting processors in signal processing and control applications. The ideal performance of a processor demands a perfect match between processor capability and program behaviour. Processor capability can be enhanced with better hardware technology, innovative architectural features and efficient resource management. From the hardware point of view, current performance varies according to whether the processor possesses a pipeline facility, is microcode/hardwired operated, has an internal cache or internal RAM, has a built-in maths coprocessor, floating-point unit etc. Program behaviour, on the other hand, is difficult to predict owing to its heavy dependence on application and run-time conditions. Other factors affecting program behaviour include algorithm design, data structure, language efficiency, programmer skill and compiler technology [11, 138]. This chapter attempts to investigate such issues within the framework of active control applications.

One of the challenging aspects of PP, as compared with sequential processing, is how to distribute the computational load across the processing elements (PEs). This requires consideration of a number of issues, including the choice of algorithm, the choice of processing topology, the relative computation and communication capabilities of the processor array and partitioning the algorithm into tasks and the scheduling of these tasks [61]. It is essential to note that in implementing an algorithm on a parallel computation platform, a consideration of:

(i) the interconnection scheme issues

(ii) the scheduling and mapping of the algorithm on the architecture

(iii) the mechanism for detecting parallelism and partitioning the algorithm into modules or subtasks

will lead to a computational speedup [7].

Many control algorithms are heterogeneous, as they usually have varying computational requirements. The implementation of an algorithm on a homogeneous architecture is constraining and can lead to inefficiencies because of the mismatch between the hardware requirements and the hardware resources. In contrast, a heterogeneous architecture having PEs of different types and features can provide a closer match with the varying hardware requirements and, thus, lead to performance enhancement. However, the relationship between algorithms and heterogeneous architectures for real-time control systems is not clearly understood [202]. The mapping of algorithms onto heterogeneous architectures is, therefore, especially challenging. To exploit the heterogeneous nature of the hardware it is required to identify the heterogeneity of the algorithm so that a close match is forged with the hardware resources available [24].

A cantilever beam system in transverse vibration is considered in this chapter. The unwanted vibrations in the structure are assumed to be due to a single point disturbance of a broadband nature. First-order central finite difference (FD) methods are used to study the behaviour of the beam and develop a suitable simulation environment as a test and verification platform. An active vibration control (AVC) system is designed to yield optimum cancellation of broadband vibration at an observation point along the beam. The controller design relations are formulated so as to allow online design and implementation and, thus, yield a self-tuning control algorithm. The beam simulation and the self-tuning AVC algorithms are implemented on several uniprocessor and multiprocessor parallel architectures involving transputers, reduced instruction set computer (RISC) processors and DSP devices. The performance in each case is assessed and comparison of the results of these implementations, on the basis of real-time interprocessor communication, computation and compiler performance, is made and discussed.

## 15.2 The cantilever beam system

Consider a cantilever beam of length $L$, clamped (fixed) at one end and free at another end. The motion of the beam in transverse vibration is governed by the well known fourth-order partial differential equation (PDE) [159]:

$$\mu^2 \frac{\partial^4 y(x,t)}{\partial x^4} + \frac{\partial^2 y(x,t)}{\partial t^2} = \frac{1}{m} U(x,t) \tag{15.1}$$

where $U(x,t)$ and $y(x,t)$ represent an applied force and the resulting deflection of the beam, from its stationary (unmoved) position, respectively at a distance $x$ from the fixed end at time $t$, $\mu$ is a beam constant given by $\mu^2 = EI/(\rho A)$, with $\rho$, $A$, $I$ and $E$ representing the mass density, cross-sectional area, moment of inertia of the beam and the Young's modulus, respectively, and $m$ is the mass of the beam. The corresponding boundary conditions at the fixed and free ends of the beam are given by:

$$y(0,t) = 0 \qquad \text{and} \qquad \frac{\partial y(0,t)}{\partial x} = 0$$

$$\tag{15.2}$$

$$\frac{\partial^2 y(L,t)}{\partial x^2} = 0 \qquad \text{and} \qquad \frac{\partial^3 y(L,t)}{\partial x^3} = 0$$

To obtain a numerical solution of the PDE in eqn. (15.1), and thus construct a suitable simulation environment characterising the behaviour of the beam, the FD method [108, 206] can be used. To do so, the partial derivative terms $\frac{\partial^4 y(x,t)}{\partial x^4}$ and $\frac{\partial^2 y(x,t)}{\partial t^2}$ in eqn. (15.1) and the boundary conditions in eqn. (15.2) are approximated using first-order central FD approximations. This involves a discretisation of the beam into a finite number of equal-length sections (segments), each of length $\Delta x$, and considering the beam motion (deflection) for the end of each section at equally-spaced time steps of duration $\Delta t$. In this manner, let $y(x,t)$ be denoted by $y_{i,j}$ representing the beam deflection at point $i$ at time step $j$ (grid point $i,j$) and $y(x+v\Delta x, t+w\Delta t)$ be denoted by $y_{i+v,j+w}$, where $v$ and $w$ are nonnegative integer numbers. Using a first-order central FD method the PDE in eqn. (15.1) and the boundary conditions in eqn. (15.2) can be expressed in terms of the FD approximations to yield the beam deflection at the grid points $i = 1, 2, \ldots, n$ as [307]:

$$Y_{j+1} = -Y_{j-1} - \lambda^2 SY_j + (\Delta t)^2 U(x,t)\frac{1}{m} \qquad (15.3)$$

where, $Y_j = [y_{1,j}\ y_{2,j}\ \cdots\ y_{n,j}]^T$ and $S$ is a pentadiagonal matrix, known as the stiffness matrix of the beam, and $\lambda^2$ is given in terms of $\mu^2$, $\Delta t$ and $\Delta x$. Eqn. (15.3) is the required relation for the simulation algorithm, characterising the behaviour of the cantilever beam system, which can be implemented on a digital computer easily. For the algorithm to be stable it is required that the iterative scheme described in eqn. (15.3), for each grid point, converges to a solution. It has been shown that a necessary and sufficient condition for stability satisfying this convergence requirement is given by [159]:

$$0 < \frac{(\Delta t)^2}{(\Delta x)^4}\,\mu^2 \le 0.25$$

## 15.3    Active vibration control

A schematic diagram of a feedforward AVC structure is shown in Figure 15.1. An unwanted (primary) point source produces broadband disturbance into the (beam) structure. This is detected by a detection sensor, processed by a controller of suitable transfer characteristics and fed to a cancelling (secondary) point source (control actuator). The secondary signal thus generated is superimposed on the primary signal so as to achieve vibration suppression along the beam. A sensor at an observation point monitors the performance of the system.

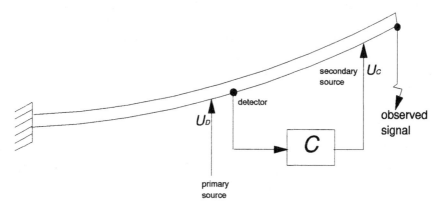

*Figure 15.1    Schematic diagram of the active vibration control structure*

The objective in Figure 15.1 is to reduce the level of vibration to zero at the observation point. This is equivalent to the minimum variance design criterion in a stochastic environment [306, 329]. This requires the primary and secondary signals at the observation point to be equal in amplitudes and have a phase difference of 180° relative to each other. Thus, using a frequency-domain representation of the structure in Figure 15.1 and synthesising the controller on the basis of this objective yields the required controller characteristics as [295]:

$$C = \left[1 - \frac{Q_1}{Q_0}\right]^{-1} \tag{15.4}$$

where $Q_0$ and $Q_1$ are the system transfer characteristics between the detection and observation points with the secondary source off and on, respectively. Thus, online design and implementation of the controller can be achieved by (a) obtaining $Q_0$ and $Q_1$ using a suitable system identification algorithm, (b) using eqn. (15.4) to calculate the controller transfer function and (c) implementing the controller on a digital processor. Moreover, to monitor system performance and update the controller characteristics upon changes in the system a supervisory level control can be utilised. This results in a self-tuning AVC mechanism composed mainly of the processes of identification and control [295].

The identification algorithm is described here as the process of estimating parameters of the required controller characteristics. In this manner, it consists of the processes of estimating the system models $Q_0$ and $Q_1$ and the controller design calculation. The RLS algorithm is used here for estimation of parameters of the system models $Q_0$ and $Q_1$. This is based on the well known least-squares method. The RLS estimation process, for an unknown plant with input $u(n)$ and output $y(n)$ described by a discrete linear model, at a time step $k$ is described by [306]:

$$\varepsilon(k) = \Psi(k)\Theta(k-1) - y(k)$$

$$\Theta(k) = \Theta(k-1) - P(k-1)\Psi^T(k)\left[1 + \Psi(k)P(k-1)\Psi^T(k)\right]^{-1}\varepsilon(k) \qquad (15.5)$$

$$P(k) = P(k-1) - P(k-1)\Psi^T(k)\left[1 + \Psi(k)P(k-1)\Psi^T(k)\right]^{-1}\Psi(k)P(k-1)$$

where $\Theta$ is the model parameter vector and $\Psi$, known as the observation matrix, is a row vector of the measured input/output signals and $P(k)$ is the covariance matrix. Thus, the RLS estimation process is to implement and execute the relations in eqn. (15.5) in the order given. The performance of the estimator can be monitored by observing the parameter set at each iteration. Once convergence has been achieved the routine can be stopped. The convergence is determined by the magnitude of the modelling error $\varepsilon(k)$ or by the estimated set of parameters reaching a steady level.

The process of calculation of parameters of the controller uses a set of design rules based on eqn. (15.4). Here, $Q_0$ and $Q_1$ are considered as second order models of a recursive form. Using these in eqn. (15.4) will lead to a set of design rules relating the parameters of the resulting fourth-order controller to those of $Q_0$ and $Q_1$. Thus, the identification algorithm, as considered in this investigation, consists in the combined implementation of eqn. (15.5), for $Q_0$ and $Q_1$, and the set of design rules for calculation of the controller parameters.

The control algorithm consists in the process of online implementation of the controller to generate the control signal. This involves the implementation of the controller, as designed through the identification algorithm above, in discrete form using the equivalent difference equation formulation. Note that in implementing this within the simulation environment, the simulation algorithm becomes an integral part of the process. Thus, the control algorithm consists of the combined implementation of the time-domain controller realisation and the simulation algorithm.

## 15.4   Hardware architectures

Three PEs, namely, an Intel 80i860 (i860) RISC processor, a Texas Instruments TMS320C40 (C40) DSP device and a T805 (T8) transputer, were utilised to develop four different heterogeneous and homogeneous parallel architectures. In general, the nature of any parallel architecture reflects the nature of its PEs and the algorithms. Therefore, to compare the performance of the parallel architectures, it is essential to explore the features of the PEs and the parallel architecture itself. The parallel architectures and their PEs are described below.

### 15.4.1   Uniprocessor architectures

The i860 is a 64-bit vector processor with 40 MHz clock speed, a peak integer performance of 40 million instructions per second (MIPS), 8 kbytes data cache and 4 kbytes instruction cache, and is capable of 80 million floating-point operations per second (MFLOPS). This is the Intel's first superscalar RISC processor possessing separate integer, floating-point, graphics, adder, multiplier and memory-management units. The i860 executes 82 instructions, including 42 RISC integer, 24 floating-point, 10 graphics, and 6 assembler pseudo operations in one clock cycle [138].

The C40 is a 32-bit DSP processor with 40 MHz clock speed, 8 kbytes on-chip RAM, and 512 bytes on-chip instructions cache, and is capable of 275 million operations per second (MOPS) and 40 MFLOPS. This DSP processor possesses six parallel high-speed communication links for interprocessor communication with 20 Mbytes/s asynchronous transfer rate at each port and 11 operations/cycle throughput. It has separate internal program, data and DMA coprocessor buses for support of massive concurrent I/O of program and data throughput, thereby maximising sustained central processing unit (CPU) performance [39, 141].

The T8 is a general-purpose medium-grained 32-bit INMOS parallel PE with 25 MHz clock speed, yielding up to 20 MIPS performance, 4 kbytes on-chip RAM and is capable of 4.3 MFLOPS. The T8 is a RISC processor possessing an onboard 64-bit floating-point unit and four serial communication links. The links operate at speeds of 20 Mbits/s achieving data rates of up to 1.7 Mbytes/sec unidirectionally or 2.3 Mbytes/s bidirectionally. Most importantly, the links allow a single transputer to be used as a node among any number of similar devices to form a powerful PP system [144, 178].

## 15.4.2   Homogeneous architectures

The homogeneous architectures considered include a network of C40s and a network of T8s. A pipeline topology is utilised for these architectures, on the basis of the algorithm structure, which is simple to realise and is well reflected as a linear farm [144]. This is shown in Figure 15.2.

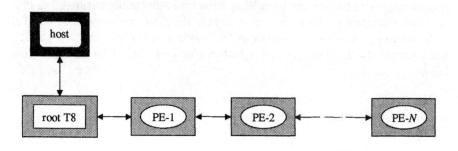

*Figure 15.2    Operational configuration of the homogeneous architectures*

The homogeneous architecture of C40s comprises a network of C40s resident on a Transtech TDM410 motherboard and a TMB08 motherboard incorporating a T8 as a root processor. The T8 possesses 1 Mbyte local memory and communicates with the TDM410 (C40s network) via a link adapter using serial-to-parallel communication links. The C40s, on the other hand, communicate with each other via parallel communication links. Each C40 processor possesses 3 Mbytes DRAM and 1 Mbyte SRAM.

The homogeneous architecture of T8s comprises a network of T8s resident on a Transtech TMB08 motherboard. The root T8 incorporates 2 Mbytes of local memory, with the rest of the T8s each having 1 Mbyte. The serial links of the processors are used for communication with one another.

## 15.4.3   Heterogeneous architectures

Three heterogeneous parallel architectures, namely, an integrated i860 and T8 (i860+T8) system, an integrated C40 and T8 (C40+T8) system and an integrated i860, T8 and C40 (i860+T8+C40) system, are considered in this study.

The operational configuration of the i860+T8 architecture is shown in Figure 15.3. This comprises a TMB16 motherboard and a TTM110 board incorporating a T8 and an i860. The TTM110 board also possesses 16 Mbytes of shared memory accessible by both the i860 and the T8, and 4 Mbytes of private memory accessible only by the T8. The i860 and the T8 processors communicate with each other via shared memory.

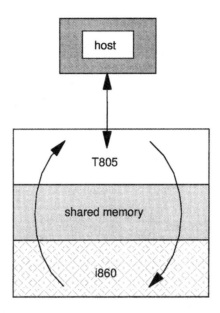

*Figure 15.3     Operational configuration of the integrated i860 and T8 architecture.*

The operational configuration of the C40+T8 architecture is shown in Figure 15.4. The T8 in this architecture is used both as the root processor providing an interface with the host, and as an active PE. The C40 and the T8 communicate with each other via serial-to-parallel or parallel-to-serial links.

The i860+T8+C40 system was developed utilising the features and facilities of the i860+T8 and the C40+T8 architectures. The architecture allows communication between the i860 and the C40 via the T8. The topology of the architecture is shown in Figure 15.5.

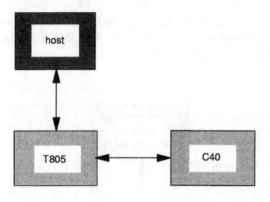

Figure 15.4    *Operational configuration of the integrated C40 and T8 architecture*

## 15.5  Software support

Software support is needed for the development of efficient programs in high-
level languages. Compilers have a significant impact on the performance of the
system. This is not to say that any particular high-level language dominates
another; most languages have advantages in certain computational domains.
The compiler itself is critical to the performance of the system as the mechanism for taking a high-level description of the application and transforming
it into hardware dependent machine language differs from one compiler to another. Identifying the foremost compiler for the application in hand is therefore
especially challenging. In control applications it is especially important to select a suitable programming language which could support highly numerical
computation. The compilers utilised include high-level languages consisting of
Portland Group and INMOS ANSI C for the i860 and the T8, 3L Parallel C
for the C40 and the T8.

## 15.6  Partitioning and mapping of algorithms

In implementing the algorithms on a parallel architecture, in addition to interprocessor communication, the issues of granularity and algorithm regularity
are required to be considered. The hardware granularity is a ratio of compu-

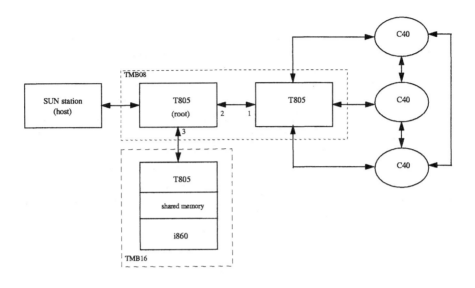

*Figure 15.5*    *Topology of the integrated i860, T8 and C40 architecture*

tational performance over the communication performance of each processor within the architecture. Similarly, task granularity is the ratio of computational demand over the communication demand of the task [183]. Regularity is a term used to describe the degree of uniformity in the execution thread of the computation. Many algorithms can be expressed by matrix computations. This leads to the so-called regular iterative (RI) type of algorithm. The simulation and control algorithms discussed above are of the RI type. In implementing these algorithms, a sequential vector processor will, principally, be expected to perform better and faster than any other processor. Moreover, since a large amount of data is to be handled for computation in these two algorithms, the performance will further be enhanced if the processor has more internal data cache, instruction cache and/or a built-in maths coprocessor. In implementing these algorithms on a PP platform, the tasks could be distributed uniformly among the PEs. However, this will require a large amount of communication between the processors. To calculate deflection of any one segment of the beam, for instance, information on the deflection of two backward and two forward segments will be required. Thus, each PE will need to send two and receive two data points in every iteration. This heavy communication time can, therefore, be a detriment to the performance of the computing platform in both homogeneous and heterogeneous architectures.

The identification algorithm is found to be of an irregular type for which sequential computation will be more suitable. The algorithm consists of two identical sections of estimation of the two models $Q_0$ and $Q_1$. Thus, a homogeneous parallel network consisting of two PEs can be used to share the load. Given the irregular nature of the algorithm a network of transputers will be expected to provide a more efficient implementation of the algorithm.

To investigate the performance of the computing platforms in the real-time implementation of the algorithms, the features discussed above were considered in the process of partitioning the algorithms such as to efficiently exploit the capabilities of the parallel hardware platforms and balance the load distribution so as to minimise interprocessor communications.

## 15.7   Implementations and results

In this investigation an aluminium type cantilever beam of length $L = 0.635$ m, mass $m = 0.037$ kg and $\mu = 1.351$ was considered and for the simulation algorithm the beam was divided into 19 segments ($n = 19$). Moreover, to cover the resonance modes of vibration of the beam and for the algorithm to be stable, a sample period of $\Delta t = 0.3$ ms was used.

### 15.7.1   Interprocessor communication

When several processors are required to work cooperatively on a single task, one expects frequent exchange of data among the subtasks which comprise the main task. The amount of data, the frequency with which the data is transmitted, the speed of data transmission, latency and the data transmission route are all significant in affecting the interprocessor communication within the architecture. The first two factors depend on the algorithm itself and how well it has been partitioned. The remaining two factors are the function of the hardware. These depend on the inter-connection strategy, whether tightly or loosely coupled. Any evaluation of the performance of the interconnection must be, to a certain extent, quantitative. However, once a few candidate networks have been tentatively selected, detailed (and expensive) evaluation including simulation can be carried out and the best one selected for a proposed application [7]. To explore the real-time performance of the parallel architectures, investigations into interprocessor communication are carried out. These interprocessor communication techniques for the different architectures are:

- T8-T8: serial communication link

- C40-C40: parallel communication link

- T8-C40: serial to parallel communication link

- T8-i860: shared memory communication

The performance of these interprocessor communication links is evaluated by utilising a similar strategy for exactly the same data block without any computation during the communication time, i.e. blocking communications. It is important to note that, although the C40 has potentially high data rates, these are often not achieved when using 3L Parallel C due to the routing of communications via the microkernel (which initialises a DMA channel). This, as noted later, incurs a significant setting up delay and becomes particularly prohibitive for small data packets.

To investigate the interprocessor communication speed using the communication links indicated above, 4000 floating-point data elements were used. The communication time was measured as the total time in sending the data from one processor to another and receiving it back. In the case of the C40-T8 and C40-C40 communications, the speed of a single line of communication was also measured using bidirectional data transmission in each of the 4000 iterations. This was achieved by changing the direction of the link at every iteration in sending and receiving the data. Figure 15.6 shows the communication times for the various links, where (1) represents a single bidirectional line of communication and (2) represents a pair of unidirectional lines of communication. Note that, as expected, the C40-C40 parallel pair of lines of communication has performed as the fastest and the C40-T8 serial to parallel single line of communication as the slowest of these communication links. Table 15.1 shows the relative communication times with respect to the C40-C40 parallel pair of lines of communication. It is noted that the C40-C40 parallel pair of lines of communication is ten times faster than the T8-T8 serial pair of lines of communication and nearly 15 times faster than the i860-T8 shared memory pair of lines of communication. The slower performance of the shared memory pair of lines of communication as compared with the T8-T8 serial pair of lines of communication is due to the extra time required in writing and reading data from the shared memory. The C40-T8 serial to parallel pair of lines of communication involves a process of transformation from serial to parallel, while transferring data from T8 to C40, and vice versa, when transferring data from C40 to T8, making the link about 17.56 times slower than the C40-C40 parallel pair of lines of communication. As noted, the C40-C40 parallel single line of communication performs about 94 times slower than the C40-C40 parallel pair of lines of communication. This is due to the utilisation of a single bidirectional line of communication in which, in addition to the sequential nature of the process of sending and receiving data, extra time is required for altering the direction of the link (data flow). Moreover, there is a setting-up delay for each communication performed. This is of the order of ten times that of the actual transmission time. These aspects are also involved in the C40-T8 serial to parallel single line of communication which performs at 115.556 times slower than the C40-C40 parallel pair of lines of communication due to the extra time required for the process of transformation from serial to parallel and vice versa.

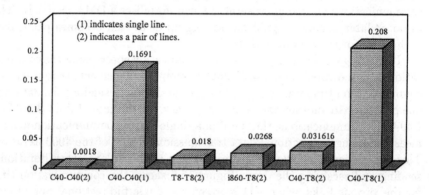

*Figure 15.6    Interprocessor communication times for various links*

To investigate the interprocessor communication issue further in a parallel processing platform the beam simulation algorithm was considered and implemented on the homogeneous architecture of transputers. It was found through numerical simulations that for the purpose of this investigation reasonable accuracy in representing the first few (dominant) modes of vibration is achieved by dividing the beam into 19 segments; thus, the beam was divided into 19 segments. Moreover, a sample period of $\Delta t = 0.3$ ms, which is sufficient to cover all the resonance modes of vibration of the beam, was selected. In this investigation, the total execution times achieved by the architectures, in implementing the simulation algorithm over 20 000 iterations, was considered in comparison with the required real time (calculated as the product of the sampling time $\Delta t$ and total number of iterations).

The algorithm considered, thus, consists in computation of deflection of nineteen equal-length segments. The computation for each segment requires information from two previous and two forward segments. Figure 15.7 shows the required communication and logical distribution process of segments into two, three and six PEs based networks. In practice, for the architecture with three PEs, the load cannot be equally distributed owing to communication overheads. The PE2 used for the computation of segments 7, 8, 9, 10, 11 and 12 requires more communication time than do the other two PEs. It was found through this investigation that the communication time for PE2 is nearly equivalent to the computation time for 2.5 segments. This has led to distributing seven segments (1,2,3,4,5,6,7) to PE1, five segments (8,9,10,11,12) to PE2 and the other seven segments to PE3 to obtain optimum performance. A similar situation occurs when using an implementation with more than three PEs. Here it is essential to note, additionally, that if the number of PEs is more than three, there will be additional communication overheads occurring in parallel to others. As a result of this the communication overhead may be bounded within a certain limit. Another important factor for reducing the communication overhead would be to pass messages in a block of two segments from one processor to another, rather than sending one segment at a time. Using such a strategy, the algorithm was implemented on networks of up to nine T8s.

To explore the computation time and the computation with communication overhead, the performance of the architecture was investigated by breaking the algorithm into fine grains (one segment as one grain). Considering the computation for one segment as a base, the grains were then computed on a single T8 increasing the number of grains from one to nineteen. The theoretical linear speedup of one to nineteen segments computation time and the actual computation time are shown in Figure 15.8. Note that the theoretical computation

Table 15.1  *Interprocessor communication times with various links (LP1) relative to the C40-C40 parallel pair of lines of communication (LP2)*

| Link | T8-T8(2) | C40-C40(1) | C40-T8(2) | C40-T8(1) | i860-T8(2) |
|---|---|---|---|---|---|
| LP1/LP2 | 10.00 | 93.889 | 17.5644 | 115.556 | 14.889 |

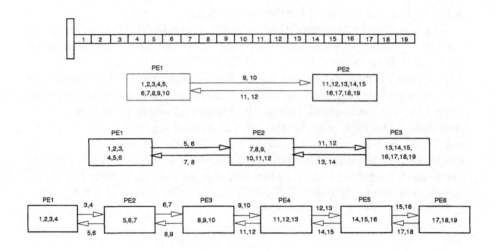

*Figure 15.7    Task allocation and communication of two, three and six proces-
sors*

time is more than the actual computation time. This implies the RISC nature
of the T8 processor. The performance for actual computation time was then
utilised to obtain the actual computation for the multiprocessor system without communication overhead. Figure 15.9 shows the real-time performance
(i.e. computation with communication overhead) and the actual computation
time with one to nine PEs. The difference between the real-time performance
and the actual computation time is the communication overhead, shown in
Figure 15.10. It is noted in Figures 15.9 and 15.10 that, due to communication overheads, the computing performance does not increase linearly with
increasing the number of Pes; the performance remains nearly at a similar
level with a network having more than six PEs. Note that the increase in
communication overhead at the beginning, with less than three PEs, is more
pronounced and remains nearly at the same level with more than five PEs.
This is due to the communication overheads among the PEs which occur in
parallel.

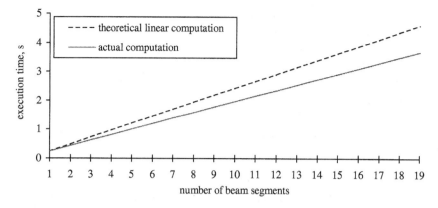

*Figure 15.8     Execution time for flexible beam segments on a single T8*

The speedup and the corresponding efficiency of the execution time for the network of up to nine T8s is shown in Table 15.2. As discussed earlier, an increase in the number of transputers has resulted in a decrease in the execution time. The relation, however, is not always linear. This implies that, with an increase in the number of processors or a transformation of the algorithm from course grain to fine grain, more and more communication is demanded. This is further evidenced in Table 15.2, which shows a nonlinear variation in speedup and efficiency. For instance, the efficiency with two T8s is 77 per cent, whereas with nine transputers the efficiency is only 24 per cent. This implies that the algorithm considered is not suitable to exploit with this hardware, due to communication overheads and run-time memory management problems.

*Table 15.2     Speedup and efficiency of the T8 network for the simulation algorithm*

| No. of T8s | Two | Three | Four | Five | Six | Seven | Eight | Nine |
|---|---|---|---|---|---|---|---|---|
| Speedup | 1.55 | 1.73 | 1.935 | 1.943 | 2.1 | 2.15 | 2.174 | 2.19 |
| Efficiency | 77% | 58% | 48% | 39% | 35% | 31% | 27% | 24% |

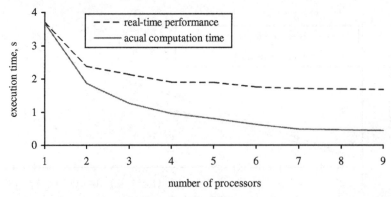

*Figure 15.9     Execution time of the simulation algorithm on the transputer network*

### 15.7.2     Compiler efficiency

In this Section results of investigations of the performance evaluation of several compilers are presented and discussed. The compilers involved are the 3L Parallel C version 2.1, INMOS ANSI C and Occam. All these compilers can be utilised with the computing platforms considered in this investigation. It has previously been reported that, although Occam is a more hardware-oriented and straight-forward programming language for parallel processing, it may not be as suitable as the Parallel C or ANSI C compilers for numerical computations [21]. To obtain a comparative performance evaluation of these compilers, the flexible beam simulation algorithm was coded, for 19 equal-length beam sections, into the three programming languages and run on a T8. Figure 15.11 shows the execution times for implementing the simulation algorithm, over 20000 iterations, using the three compilers. It is noted that the performances with Parallel C and ANSI C are nearly at a similar level and at about 1.5 times faster than those with Occam.

### 15.7.3     Code optimisation

The code optimisation facility of compilers for hardware is another important component affecting the real-time performance of a processor. Almost always optimisation facilities enhance the real-time performance of a processor. The i860 and the C40 have many optimisation features [140, 141, 142]. The TMS320 floating-point DSP optimising C compiler is the TMS320 version of

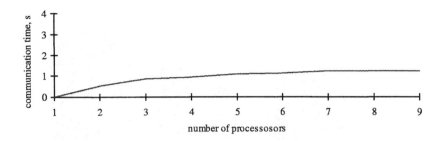

*Figure 15.10   Communication overhead of the transputer network*

the 3L Parallel C compiler [143]; it has many options, constituting three levels of optimisation, which aid the successful optimisation of C source code files on the C40. The Portland Group (PG) C compiler is an optimising compiler for the i860 [140]. It incorporates four levels of optimisation.

To measure the performance attainable from the compiler optimisers, so as to fully utilise the available features, experiments were conducted to compile and run the beam simulation algorithm on the i860 and the C40. To study the effect of the PG optimising compiler, the beam simulation algorithm was compiled and implemented on the i860 with five beam segments and $\Delta t = 0.055$ ms. The execution time of the processor in implementing the algorithm over 20 000 iterations was recorded with each of the four levels of optimisation. Figure 15.12 shows the execution times achieved in implementing the beam simulation algorithm, where level 0 corresponds to no optimisation. The corresponding execution time speedups achieved with each optimisation level in implementing the beam simulation algorithm are shown in Figure 15.13. It is noted in Figures 15.12 and 15.13 that the performance of the processor in implementing the beam simulation algorithm has enhanced. However, the enhancement is not significant beyond the first level. This is because the beam simulation algorithm is in a matrix format and thus does not need as much improvement [140]. With algorithms of irregular nature, however, significant enhancement is achieved in the performance of an architecture with code optimisation.

To study the effect of the optimisers on the performance of the system further, optimisation level 0 (no optimisation) and level 2 were used with the

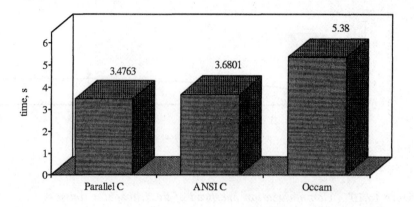

*Figure 15.11     Performance of the compilers in implementing the simulation
algorithm on the T8*

3L optimiser in implementing the algorithm on the C40. Similarly, with the
i860, using the PG compiler, optimisation level 0 and level 4 were utilised.
The beam simulation algorithm was coded for various task sizes, by changing
the number of beam segments. The algorithm was thus implemented on the
i860 and the C40. Figure 15.14 shows the execution times achieved by the
processors in implementing the algorithm over 20 000 iterations. It is noted
that the execution time varies approximately linearly as a function of the task
size. The slight variation in gradient noted in Figure 15.14*b* with the algorithm
implemented on the C40 is more likely due to computational error.

It is noted in Figure 15.14 that the performance of the processors in implementing the algorithm has been enhanced with code optimisation. Figure
15.15 shows the speedups achieved with code optimisation in implementing the
beam simulation algorithm on the processors. It is noted that the execution
time speedups achieved with code optimisation reach a similar level with both
the i860 and the C40. At the lower end of task sizes the speedup with the
i860 is relatively larger and continues to decrease with an increase in the task
size. The speedup with the implementation on the C40, on the other hand,
increases with the task size rapidly at the lower end and slowly beyond 20
beam segments. This suggests that the optimisation for the C40 is performing
better than that for the i860 in this type of application.

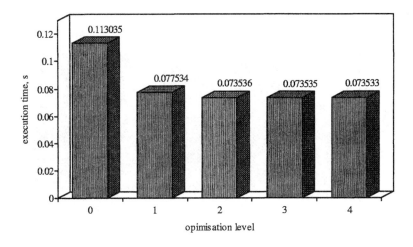

*Figure 15.12    Execution times of the i860 in implementing the beam simulation algorithm with the Portland Group C compiler optimiser*

### 15.7.4    Simulation algorithm

The beam simulation algorithm was implemented as a sequential process in the case of the uniprocessor-based architectures and in the case of the multiprocessor-based architectures it was partitioned for equal load distribution among the processors. The execution times achieved by the architectures, in implementing the algorithm over 20 000 iterations, are shown in Figure 15.16 where the required real-time is calculated as the product of the sample time $\Delta t$ and total number of iterations. The simulation algorithm, as discussed earlier, is mainly of a matrix-based computational type for which the powerful vector-processing resources of the i860 are exploited and utilised to achieve the shortest execution time among the uniprocessor architectures, and with the i860+T8 among the parallel architectures. The C40 and the T8 do not have such vector-processing resources, making them 6.053 and 9.864 times slower than the i860, respectively. This implies that the C40 and the T8 are not performing well in situations where the algorithm is of matrix type, and extensive run-time memory management is involved. The speedup achieved with two C40s as compared to a single C40 is only 1.42. Although the program has reduced to a half for a single C40, due to the nature of the algorithm, the C40+C40 has

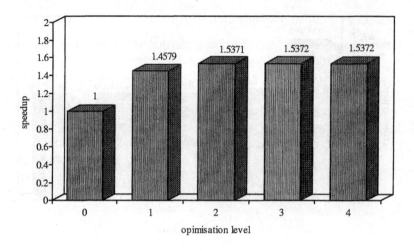

Figure 15.13    *Speedup with the Portland Group C compiler optimiser in implementing the beam simulation algorithm on the i860*

not achieved a performance better than that of a single i860 vector processor. The speedup achieved with two T8s as compared to a single T8 is 1.23. This speedup is similar to that achieved with two C40s in comparison to a single C40, although a serial communication link is utilised in the case of the T8s. Thus, the main factor influencing the C40+T8 to perform slightly slower than the C40+C40 is the serial-to-parallel communication link utilised, and the slower T8 processor incorporated in this architecture.

### 15.7.5 The identification algorithm

As discussed earlier, the identification algorithm is composed of two components of similar nature and length, while estimating parameters of $Q_0$ and $Q_1$, and a process of controller design calculation. In case of the uniprocessor architectures the algorithm was implemented sequentially. In case of the parallel architectures, on the other hand, the algorithm was partitioned so that the load at estimating parameters of $Q_0$ and $Q_1$ was equally distributed among the two PEs, with one of the PEs further carrying out the calculation of the parameters of the controller. The execution times of the computing platforms, in implementing the identification algorithm over 1000 iterations using second-order models for $Q_0$ and $Q_1$, are shown in Figure 15.17. It is noted that the C40 and the C40+C40 architectures have performed the fastest among the uniprocessor and parallel architectures, respectively. However, the speedup achieved with two C40s as compared to a single C40 is only 1.35. This could be due to the nature of the identification algorithm for which the pipeline nature of the C40 DSP device is not exploited much, even after a reduction of the program into two segments. The algorithm does incorporate some matrix manipulation. However, as a result of the irregular nature of the algorithm, the i860 has performed slower than the C40. This is further noted in the performance of the i860+T8 as compared to that of the C40+T8. In contrast, the T8 has performed significantly well; the speedup achieved with two T8s as compared to a single T8 is 2.22. This super-linear speedup results from a significant reduction in the data handling, for which the available internal memory of the T8 is sufficient, thus reducing the run-time memory-management load.

### 15.7.6 The control algorithm

The control algorithm, as outlined earlier, is essentially composed of the simulation algorithm and the realisation of a recursive filter structure. Thus, in the case of the uniprocessor-based architectures, the algorithm was implemented as a sequential process, whereas in the case of the two processor-based architectures, the algorithm was partitioned for the beam simulation part in a similar manner to the case of the simulation algorithm, with one of the PEs additionally carrying out calculation of the control signal.

The execution times of the computing platforms, in implementing the control algorithm over 20 000 iterations, are shown in Figure 15.18. Note that the beam simulation forms a large proportion of the control algorithm. This makes the algorithm mainly an RI type. Thus, as in the case of the simulation algorithm, the powerful vector-processing resources of the i860 are utilised to achieve the shortest execution time among the uniprocessor architectures and with the i860+T8 among the parallel architectures. The execution times achieved with the architectures are similar to those of the beam simulation algorithm. The speedup achieved with two C40s as compared to a single C40 is 1.4. Similarly, the speedup achieved with two T8s as compared to a single T8 is 1.23. These are consistent with the performance of these architectures in the case of the beam simulation algorithm. Therefore, a similar interpretation and explanation can be made with regard to the speedups and performances of the architectures in the case of the beam control algorithm.

### 15.7.7    The combined simulation, identification and control algorithm

It is noted in the above investigations that the i860 achieved best performance in implementing the beam simulation and the beam control algorithms. In contrast, the C40 achieved best performance in implementing the beam identification algorithm. Thus, for best solution in implementing the beam AVC (combined simulation, identification and control) algorithm, an integrated i860+T8+C40 system was considered. The T8 in the architecture was used for communication between the i860 and the C40. Thus, the simulation and control algorithms were allocated to the i860 and the identification algorithm was allocated to the C40 processor. The execution time achieved with the i860+T8+C40 architecture is shown in Figure 15.19 along with those of the i860, C40 and T8 uniprocessors. It is noted that the i860+T8+C40 architecture has performed 2.64 and 7.23 times faster than a single i860 and a single C40, respectively. As compared to a single T8, it performed 12 times faster. This demonstrates that matching the computing requirements of the algorithm with the computing resources of the architecture leads to significant performance enhancement.

### 15.7.8    Comparative performance of the architectures

In this Section a summary of the performances of the computing platforms in implementing the algorithms is presented. The execution times achieved with the uniprocessor and parallel architectures, in implementing the algorithms, are shown in Table 15.3. The algorithms have been listed according to their degree of regularity relative to one another; the simulation at the top is of the most regular and the identification algorithm at the bottom is the most irregular among the algorithms. It is noted that, among the uniprocessor architectures, the i860 performs as the fastest in implementing regular and matrix-based algorithms. In contrast, the C40 performs the fastest of the uniprocessors in implementing algorithms of irregular nature. In a similar manner, the performance of the T8 is enhanced by an increasing degree of irregularity in an algorithm.

It is noted in Table 15.3 that among the parallel architectures the suitability of the i860 for regular and matrix-based algorithms is well reflected in the shortest execution times achieved with the i860+T8 in implementing the beam simulation and control algorithms. In contrast, the suitability of the C40 is reflected in achieving the shortest execution times with C40+C40 in implementing the identification algorithm.

Table 15.3 also shows the execution times of the architectures in implementing the combined flexible beam simulation, identification and control algorithms. This shows that suitable matching of the computing requirements of the algorithm with the computing capabilities of the architecture enhances the performance of the architecture significantly.

*Table 15.3    Performance of the hardware architectures (s)*

| Algorithm | i860 | C40 | T8 | i860+T8 | C40+ C40 | C40+ T8 | T8+T8 | i860+T8 +C40 |
|---|---|---|---|---|---|---|---|---|
| Simulation | 0.38 | 2.3 | 3.68 | 0.99 | 1.618 | 1.7669 | 3.008 | - |
| Control | 0.41 | 2.68 | 3.695 | 1.04 | 1.92 | 1.963 | 3.009 | - |
| Identification | 0.35 | 0.179 | 0.674 | 0.6118 | 0.2644 | 0.6114 | 0.6066 | - |
| AVC | 1.11 | 3.04 | 5.04 | - | - | - | - | 0.42 |

## 15.8    Conclusions

This chapter has explored the real-time implementation of DSP and control algorithms within an AVC framework on a number of parallel and sequential hardware architectures. Special features, such as vector processing resources in an i860, are exploited to give even better performance in applications involving matrix manipulations. Utilisation of these devices in a parallel architecture leads to penalties due to excessive communication overheads. The C40 DSP device, for example, uses a pair of (fast) parallel lines of communication and yet is not performing impressively within a parallel architecture in implementing the algorithms used. Transputers, on the other hand, are found to be performing efficiently in carrying out tasks of an irregular nature. Moreover, the communication overhead of transputers in a parallel network is smaller than that due to the DSP devices. It has been demonstrated that there is, generally, a mismatch between the hardware requirements of an algorithm and the hardware resources of the architecture leading to a disparity in their relative performance. Therefore, to fully exploit the architectures a close match needs to be forged between the algorithm and the underlying hardware, with due consideration of the suitable programming language for the application, and issues such as algorithmic regularity and granularity. It has been demonstrated through the investigations carried out that, generally, there is no one PP architecture for the solution of the best real-time performance in terms of computation, communication and cost. Computational performance of processors varies with granularity of hardware and granularity and regularity of an algorithm. Thus, in designing a parallel architecture, identification of hardware heterogeneity to allow exploitation of capabilities of PEs and algorithm heterogeneity to provide suitable distribution of tasks among the PEs accordingly are required.

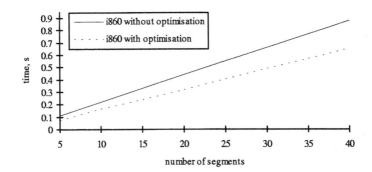

(a) With the i860.

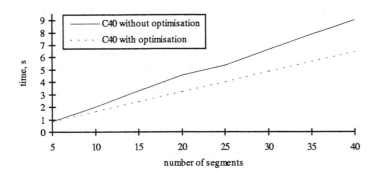

(b) With the C40.

Figure 15.14    *Execution times of the processors in implementing the beam simulation algorithm*

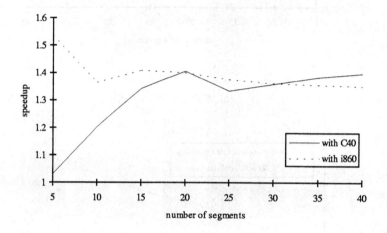

Figure 15.15    *Optimisation speedup with the processors in implementing the beam simulation algorithm*

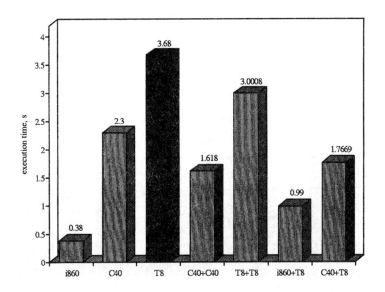

Figure 15.16    *Execution times of the computing platforms in implementing the flexible beam simulation algorithm*

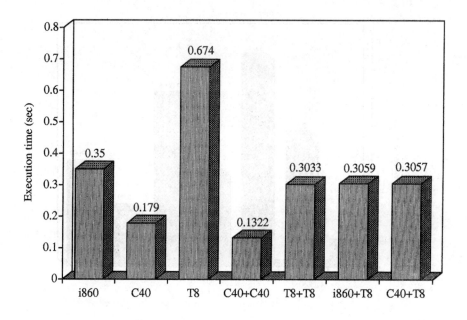

Figure 15.17    *Execution times of the computing platforms in implementing the flexible beam identification algorithm*

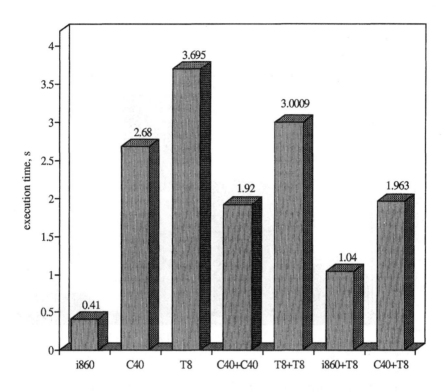

Figure 15.18    *Execution times of the computing platforms in implementing the flexible beam control algorithm*

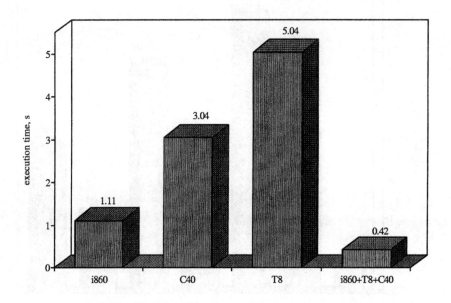

Figure 15.19    *Execution times of the computing platforms in implementing the flexible beam AVC algorithm*

# Bibliography

[1] K. J. Å ström and B. Wittenmark. *Adaptive Control.* Addison Wesley, Reading, Massachusetts, 1989, 1995.

[2] M. Jessel adn G. A. Mangiante. Active sound absorbers in an air duct. *Journal of Sound and Vibration*, 23(3):383–390, 1972.

[3] G. S. Aglietti. *Active Control of Microvibrations for Equipment Loaded Spacecraft Panel.* PhD thesis, ISVR, University of Southampton, UK, 1999.

[4] G. S. Aglietti, S. B. Gabriel, R. S. Langley, and E. Rogers. A modeling technique for active control design studies with application to spacecraft microvibration. *The Journal of The Acoustical Society of America*, 102(4):2158–2166, 1997.

[5] G. S. Aglietti, R. S. Langley, E. Rogers, and S. B. Gabriel. An efficient model of an equipment loaded panel for active control design studies. *Journal of The Acoustical Society of America*, (Accepted for publication), 2000.

[6] G. S. Aglietti, J. Stoustrup, E. Rogers, R. S. Langley, and S. B. Gabriel. LTR control of microvibration. *Proc: IEEE International Conference on Control Applications*, pages 624–628, 1998.

[7] D. P. Agrawal, V. K. Janakiram, and G. C. Pathak. Evaluating the performance of multicomputer configuration. *IEEE Computer*, 19(5):23–37, 1986.

[8] T. E. Alberts, L. J. Love, E. Bayo, and H. Moulin. Experiments with endpoint control of a flexible link using the inverse dynamics approach and

passive damping. *Proceedings of American Control Conference*, (1):350–355, 1990.

[9] J. F. Allan. The stabilization of ships by activated fins. *Trans. Inst. Naval Architects*, 87:123–159, 1945.

[10] R. B. Allen. Several studies on natural language and backpropagation. *Proceedings of IEEE First International Conference on Neural Networks, San Diego*, 2:335–341, 1987.

[11] A. J. Anderson. A performance evaluation of microprocessors, dsps and the transputer for recursive parameter estimation. *Microprocessors and Microsystems*, 15:131–136, 1991.

[12] B. D. O. Anderson and R. E. Skelton. The generation of all $q$-Markov covers. *IEEE Trans. Circuits and Systems*, 35:375–384, 1988.

[13] O. L. Angevine. Active systems for attenuation of noise. *International J. of Active Control*, 1:302–310, 1995.

[14] Anon. Perfectionnements apportés aux paliers pour corps tournants, notamment pour ensembles devant tourner à l'interieur d'une enceinte étanche. *French Patent FR 1 186 527.*, 1959. Filed: Nov. 18, 1957. Patented: Feb. 23.

[15] Anon. Intelligent Solutions. Engineering materials ease the burden on the environment, April, 1995. In: *Chemistry with Chlorine*, Bayer AG, Public Relations Department, Leverkusen, Germany.

[16] D. M. Aspinwall. Acceleration profiles for minimising residual response. *ASME Journal of Dynamic Systems, Measurement and Control*, 102(1):3–6, 1980.

[17] Rafaely B. *Feedback control of sound*. PhD thesis, ISVR, University of Southampton, 1997.

[18] Rafaely B. and Elliott S.J. Adaptive internal model controller stability analysis. *Proc. INTERNOISE96*, pages 1021–1024, 1996.

[19] Rafaely B. and Elliott S.J. $H_2/H_\infty$ active control of sound in a headrest: design and implementation. *IEEE Trans. on Control System Technology*, 1999.

[20] W. R. Babcock and A. G. Cattaneo. Method and Apparatus for Generating an Acoustic Output from an Ionized Gas Stream, 1971. *U.S. Patent US 3,565,209.* Filed: Feb. 28, 1968. Patented: Feb. 23.

[21] G. Bader and E. Gehrke. On the performance of transputer networks for solving linear systems of equation. *Parallel Computing,* 17:1397–1407, 1991.

[22] K.H. Baek and S.J. Elliott. Genetic algorithms for choosing source locations in active control systems. *Proceedings of the Institute of Acoustic,* 15(3):437–445, 1993.

[23] K.H. Baek and S.J. Elliott. Natural algorithms for choosing source location in active control systems. *Journal of Sound and Vibration,* 186:245–267, 1995.

[24] M. J. Baxter, M. O. Tokhi, and P. J. Fleming. Parallelising algorithms to exploit heterogeneous architectures for real-time control systems. *Proceedings of IEE Control-94 Conference, Coventry, 21-24 March 1994, 2, 1266-1271,* 1994.

[25] E. Bayo. Computed torque for the position control of open-loop flexible robots. *Proceedings of IEEE International Conference on Robotics and Automation, Philadelphia, 25-29 April, 316-321,* 1988.

[26] J. Bell. Adaptive optics clears the view for industry. *Opto & Laser Europe,* 45:17–20, 1997.

[27] S. A. Billings and W. S. F. Voon. Correlation based model validity tests for non-linear systems. *International Journal of Control,* 15(6):601–615, 1986.

[28] R. R. Bitmead. Iterative identification and control – a survey. *Proc. IFAC World Congress (Sidney, Australia),* 1993.

[29] E. Bjarnason. Analysis of the filtered-x lms algorithm. *IEEE Trans. Speech and Audio Processing,* 3(3):504–514, 1995.

[30] J. C. Blcazy. Electronic sound absorber. *Journal of Audio Engineering Society,* 10(2):135–139, 1962.

[31] V. Blondel, M. Gevers, and R. R. Bitmead. When is a model good for control design? *CDC'97 San Diego,* 1997.

392   *Bibliography*

[32] W. Böhm. Untersuchungen zur breitbandig wirksamen aktiven Kompensation instationär angeregter Schallfelder, 1992. Ph. D. Thesis, Göttingen.

[33] A. V. Boiko and V. V. Kozlov. Strategy of the flow mems control at laminar–turbulent transition in a boundary layer. In: G.E.A. Meier, P.R. Viswanath (eds.): IUTAM Symposium on Mechanics of Passive and Active Flow Control:203–208, 1999.

[34] W. J. Book. Recursive lagrangian dynamics of flexible manipulator arms. *International Journal of Robotics Research*, 3(3):87–101, 1984.

[35] S. et al. Boyd. A new cad method and associated architectures for linear controllers. *IEEE Trans. Automatic Control*, AC33:268–283, 1988.

[36] M. J. Brennan, M. J. Day, and R. Pinnington S. J. Elliott. Piezoelectric actuators and sensors. *Proc: IUTAM Symposium on The Active Control of Vibration (Eds C. R. Burrows and P. S. Keogh)*, pages 263–274, 1994.

[37] M. Bronzel. Aktive Schallfeldbeeinflussung nicht-stationärer Schallfelder mit adaptiven Digitalfiltern, 1993. Ph. D. Thesis, Göttingen.

[38] D. S. Broomhead and D. Lowe. Multivariable functional interpolation and adaptive networks. *Complex Systems*, 2:321–355, 1988.

[39] A. Brown. Dsp chip with no parallel? *Electronics World + Wireless World*, pages 878–879, 1991.

[40] A.J. Bullmore, P.A. Nelson, and S.J. Elliott. Active minimisation of acoustic potential energy in harmonically excited cylindrical enclosed sound fields. *AIAA Paper*, 86-1958, 1986.

[41] A. V. Bykhovskii. Sposob polawlenija shuma w sluchowom organe (Technique for noise suppression in the ear), 1960. Patent UdSSR SU 133 631. Filed: Aug. 24, 1949. Published: Patent Bulletin No. 22.

[42] R. H. Cannon and E. Schmitz. Initial experiments on the end-point control of a flexible one-link robot. *International Journal of Robotics Research*, 3(3):62–75, 1984.

[43] A. B. Carlson. *Communication systems*, volume 2nd edn. McGraw-Hill, 1975.

[44] C. Carme et al. Haut-parleur linéaire, 1999. French Patent Application FR 2 766 650 A1. Filed: July 23, 1997. Published: Jan. 29.

[45] Ross C.F. and Purver M.R.J. Active cabin noise control. *Proc. ACTIVE97*, pages xxxix–xlvi, 1997.

[46] J. C. H. Chang and T. T. Soong. The use of aerodynamic appendages for tall building control, 1980. In: H. H. E. Leipholz (ed.): Structural Control, North-Holland Publ. Co.

[47] G. B. B. Chaplin. Anti-sound — The Essex breakthrough, 1983. Chartered Mechanical Engineer (CME) 30.

[48] G. B. B. Chaplin, A. Jones, and O. Jones. Method and apparatus for low frequency active attenuation. *US Patent No. 4527282*, 1985.

[49] G. B. B. Chaplin et al. The Cancelling of Vibrations Transmitted through a Fluid in a Containing Vessel, 1979. International Patent Application WO 81/01479. Published: May 28, 1981. Priority (GB): Nov. 10.

[50] B. D. Charles et al. Blade Vortex Interaction Noise Reduction Techniques for a Rotorcraft, 1996. U.S. Patent US 5,588,800. Filed: May 31, 1994. Patented: Dec. 31.

[51] Y. P. Cheng. *Sliding Mode Controller Design for flexible multi-link manipulators*. PhD thesis, University of Texas at Arlington. Ann Arbor, Michigan, USA, 1989.

[52] P. C. Ching and S. W. Wu. Real-time digital signal processing system using a parallel processing architecture. *Microprocessors and Microsystems*, 13(10):653–658, 1989.

[53] R. L. Clark and C. R. Fuller. Active Structural Acoustic Control with Adaptive Structures Including Wavenumber Considerations, April 1991. In: C. A. Rogers and C. R. Fuller (eds.): Proc. of the 1st Conference on Recent Advances in Active Control of Sound and Vibration. Blacksburg, VA.

[54] R. L. Clark and G. P. Gibbs. A novel approach to feedforward higher-harmonic control, 1994. J. Acoust. Soc. Am. 96.

[55] H. Coanda. Procédé de protection contre les bruits, 1931. Brevet d'Invention (French Patent Application) FR 722.274. Filed: Oct. 21, 1930. Patented: Dec. 29.

[56] H. Coanda. Procédé et dispositif de protection contre les bruits, 1934. Brevet d'Invention (French Patent Application) FR 762.121. Filed: Dec. 31, 1932. Patented: Jan. 18.

[57] G. P. Collins. Making Stars to See Stars: DOD Adaptive Optics Work is Declassified, 1992. Physics Today 45.

[58] W. B. Conover. Fighting noise with noise. *Noise Control*, 92:78–82, 1956.

[59] W. B. Conover. Noise reducing system for transformers. *US Patent No. 2776020*, 1957.

[60] W. B. Conover and W. F. M. Gray. Noise Reducing System for Transformers, 1957. U.S. Patent US 2,776,020. Filed: Feb. 9, 1955. Patented: Jan. 1.

[61] T. P. Crummey, D. J. Jones, P. J. Fleming, and W. P. Marnane. A hardware scheduler for parallel processing in control applications. *Proceedings of IEE Control-94 Conference, Coventry, 21-24 March 1994, 2, 1098-1103*, 1994.

[62] M. A. Daniels. Loudspeaker Phase Distortion Control Using Velocity Feedback, 1998. U.S. Patent US 5,771,300. Filed: Sept. 25, 1996. Patented: June 23.

[63] J. J. D'Azzo and C. H. Houpis. *Feedback control system analysis and synthesis*, volume 2nd edn. McGraw-Hill, 1966.

[64] P. DeFonseca, P. Sas, and H. Brussel. Optimisation methods for choosing sensor and actuator locations in an actively controlled double-panel partition. In *Proceedings of the SPIE*, volume 3041, pages 124–135, 1997.

[65] R. Dellman, I. Glicksber, and O. Gross. On the bang-bang control problem. *Quarterly of Applied Mechanics*, 14(1):11–18, 1956.

[66] S. Deus. Aktive Schalldämpfung im Ansaugkanal von Gebläsen, 1988. In: Fortschritte der Akustik – DAGA 98 , DEGA e.V., Oldenburg, Germany.

[67] P. J. Dines. Active control of flame noise, 1984. Ph. D. Thesis, Cambridge, England.

[68] P. Doel. A revolution in telescope design. *Opto & Laser Europe*, 69, 1999.

[69] J. C. Doyle, B. A. Francis, and A. R. Tannenbaum. *Feedback Control Theory.* Mcmillan, New York, 1992.

[70] Morgan D.R. An analysis of multiple cancellation loops with a filter in the auxiliary path. *IEEE Trans. Acoustics and Signal Processing*, ASSP-28:454–467, 1980.

[71] G. P. Eatwell. The Use of the Silentseat in Aircraft Cabins, 1991. In: C.A. Rogers and C.R. Fuller (eds.): Proceedings of the 1st Conference on Recent Advances in Active Control of Sound and Vibration. Blacksburg, VA, USA, April 15–17.

[72] G. P. Eatwell et al. Piezo Speaker for Improved Passenger Cabin Audio Systems, 1995. International Patent Application WO 97/17818 A1. Published: May 15, 1997. Priority (US): Sept. 25.

[73] D. L. Edberg and A. H. von Flotow. Progress Toward a Flight Demonstration of Microgravity Isolation of Transient Events, 1992. World Space Congress, 43rd Congress of the International Astronautical Federation, Paper IAF-92-0781.

[74] D. Egelhof. Einrichtung zur Schwingungsdämpfung, 1987. German Patent Application DE 35 41 201. Filed: Nov. 21, 1985. Published: May 27.

[75] Kh. Eghtesadi and H. G. Leventhall. Comparison of active attenuators of noise in ducts. *Acoustics Letters*, 4(10):204–209, 1981.

[76] Kh. Eghtesadi et al. Industrial Applications of Active Noise Control, 1993. NOISE-93. International Noise and Vibration Control Conference, St. Petersburg, Russia, May 31 – June 3, 1993. Proceedings: Vol. 2.

[77] S. J. Elliot and C. C. Boucher. Interaction between multiple feedforward active control systems. *IEEE Trans. Speech and Audio Processing*, 2:521–530, 1994.

[78] S. J. Elliot, I. M. Stothers, and P. A. Nelson. A multiple error lms algorithm and its application to the active control of sound and vibration. *IEEE Trans. Acoust. Speech, and Signal Processing*, ASSP-35:1423–1434, 1987.

[79] S. J. Elliott and P. A. Nelson. Multichannel active sound control using adaptive filtering, 1988. ICASSP '88, Paper A3.4, Proceedings.

[80] S. J. Elliott and P. A. Nelson. *Active Noise Control.* Academic Press, New York, 1993.

[81] S. J. Elliott and B. Rafaely. Frequency-domain adaptation of feedforward and feedback controllers. *Proc. Int. Symposium on Active Control and Vibration, ACTIVE'97, Budapest*, pages 771–788, 1997.

[82] S. J. Elliott, I. M. Stothers, and P. A. Nelson. A multiple error lms algorithm and its application to the active control of sound and vibration. *IEEE Transactions on Acoustics, Speech, and Signal Processing*, 35(10):1423–1434, 1987.

[83] Stothers I.M. Elliott S.J., Nelson P.A. and Boucher C.C. In-flight experiments on the active control of propeller-induced cabin noise. *Journal of Sound and Vibration*, pages 219–238, 1990.

[84] L. J. Eriksson. Active Attenuation System with On-line Modeling of Speaker, Error Path and Feedback Path, 1987. U.S. Patent US 4,677,676. Filed: Feb. 11, 1986. Patented: June 30.

[85] L. J. Eriksson. Development of the filtered-u algorithm for active noise control. *J. Acoust. Soc. America*, 89:257–265, 1991.

[86] L. J. Eriksson, M. C. Allie, C. D. Bremigan, and R. A. Greiner. Active noise control using adaptive digital signal processing. *Proceedings of the IEEE International Conference on Acoustics, Speech, and Signal Processing, New York, 2594-2597*, 1988.

[87] L. J. Eriksson, M. C. Allie, and R. A. Greiner. The selection and application of an iir adaptive filter for use in active sound attenuation. *IEEE Transactions on Acoustics, Speech, and Signal Processing*, 35(4):433–437, 1987.

[88] L. J. Eriksson, M. C. Allie, and R. H. Hoops. Active Acoustic Attenuation System for Higher Order Mode Non-Uniform Sound Field in a

Duct, 1989. U.S. Patent US 4,815,139. Filed: March 16, 1988. Patented: March 21.

[89] S. M. Kuo *et al. Active Noise Control Systems -Algorithms and DSP Implementation.* Wiley-Interscience, 1996.

[90] F. Evert, D. Ronneberger, and F.-R. Grosche. Application of linear and nonlinear adaptive filters for the compensation of disturbances in the laminar boundary layer. *Z. angew. Math. und Mech. (ZAMM), to be published in 2000,* 2000.

[91] P. L. Feintuch. An Adaptive Recursive LMS Filter, 1976. Proc. IEEE 64, Comments: 65 (1977) 1399 and 1402.

[92] P. L. Feintuch, N. J. Bershad, and A. K. Lo. A frequency domain model for filtered lms algorithm - stability analysis, design, and elimination of the training mode. *IEEE Trans. Acoust. Speech, and Signal Processing,* ASSP-41:1518–1531, 1993.

[93] W. B. Ferren and Bernard. Active control of simulated road noise. *SAE Proc. Noise and Vibration Conference,* pages 69–82, 1991.

[94] J. E. FfowcsWilliams and W. Möhring. *Active Control of Kelvin-Helmholtz Waves,* volume In: G.E.A. Meier, P.R. Viswanath (eds.): IU-TAM Symposium on Mechanics of Passive and Active Flow Control, pages 343–348. Kluwer Academic Publishers, 1999.

[95] Powell J.D. Franklin G.F. and Emani-Naeini A. *Feedback control of dynamic systems.* Addison-Wesley, New Jersey, 1994.

[96] R. Freymann. Dynamic Interactions Between Active Control Systems and a Flexible Aircraft Structure, 1986. Proc. 27th AIAA/ASME/SAE SDM Conference, AIAA Paper 86-0960.

[97] R. Freymann. Von der Pegelakustik zum Sounddesign, 1996. In: Fortschritte der Akustik – DAGA '96, DEGA e.V., Oldenburg, Germany.

[98] R. Q. Fugate et al. Measurement of atmospheric wavefront distortion using scattered light from a laser guide-star, 1991. Nature 353 (Sept. 12.

[99] C. R. Fuller. Analytical model for investigation of interior noise characteristics in aircraft with multiple propellers including synchrophasing, 1986. J. Sound Vib. 109.

[100] C. R. Fuller and C. A. Rogers adn H. H. Robertshaw. Control of sound radiation with active/adaptive structures. *Journal of Sound and Vibration*, 157(1):19–39, 1992.

[101] C. R. Fuller, S. J. Elliott, and P. A. Nelson. Active Control of Vibration, 1995. Academic Press, London etc.

[102] C. R. Fuller, C. H. Hansen, and S. D. Snyder. Active Control of Sound Radiation from a Vibrating Rectangular Panel by Sound Sources and Vibration Inputs: An Experimental Comparison, 1991. J. Sound Vib. 145.

[103] C. R. Fuller and A. H. von Flotow. Active control of sound and vibration. *IEEE Control Systems Magazine*, 15(6):9–19, 1995.

[104] S. Ganesan. A dual-dsp microprocessor system for real-time digital correlation. *Microprocessors and Microsystems*, 15(7):379–384, 1991.

[105] Garcia-Bonito, S.J.Elliot, and C.C.Boucher. Novel secondary source for a local active noise contorl system. *ACTIVE 97*, pages 405–418, 1997.

[106] F. Garcia-Nocetti and P. J. Fleming. *Parallel processing in digital control*. Springer-Verlag, 1992.

[107] L. Gaul. Aktive Beeinflussung von Fügestellen in mechanischen Konstruktionselementen und Strukturen, 1997. German Patent Application DE 197 02 518 A1. Filed: Jan. 24, Published: June 12.

[108] T. H. Gawain and R. E. Ball. Improved finite difference formulas for boundary value problems. *International Journal for Numerical Methods in Engineering*, 12:1151–1160, 1978.

[109] S¿ S. Ge, T. H. Lee, and G. Zhu. Improving regulation of a single-link flexible manipulator with strain feedback. *IEEE Transaction on Robotics and Automation*, 14(1):179–185, 1998.

[110] D. H. Gilbert. Echo Cancellation System, 1989. U.S. Patent US 4,875,372. Filed: May 3, 1988. Patented: Oct. 24.

[111] D.E. Goldberg. *Genetic algorithms in search, optimisation and machine learning*. Addison-Wesley Publishing Company, 1989.

[112] D.E. Goldberg and J. Richardson. Genetic algorithms with sharing for multi-modal function optimisation. In *Proceedings of the Second International Conference on Genetic Algorithms*, pages 44–49, 1987.

[113] W. Gossman and G. P. Eatwell. Active High Transmission Loss Panel, 1994. U.S. Patent US 5,315,661. Filed: Aug. 12, 1992. Patented: May 24.

[114] M. Green and D. J. N. Limebeer. *Linear Robust Control*. Prentice Hall, Englewood Cliffs, 1995.

[115] G. Gu and P. P. Khargonekar. A class of algorithms for identification in $h_\infty$. *Automatica*, 28:299–312, 1992.

[116] D. Guicking. Active noise and vibration control. *Annotated Reference Bibliography with Keyword and Author Index. 3rd Edition (Feb. 1988), 1708 refs. 1st Suppl.(Aug.1991)*, page 2193, 1995.

[117] D. Guicking. Active control of vibration and sound – an overview of the patent literature. *ISMA 21, International Conference on Noise and Vibration Engineering, Leuven, Belgium, Sept. 18-20*, pages 199–220, 1996.

[118] D. Guicking and H. Freienstein. Broadband Active Sound Absorption in Ducts with Thinned Loudspeaker Arrays, 1995. In: ACTIVE 95, Newport Beach, CA, USA, July 6–8.

[119] D. Guicking and K. Karcher. Active Impedance Control for One-Dimensional Sound, 1984. ASME J. Vib. Acoust. Stress Rel. in Design 106.

[120] Hareo Hamada, Kazuaki Nimura, and Seigo Uto. Active noise control using adaptive prediction -implementation by adaptive filter. *Proc. Acoust. Soc. Japan Symposium*, pages 531–532, 1992.

[121] C. H. Hansen and S. D. Snyder. Design considerations for active noise control systems implementing the multiple input, multiple output lms algorithm. *Journal of Sound and Vibration*, 159(1):157–174, 1992.

[122] C.H. Hansen and S.D. Snyder. *Active control of noise and vibration*. E&FN Spon, London, 1997.

[123] J. Hansen. Eine anwendungsreife lösung für die aktive minderung von abgasgeräuschen industrieller dieselmotoren und drehkolbenpumpen. *VDI Bericht Nr. 1491*, page 199, 1999.

[124] E. Hänsler. The hands-free telephone problem - an annotated bibliography. *Signal Processing*, 27:259–271, 1992.

[125] T. R. Harley.    Active noise control stethoscope.    *U.S. Patent US 5,610,987. First filed: Aug. 16, 1993; Patented: March 11*, 1997.

[126] M. Harper. Active control of surge in a gas turbine engine. *In: C. A. Rogers and C. R. Fuller (eds.): Proc. of the 1st Conference on Recent Advances in Active Control of Sound and Vibration. Blacksburg, VA, April*, pages 133–149, 1991.

[127] D. Hassler. Apparatus and method for suppressing reflections at an ultrasound transducer. *U.S. Patent US 5,245,586. Patented: Sept. 14, 1993. Priority (EP): Nov. 15*, 1991.

[128] C. M. Heatwole and R. J. Bernhard. The selection of active noise control reference transducers based on the convergence speed of the lms algorithm. *Proc. INTER-NOISE 94*, pages 1377–1382, 1994.

[129] W. von Heesen. Practical experience with an active noise control installation in the exhaust gas line of a co-generator plant engine. *ACUSTICA/acta acustica 82*, pages Suppl. 1, p. S 195, 1996.

[130] H. Heller, W. Splettstoesser, and K.-J. Schultz. Helicopter rotor noise research in aeroacoustic wind tunnels – state of the art and perspectives. *NOISE-93. International Noise and Vibration Control Conference, St. Petersburg, May 31 – June 3*, pages p. 39–60, 1993.

[131] A. J. Helmicki, C. A. Jacobson, and C. N. Nett. Control oriented system identification: a worst-case\deterministic approach in $h_\infty$'. *IEEE Trans. Automatic Control*, AC-36:1163–1178, 1991.

[132] N. Hesselman. Investigation of noise reduction on a 100 kva transformer tank by means of active methods. *Applied Acoustics*, 11(1):27–34, 1978.

[133] S. Hildebrand and Z. Q. Hu. Global quieting system for stationary induction apparatus. *U.S. Patent US 5,617,479. First filed: Sept. 3 1993*, Patented: April 1, 1997.

[134] Chih-Ming Ho et al. Active flow control by micro systems. In: G.E.A. Meier, P.R. Viswanath (eds.): IUTAM Symposium on Mechanics of Passive and Active Flow Control:195–202, 1999.

[135] Y. Hori et al. Vibration/noise reduction device for electrical apparatus. *U.S. Patent US 4,435,751. Patented: March 6, 1984. Priority (JP): July 3*, 1980.

[136] H. Hort. Beschreibung und versuchsergebnisse ausgeführter schiffsstabilisierungsanlagen. *Jahrbuch der Schiffbautechnischen Gesellschaft 35*, pages 292–312, 1934.

[137] C. G. Hutchens and S. A. Morris. Method for acoustic reverberation removal. *U.S. Patent US 4,796,237. Filed: Jan. 28, 1987. Patented: Jan. 3*, 1989.

[138] K. Hwang. *Advanced computer architecture - parallelism scalability programmability*. McGraw-Hill, California, 1993.

[139] Y. Ichikawa and T. Sawa. Neural network application for direct feedback controllers. *IEEE - Transactions on Neural Networks*, 3:224–231, 1992.

[140] Portland Group Inc. *PG tools user manual*. Portland Group Inc., USA, 1991.

[141] Texas Instruments. *TMS320C40 user's guide*. Texas Instruments, USA, 1991a.

[142] Texas Instruments. *TMS320C4x user's guide*. Texas Instruments, USA, 1991b.

[143] Texas Instruments. *TMS320 floating-point DSP optimising C compiler user's guide*. Texas Instruments, USA, 1991c.

[144] G. W. Irwin and P. J. Fleming (eds.). *Transputers for real-time control*. Research Studies Press, 1992.

[145] Y. Ishida and N. Adachi. Active noise control by an immune algorithm: adaptation inimmune system as evolution. In *Proceedings of IEEE International Conference on Evolutionary Computing*, volume xxii+891, pages 150–153, 1996.

[146] M. Izumi. Control of structural vibration — past, present and future. *International Symposium on Active Control of Sound and Vibration, Tokio, April*, pages 195–200, 1991.

[147] L. B. Jackson. *Digital filters and Signal Processing*. Kluwer Academic Publishers, London, 1989.

[148] H. B. Jamaluddin. *Nonlinear system identification using neural networks*. PhD thesis, Department of Automatic Control and Systems Engineering, The University of Sheffield, UK, 1991.

[149] M. J. M. Jessel. La question des absorbeurs actifs. *Revue d'Acoustique* 5, pages 37–42, 1972.

[150] F. Jiang, N. Ojiro, H. Ohmori, and A. Sano. Adaptive active noise control schemes in time domain and transform domains. *Proc. 34th IEEE Conf. Decision and Control*, New Orleans, USA:2165–2172, 1995.

[151] F. Jiang, H. Tsuji, H. Ohmori, and A. Sano. Adaptation for active noise control. *IEEE Control Systems Magazine*, 17(6):36–47, 1997.

[152] S. Johansson et al. Performance of a multiple versus a single reference mimo anc algorithm based on a dornier 328 test data set. *The 1997 International Symposium on Active Control of Sound and Vibration, Budapest, Hungary, Aug. 21–23*, pages 521–528, 1997.

[153] T. Kailath. *Linear Systems*. Prentice-Hall, Englewood Cliffs, New Jersey, 1980.

[154] M. Kallergis. Experimental results on propeller noise attenuation using an 'active noise control' technique. *14th DGLR/AIAA Aeroacoustics Conference, Aachen, Germany, May 11-14, AIAA Paper 92-02-155*, pages 907–918, 1992.

[155] S.K. Katsikas, D. Tsahalis, D. Manolas, and S. Xanthakis. Genetic algorithms for active noise control. In *Proceedings of Noise-93*, pages 167–171, St. Petersburg, Russia, 1993.

[156] S.K. Katsikas, D. Tsahalis, D. Manolas, and S. Xanthakis. A genetic algorithm for active noise control actuator positioning. *Mechanical Systems and Signal Processing*, 9:697–705, 1995.

[157] G. Kim and R. Singh. A study of passive and adaptive hydraulic engine mount systems with emphasis on non-linear characteristics. *J. Sound and Vibration*, 179:427–453, 1995.

[158] M. J. Korenberg, S. A. Billings adn Y. P. Liu, and P. J. McIlroy. Orthogonal parameter estimation algorithm for nonlinear stochastic systems. *International Journal of Control*, 48(1):193–210, 1988.

[159] P. K. Kourmoulis. *Parallel processing in the simulation and control of flexible beam structure system.* PhD thesis, Department of Automatic Control and Systems Engineering, The University of Sheffield, UK, 1990.

[160] S. M. Kuo, Y. C. Huang, and Z. Pan. Acoustic noise and echo cancellation microphone system for videoconferencing. *IEEE Transactions on Consumer Electronics*, 41:1150–1158, 1995.

[161] S. M. Kuo and D. R. Morgan. *Active Noise Control Systems.* John Wiley and Sons, New Jersey, 1996.

[162] S.M. Kuo and D.R. Morgan. *Active Noise Control Systems.* John Wiley & Sons, Inc., New York, 1996.

[163] Heinrich Kuttruff. *Room Acoustics Second Edition.* Applied Science Publish. Ltd, 1973-79.

[164] B. Lange and D. Ronneberger. Control of pipe flow by use of an aeroacoustic instability. In: G.E.A. Meier, P.R. Viswanath (eds.): IUTAM Symposium on Mechanics of Passive and Active Flow Control:305–310, 1999.

[165] A. Lapedes and R. Farber. Nonlinear signal processing using neural networks: Prediction and system modelling. *Preprint LA-UR-87-2662, Los Alamos National Laboratory, Los Alamos*, 1987.

[166] J. D. Leatherwood et al. Active vibration isolator for flexible bodies. *U.S. Patent US 3,566,993. Filed: March 26, 1969. Patented: March 2, 1971.*

[167] L. Lecce, A. Ovallesco, A. Concilio, and A. Sorrentino. Optimal positioning of sensors using genetic algorithms in an active noise control system with piezoelectric actuators. In *Proceedings of Topical Symposium VI on Intelligent Materials and Systems of the 8th CIMTEC world ceramics conference and forum on new materials*, pages 307–314, Faenza, Italy, 1995.

[168] H.-J. Lee, C.-H. Yoo, J.-H. Yun, and D.-H. Youn. An active noise control system for controlling humming noise generated by a transformer. *Internoise 97, Budapest, Aug. 25–27, 1997. Proceedings: Vol. I*, pages 517–520, 1997.

[169] S. Lee. Noise killing system of fans. *U.S. Patent US 5,791,869. Patented: Aug. 11, 1998. Priority (KR): Sept. 18*, 1995.

[170] R. R. Leitch and M. O. Tokhi. The implementation of active noise control systems using digital signal processing techniques. *Proceedings of the Institute of Acoustics*, 8(Part 1):149–157, 1986.

[171] R. R. Leitch and M. O. Tokhi. Active noise control system. *Proc. IEE*, 134A:525–546, 1987.

[172] R. R. Leitch and M. O. Tokhi. Active noise control systems. *IEE Proceedings-A*, 134(6):525–546, 1987.

[173] I. J. Leontaritis and S. A. Billings. Input-output parametric models for nonlinear systems, part i: Deterministic nonlinear systems; part ii: Stochastic nonlinear systems. *International Journal of Control*, 41(2):303–344, 1985.

[174] H. G. Leventhall. Developments in active attenuators. *Proceedings of Noise Control Conference, Warsaw, 13-15 October 1976, 33-42*, 1976.

[175] H. G. Leventhall and Kh. Eghtesadi. Active attenuation of noise: Monopole and dipole systems. *Proceedings of Inter-noise 79: International Conference on Noise Control Engineering, Warsaw, 11-13 September 1979, I, 175-180*, 1979.

[176] H. W. Liepmann and D. M. Nosenchuk. Active control of laminar–turbulent transition. *J. Fluid Dynamics 118*, pages 201–204, 1982.

[177] L. Ljung. *System Identification. Theory for the User.* Prentice-Hall, Englewood Cliffs, 1987.

[178] Transtech Parallel Systems Ltd. *Transtech parallel technology.* Transtech Parallel Systems Ltd, UK, 1991.

[179] P. Lueg. Process of silencing sound oscillations. *U.S. Patent US 2,043,416. Filed: March 8, 1934. Patented: June 9*, 1936.

[180] P. Lueg. Process of silencing sound oscillations. *US Patent 2043416*, 1936.

[181] P. Lueg. Verfahren zur dämpfung von schallschwingungen. *German Patent DE 655 508. Filed: Jan. 27, 1933. Patented: Dec. 30*, 1937.

[182] H. O. Madsen, S. Krenk, and N. C. Lind. *Methods of Structural Safety.* Prentice-Hall, New Jersey, 1996.

[183] L. P. Maguire. *Parallel architecture for Kalman filtering and self-tuning control.* PhD thesis, The Queen's University of Belfast, UK, 1991.

[184] A. Mallock. A method of preventing vibration in certain classes of steamships. *Trans. Inst. Naval Architects 47*, pages 227–230, 1905.

[185] G. Mangiante. Active sound absorption. *J. Acoust. Soc. Am. 61*, pages 1516–1523, 1977.

[186] G. Mangiante and J. P. Vian. Application du principe de huygens aux absorbeurs acoustiques actifs. ii: Approximations du principe de huygens. *Acustica 37*, pages 175–182, 1977.

[187] R. A. Mangiarotty. Control of laminar flow in fluids by means of acoustic energy. *U.S. Patent US 4,802,642. Filed: Oct. 14, 1986. Patented: Feb. 7*, 1989.

[188] M. Mano. Ship design considerations for minimal vibration. *Ship Technology and Research (STAR) 10th Symposium of the Society of Naval Architects and Marine Engineers (SNAME)*, pages 143–156, 1985.

[189] D.A. Manolas, T. Gialamas, and D.T. Tsahalis. A genetic algorithm for the simultaneous optimization of the sensor and actuator positions for an active noise and/or vibration control system. In *Proceedings of Inter Noise 96*, pages 1187–1191, 1996.

[190] T. Martin and A. Roare. Active noise control of acoustic sources using spherical harmonic expansion and a genetic algorithm: simulation and experiment. *Journal of Sound and Vibration*, 212:511–523, 1998.

[191] Nobuo KOIZUMI Masato MIYOSHI, Junko SHIMIZU. N arrangements of noise-controlled points for producing lager quiet zones with multipoint active noise control. *Inter-Noise'94 Congress*, pages 1299–1304, 1994.

[192] W. P. Mason. Piezoelectric damping means for mechanical vibrations. *U.S. Patent US 2,443,417. Filed: March 29, 1945. Patented: June 15*, 1948.

[193] The MathWorks. *MATLAB Version 5.3: High Performance Numeric Computation and Visualisation Package*, volume www.mathworks.com of *ftp.mathworks.com*. The MathWorks, Inc., Natick, MA 01760-1500, 1999.

[194] J. M. McCool et al. Adaptive detector. *U.S. Patent US 4,243,935. Filed: May 18, 1979. Patented: Jan. 6*, 1981.

[195] T. McKelvey. Frequency domain identification. *IFAC Symposium on System Identification, Santa Barbara*, CD, Plenary Talk:1–12, 2000.

[196] R. L. McKinley. Development of active noise reduction earcups for military applications. *ASME Winter Annual Meeting, Anaheim, CA, Dec.*, pages NCA–8B, 1986.

[197] P. H. Meckl and W. P. Seering. Active damping in a three-axis robotics manipulator. *Journal of Vibration, Acoustics, Stress and Reliability in Design*, 107(1):38–46, 1985a.

[198] P. H. Meckl and W. P. Seering. Minimising residual vibration for point-to-point motion. *Journal of Vibration, Acoustics, Stress and Reliability in Design*, 107(4):378–382, 1985b.

[199] P. H. Meckl and W. P. Seering. Reducing residual vibration in systems with time-varying resonances. *Proceedings of IEEE International Conference on Robotics and Automation, 1690-1695*, 1987.

[200] P. H. Meckl and W. P. Seering. Controlling velocity limited system to reduce residual vibration. *Proceedings of IEEE International Conference on Robotics and Automation, Philadelphia, USA, 1428-1433*, 1988.

[201] P. H. Meckl and W. P. Seering. Experimental evaluation of shaped inputs to reduce vibration of a cartesian robot. *Transaction of the ASME Journal of Dynamic Systems, Measurement and Control*, 112(6):159–165, 1990.

[202] G. M. Megson. Practical steps towards algorithmic engineering. *IEE Digest No. 1992/204: Colloquium on Applications of Parallel and Distributed Processing in Automation and Control, London*, 1992.

[203] L. Meirovitch. Dynamics and control of structures. *John Wiley & Sons, New York etc. 1990*, 1990.

[204] J. Melcher and A. Büter. Adaptive structures technology for structural acoustic problems. *Proc. 1st Joint CEAS/AIAA Aeronautics Conference (16th AIAA Aeroacoustics Conference), Munich, Germany, June 12-15*, pages 1213–1220, 1995.

[205] T. Meurers and S. M. Veres. Iterative design for disturbance attenuation. *Int. J. Acoustics and Vibration*, 4:76–83, 1999.

[206] A. R. Mitchell and D. F. GRiffiths. *The finite difference method in partial differential equations*. John Wiley and Sons, New York, 1980.

[207] H. Miyamoto, H. Ohmori, and A. Sano. Parametrization of all plug-in adaptive controllers for sinusoidal disturbance rejection. *Proc. Amer. Control. Conf. (ACC99)*, San Diego, USA, 1999.

[208] I. Moore, T. J. Sutton, and S. J. Elliott. Use of nonlinear controllers in the active attenuation of road noise inside cars. *Proc. Recent Advance in Active Control of Sound and Vibration*, pages 682–690, 1991.

[209] D. R. Morgan. An analysis of multiple correlation cancellation loops with a filter in the auxiliary path. *IEEE Trans. Acoust. Speech, and Signal Processing*, ASSP-28:454–467, 1980.

[210] S. Morishita and J. Mitsui. An electronically controlled engine mount using electro-rheological fluid. *SAE Special Publication 936*, pages 97–103, 1992.

[211] P. M. Morse and K. U. Ingard. Theoretical acoustics. *McGraw-Hill Book Co., New York etc. Chapter 7.1*, 1968.

[212] E. Mosca and G. Zappa. ARX modelling of controlled ARMAX plants and LQ adaptive controllers. *IEEE Trans. Automatic Control*, AC-34:371–375, 1989.

[213] H. Moulin and E. Bayo. On the accuracy of end-point trajectory tracking for flexible arms by non-causal inverse dynamic solution. *Transaction of the ASME Journal of Dynamic Systems, Measurement and Control*, 113(2):320–324, 1991.

[214] G. V. Moustakides. Correcting the instability due to finite precision of the fast kalman identification algorithms. *Signal Processing 18*, pages 33–42, 1989.

[215] M. L. Munjal and L. J. Eriksson. An analytical, one-dimensional, standing-wave model of a linear active noise control system in a duct. *The Journal of the Acoustical Society of America*, 84(3):1086–1093, 1988.

[216] K. S. Narendra and K. Parthasarathy. Identification and control of dynamical systems using neural networks. *IEEE Transactions on Neural Networks*, 1(1):4–27, 1990.

[217] K.S. Narendra and D.N. Steeter. An adaptive procedure for controlling undefined linear processes. *IEEE Trans. Automatic Control*, pages 545–548, 1964.

[218] P. A. Nelson, A. R. D. Curtis, S. J. Elliott, and A. J. Bullmore. The active minimization of harmonic enclosed sound fields, part i: Theory. *Journal of Sound and Vibration*, 117(1):1–13, 1987.

[219] P. A. Nelson and S. J. Elliott. *Active Control of Sound*. Academic Press, London, 1995.

[220] P.A. Nelson, A.R.D. Curtis, and S.J. Elliott. Quadratic optimisation problems in the active control of free and enclosed sound fields. *Proceedings of the Institute of Acoustics*, 7:55–64, 1985.

[221] H. T. Ngo. Tip vortex reduction system. *U.S. Patent US 5,791,875. Filed: Sept. 10, 1996. Patented: Aug. 11, 1998*, 1998.

[222] D. Nguyen and B. Widrow. Improving the learning speed of 2-layer neural networks by choosing initial values of the adaptive weights. *Proceedings of International Joint Conference on Neural Networks, Washington DC, 3, 21-26*, 1990.

[223] B. Ninnness. A stochastic approach to linear estimation in $h_{infty}$. *Automatica*, 34(4):405–414, 1998.

[224] J. P. Norton. *An Introduction to Identification*. Academic Press, London, 1986.

[225] H. Ohmori, N. Narita, and A. Sano. A new design of plug-in adaptive controller for rejection of periodic disturbances. *Proc. IFAC Symp. Adaptive Systems in Control and Signal Processing, Glasgow, UK:297–302*, 1998.

[226] N. Ojiro, M. Kajiki, H. Ohmori, H. Tsuji, and A. Sano. Adaptive algorithm considering stability and application to adaptive active noise canceling. *IEICE Trans. Fundamentals*, J78-A(10):1289–1297, 1995.

[227] H. F. Olson. Electronic control of noise, vibration, and reverberation. *J. Acoust. Soc. America*, 28:966–972, 1956.

[228] H. F. Olson. Electronic sound absorber. *U.S. Patent US 2,983,790. Filed: April 30, 1953. Patented: May 9, 1961*, 1961.

[229] H. F. Olson. Electronic sound absorber. *US Patent No. 2983790*, 1961.

[230] H. F. Olson and E. G. May. Electronic sound absorbers. *Journal of the Acoustical Society of America*, 25:1130–1136, 1953.

[231] H.F. Olson. Electronic Control of Noise, Vibration and Reverberation. *Journal of the Acoustical Society of America 28*, pages 966–972, 1956.

[232] T. Onsay and A. Akay. Vibration reduction of a flexible arm by time optimal open-loop control. *Journal of Sound and Vibration*, 142(2):283–300, 1991.

[233] A V Oppenheim, K C Zangi, M Feder, and D Gauger. Single sensor active noise cancellation based on the em algorithm. *ICASSP*, 1992.

[234] R. Ortega and R. Lozano. A journal on direct adaptive control of systems with bounded disturbances. *Automatica*, 23(2):253–, 1987.

[235] Nelson P.A. and Elliott S.J. *Active Control of Sound*. Academic Press, New York, 1992.

[236] N. R. Petersen. Design of large scale tuned mass dampers. *In: H. H. E. Leipholz (ed.): Structural Control, North-Holland Publ. Co.*, pages 581–596, 1980.

[237] J. Piraux and S. Mazzanti. Broadband active noise attenuation in three-dimensional space. *Internoise 85, München, Proceedings*, pages 485–488, 1985.

[238] H. Poerwanto. *Dynamic simulation and control of flexible manipulator systems*. PhD thesis, Department of Automatic Control and Systems Engineering, The University of Sheffield, UK, 1998.

[239] L.D. Pope. On the transmission of sound through finite closed shells: Statistical energy analysis, modal coupling, and non-resonant transmission. *Journal of the Acoustical Society of America*, 50(3):1004–1018, 1971.

[240] S. R. Popovich. Fast adapting control system and method. *U.S. Patent US 5,602,929. Filed: Jan. 30, 1995. Patented: Feb. 11, 1997*, 1997.

[241] S. Pottie and D. Botteldooren. Optimal placement of secondary sources for active noise control using a genetic algorithm. In *Proceedings of Inter Noise 96*, pages 1101–1104, 1996.

[242] M. J. D. Powell. Radial basis functions for multivariable interpolation: A review. *Proceedings of the IMA Conference on Algorithms for the Approximation of Functions and Data, RMCA, Shrivenham*, 1985.

[243] H. Preckel and D. Ronneberger. Dynamic control of the jet-edge-flow. pages 349–354, 1999.

[244] S.S. Rao, T.S. Pan, and V.B. Venkavya. Optimal placement of actuators in actively controlled structures using genetic algorithms. *AIAA Journal*, 29:942–943, 1991.

[245] K. S. Rattan, V. Feliu, and Jr. H. B. Brown. Tip position control of flexible arms. *IEEE International Conference on Robotics and Automation, 3, 1803-1808*, 1990.

[246] Lord Rayleigh. The theory of sound. *Vol. II, Chapter XIV, § 282: Two Sources of Like Pitch; Points of Silence; Experimental Methods. MacMillan & Co, London etc., 1st ed. 1877/78: p. 104-106; 2nd ed. 1894/96 and Reprints (Dover, New York)*, pages 116–118, 1978.

[247] D. Rees and P. Witting. *Controller implementations using novel processors, in K. Warwick and D. Rees (eds.): Industrial digital control systems, 2nd edn.* Peter Peregrinus, London, 1988.

[248] R.E.Skelton and G. Shi. Iterative identification and control using a weighted q-Markov cover with measurement noise. *Signal Processing*, 52:217–234, 1996.

[249] C. F. Ross. Experiments on the active control of transformer noise. *Journal of Sound and Vibration*, 61(4):473–480, 1978.

[250] C. F. Ross. An adaptive digital filter for broadband active sound control. *Journal of Sound and Vibration*, 80(3):381–388, 1982.

[251] A. Roure. Self-adaptive broadband active sound control system. *Journal of Sound and Vibration*, 101(3):429–441, 1985.

[252] C.E. Ruckman and C.R. Fuller. Optimising actuator locations in active noise control systems using subset selection. *Journal of Sound and Vibration*, 186:395–406, 1995.

[253] A. Sano. Wavelet transform and its application to system identification. *System/Control/Information*, 42(2):103–110, 1998.

[254] K. Schaaf. Private communication from the research department. *Volkswagen AG, Wolfsburg, Germany*, 1988.

[255] J. Scheuren, U. Widmann, and J. Winkler. Active noise control and sound quality design in motor vehicles. *SAE Technical Paper Series No. 1999-01-1846. Proc. of the 1999 Noise and Vibration Conference and Exposition, Traverse City, MI, USA, May 17-20*, 1999.

[256] R. Schirmacher. Schnelle algorithmen für adaptive iir-filter und ihre anwendung in der aktiven schallfeldbeeinflussung. *Ph. D. Thesis, Göttingen 1996. Abstract: ACUSTICA – acta acustica 82 (1996) 384*, 1996.

[257] R. Schirmacher and D. Guicking. Theory and implementation of a broadband active noise control system using a fast rls algorithm. *Acta Acustica 2 (1994) 291-300*, 1994.

[258] R. J. P. Schrama. Accurate models for control design: the necessity of an iterative scheme. *IEEE Trans. Automatic Control*, AC-37:991–994, 1992.

[259] M. R. Schroeder, D. Gottlob, and K. F. Siebrasse. Comparative study of european concert halls. correlation of subjective preference with geometric and acoustic parameters. *J. Acoust. Soc. Am. 56 (1974) 1195 1201*, 1974.

[260] B. Schwarzschild. First of the twin 10-meter keck telescopes starts doing astronomy. *Physics Today*, 46:17–18, 1993.

[261] T. J. Sejnowski and C. R. Rosenberg. Nettalk: A parallel network that learns to read aloud. Technical Report JHU/EECS-86/01, Johns Hopkins University USA, 1986.

[262] W. R. Short. Global low frequency active noise attenuation. *Proceedings of Inter-noise 80: International Conference on Noise Control Engineering, Florida, 8-10 December 1980, II*, pages 695–698, 1980.

[263] B. Siciliano and W. J. Book. A singular perturbation approach to control of lightweight flexible manipulators. *International Journal of Robotics Research*, 7(4):79–90, 1988.

[264] M. H. Silverberg et al. Outbound noise cancellation for telephonic handset. *U.S. Patent US 5,406,622. Filed: Sept. 2, 1993. Patented: April 11, 1995*, 1995.

[265] M.T. Simpson and C.H. Hansen. Use of genetic algorithms to optimise actuator placement for active control of interior noise in a cylinder with floor structure. *Noise Control Engineering Journal*, 44:169–184, 1996.

[266] N. C. Singer and W. P. Seering. Using causal shaping techniques to reduce residual vibration. *Proceedings of IEEE International Conference on Robotics and Automation, Philadelphia, 1434-1439*, 1988.

[267] N. C. Singer and W. P. Seering. Pre-shaping command inputs to reduce systems vibration. *Transaction of the ASME Journal of Dynamic Systems, Measurement and Control*, 112(1):76–82, 1990.

[268] N. C. Singer and W. P. Seering. An extension of command shaping methods for controlling residual vibration using frequency sampling. *Proceedings of IEEE International Conference on Robotics and Automation, 800-805*, 1992.

[269] Elliott S.J. and Nelson P.A. Active noise control. 28:12–35, 1993.

[270] R. E. Skelton. Model error concept in control design. *Int. J. Control*, 49:1725–53, 1989.

[271] R. E. Skelton and J. Lu. Iterative identification and control design using finite-signal-to-noise models. *Mathematical Modelling of Systems*, 3(1):102–135, 1997.

[272] S.D. Snyder and C.H. Hansen. The design of systems to actively control periodic sound transmission into enclosed spaces, Part 1: Analytical models. *Journal of Sound and Vibration*, 170(4):433–449, 1994.

[273] S.D. Snyder and C.H. Hansen. The design of systems to actively control periodic sound transmission into enclosed spaces, Part 2: Mechanisms and trends. *Journal of Sound and Vibration*, 170(4):451–472, 1994.

[274] M. M. Sondhi. Closed loop vibration echo canceller using generalized filter networks. *U.S. Patent US 3,499,999. Filed: Oct. 31, 1966. Patented: March 10, 1970*, 1970.

[275] T. T. Soong and H. G. Natke. From active control to active structures. *VDI-Berichte 695 (1988) 1–18*, 1988.

[276] W. D. Stanley. *Electronic communications systems*. Reston Publishing Company, Virginia, 1982.

[277] H. R. Stark and C. Stavrinidis. Esa microgravity and microdynamics activities - an overview. *Acta Astronautica*, 34:205–221, 1994.

[278] G. J. Stein. A driver's seat with active suspension of electropneumatic type. *ASME J. of Vibration and Acoustics 119 (1997) 230–235*, 1997.

[279] Trevor J. Sutto Stephen J. Elliott. Performance of feedforward and feedback systems for active control. *IEEE Transactions on speech and audio processing*, 4(3):214–223, 1996.

[280] J. C. Stevens and K. K. Ahuja. Recent advances in active noise control. *AIAA Journal*, 29:1058–1067, 1991.

[281] Hyoun suk Kim, Youngjin Park, and Kang hyug Sur. Active noise control of road booming noise with constraint multiple filtered-x lms algorithm. *Proc. INTER-NOISE 96*, pages 1155–1158, 1996.

[282] T. J Sutton, S. J. Elliott, P. A. Nelson, and I. Moore. The active control of road noise inside vehicles. *Proc. INTER-NOISE;90*, pages 1247–1250, 1990.

[283] T. J Sutton, S. J. Elliott, P. A. Nelson, and I. Moore. The active control of road noise inside vehicles. *Proc. INTER-NOISE 90*, pages 1247–1250, 1990.

[284] J. C. Swigert. Shaped torque techniques. *Journal of Guidance and Control*, 3(5):460–467, 1980.

[285] M. A. Swinbanks. The active control of sound propagation in long ducts. *J. Sound Vib. 27 (1973) 411–436*, 1973.

[286] M. A. Swinbanks. Active control of sound propagation in long ducts. *Journal of Sound and Vibration*, 27(3):411–436, 1973.

[287] M. A. Swinbanks. The active control of noise and vibration and some applications in industry. *Proc. IMechE 198 A (1984), No. 13, p. 281–288*, 1984.

[288] G. Syswerda. Uniform crossover in genetic algorithms. In *Proceedings of the Third International Conference on Genetic Algorithms*, pages 2–9, George Mason Univ., Washington DC, 1989.

[289] M. Taki et al. Sound attenuating system. *U.S. Patent US 5,347,585. Filed: Sept. 10, 1991. Patented: Sept. 13, 1994*, 1994.

[290] K.S. Tang, K.F. Man, S. Kwong, C.Y. Chan, and C.Y. Chu. Application of the genetic algorithm to real time active noise control. *Real Time Systems*, 11:289–302, 1996.

[291] K.S. Tang, K.F. Man, S. Kwong, and Q. He. Genetic algorithms and their applications. *IEEE Signal Processing Magazine*, 13:22–37, 1996.

[292] D.R. Thomas, P.A. Nelson, and S.J. Elliott. Active control of the transmission of sound through a thin cylindrical shell, Part II: The minimisation of acoustic potential energy. *Journal of Sound and Vibration*, 167(1):113–128, 1993.

[293] M. O. Tokhi and A. K. M. Azad. Active vibration suppression of flexible manipulator systems: Open-loop control methods. *International Journal of Active Control*, 1(1):15–43, 1995.

[294] M. O. Tokhi and A. K. M. Azad. Control of flexible manipulator systems. *Proceedings of IMechE-I: Journal of Systems and Control Engineering*, 210(I2):113–130, 996.

[295] M. O. Tokhi and M. A. Hossain. Adaptive active control of noise and vibration. *Proceedings of the Institute of Acoustics*, 16(Part 2):245–253, 1994.

[296] M. O. Tokhi and M. A. Hossain. Cisc, risc and dsp processors in real-time signal processing and control. *Microprocessors and Microsystems*, 19(5):291–300, 1995.

[297] M. O. Tokhi and M. A. Hossain. Real-time active control using sequential and parallel processing methods. *Proceedings of the fourth International Congress on Sound and Vibration, St Petersburg, 24-27 June 1996, 1, 391-398*, 1996.

[298] M. O. Tokhi, M. A. Hossain, M. J. Baxter, and P. J. Fleming. Heterogeneous and homogeneous parallel architectures for real-time active vibration control. *IEE Proceedings-D: Control Theory and Applications*, 142(6):1–8, 1995.

[299] M. O. Tokhi and R. R. Leitch. Practical limitations in the controller design for active noise control systems in three-dimensions. *Proceedings of Inter-noise 88: International Conference on Noise Control Engineering, Avignon, 30 August - 01 September 1988, 1037-1040*, 1988.

[300] M. O. Tokhi and R. R. Leitch. Self-tuning active noise control. *IEE Digest No 1989/46: Colloquium on Adaptive Filters, London, 22 March 1989, 9/1-9/4*, 1989.

[301] M. O. Tokhi and R. R. Leitch. Design and implementation of self-tuning active noise control systems. *IEE Proceedings-D: Control Theory and Applications*, 138(4):421–430, 1991a.

[302] M. O. Tokhi and R. R. Leitch. The robust design of active noise control systems based on relative stability measures. *Journal of the Acoustical Society of America*, 90(1):334–345, 1991b.

[303] M. O. Tokhi and R. R. Leitch. Design of active noise control systems operating in three-dimensional dispersive propagation medium. *Noise Control Engineering Journal*, 36(1):41–53, 1991c.

[304] M. O. Tokhi and R. R. Leitch. Active noise control. *The Oxford Engineering Sciences Series, Vol. 29. Clarendon Press (Oxford University Press), Oxford, UK 1992*, 1992.

[305] M. O. Tokhi and R. R. Leitch. Parametric analysis of field cancellation in a three-dimensional propagation medium. *Journal of Sound and Vibration*, 155(3):497–514, 1992a.

[306] M. O. Tokhi and R. R. Leitch. *Active noise control*. Clarendon Press, Oxford, 1992b.

[307] M. O. Tokhi, G. S. Virk, and M. A. Hossain. Integrated dsp$^3$ systems for adaptive active control. *IEE Digest No. 1992/185: Colloquium on Active Techniques for Vibration Control - Sources, Isolation and Damping, London, 28 October 1992, 6/1-6/4*, 1992.

[308] M. O. Tokhi and R. Wood. Active control of noise using neural networks. *Proceedings of the Institute of Acoustics*, 17(Part 4):209–216, 1995.

[309] M. O. Tokhi and R. Wood. Radial basis function neuro-active noise control. *Preprints of IFAC-96: 13th World Congress, San Francisco, 30 June-05 July 1996, K, 97-102*, 1996.

[310] M. O. Tokhi and R. Wood. Active noise control using radial basis function networks. *Control Engineering Practice*, 5(9):1311–1322, 1997a.

[311] M. O. Tokhi and R. Wood. Active noise control using multi-layered perceptron neural networks. *International Journal of Low Frequency Noise, Vibration and Active Control*, 16(2):109–144, 1997b.

[312] Hironori Tokuno, Yuko Watanabe, Hareo Hamada, Ole Kirkeby, and Philip A Nelson. Binaural reproduction in stereo dipole. *ASA and ASJ, Third Joint Meeting*, pages 1317–1322, 1996.

[313] Hironori Tokuno, Yuko Watanabe, Hareo Hamada, and Philip A Nelson. Virtual source image in stereo dipole system (in Japanese). *Proc. Acoust. Soc. Japan Symposium*, pages 591–592, 1997.

[314] M. C. J. Trinder and P. A. Nelson. The acoustical virtual earth and its application to ducts with reflecting terminations. *Proceedings of Internoise 83:International Conference on Noise Control Engineering, Edinburgh, 13-15 July 1983, I, 447-450*, 1983a.

[315] M. C. J. Trinder and P. A. Nelson. Active noise control in finite length ducts. *Journal of Sound and Vibration*, 89(1):95–106, 1983b.

[316] D.T. Tsahalis, S.K. Katsikas, and D.A. Manolas. A genetic algorithm for optimal positioning of actuators in active noise control: Results from the ASANCA project. *Proceedings of Inter Noise 93*, pages 83–88, 1993.

[317] H. Unbehauen and G. P. Rao. *Identification of Continuous Systems*. North-Holland, Amsterdam, 1987.

[318] H. Unbehauen and G. P. Rao. Continous-time approaches to system identification: - a survey. *Automatica*, 26(1):23–35, 1990.

[319] A. Vang. Vibration dampening. *U.S. Patent US 2,361,071. Filed: Sept. 23, 1942. Patented: Oct. 24, 1944*, 1944.

[320] I. Veit. Gehörschutz-kopfhörer. elektronik kontra lärm. *Funkschau 23 (1988) 50–52*, 1988.

[321] I. Veit. Anordnung und verfahren zur aktiven reduzierung von reifenschwingungen. *German Patent DE 197 23 516 C1. Filed: June 5, Patented: Oct. 29, 1998*, 997.

[322] S. M. Veres. Iterative identification and control redesign via unfalsified sets of models: a basic scheme. *Int. J. Control*, 72(10):887–903, 1999.

[323] S. M. Veres and D. S. Wall. *Synergy and Duality of Identification and Control*. Taylor & Francis, London, 2000.

[324] M. Vogt. General conditions of phase cancellation in an acoustic field. *Archives of Acoustics*, 1(1):109–125, 1976.

[325] Ren W. and Kumar P.R. Adaptive active noise control: structures algorithms and convergence analysis. *Proc. InterNoise89*, pages 435–441, 1989.

[326] B. Wang. Optimal placement of microphones and piezoelectric transducer actuators for far-field sound radiation control. *Journal of the Acoustical Society of America*, 99(5):2975–2984, 1996.

[327] S. Wang. Open-loop control of a flexible robot manipulator. *International Journal of Robotics and Automation*, 1(2):54–57, 1986.

[328] C.T. Wangler and C.H. Hansen. Genetic algorithm adaptation of nonlinear filter structures for active sound and vibration control. *ICASSP*, 1994.

[329] P. E. Wellstead and M. B. Zarrop. *Self-tuning systems - Control and signal processing*. John Wiley, Chichester, 1991.

[330] U. Weltin. Aktive schwingungskompensation bei verbrennungsmotoren. *Fortschrittsberichte VDI, Reihe 12, No. 179 (1993)*, III:505–508, 1993.

[331] M. Wenzel. Untersuchungen zur breitbandigen messung und regelung der akustischen wandimpedanz an einer aktiven schallwand mit adaptiven filtern. *Ph. D. Thesis, Göttingen 1992*, 1992.

[332] D. Whitley. The genitor algorithm and selection pressure: Why rank-based allocation of reproductive trials is best. *The 3rd International Conference on Genetic Algorithms*, pages 116–121, 1989.

[333] D. Whitley and T. Hanson. Optimising neural networks using faster, more accurate genetic search. *The 3rd International Conference on Genetic Algorithms*, 1989.

[334] G. Wickern. Windkanal. *German Patent Application DE 19702390 A1. Filed: Jan. 24, 1997. Published: July 30, 1998*, 1976.

[335] B. Widrow. Ecg enhancement by adaptive cancellation of electrosurgical interference. *U.S. Patent US 4,537,200. Filed: July 7, 1983. Patented: Aug. 27, 1985*, pages 391–396, 1985.

[336] B. Widrow. Seismic exploration method and apparatus for cancelling interference from seismic vibration source. *U.S. Patent US 4,556,962. Filed: April 21, 1983. Patented: Dec. 3, 1985*, 1985.

[337] B. Widrow and S. D. Stearns. *Adaptive Signal Processing*. Prentice Hall. Prentice Hall, Englewood Cliffs, 1985.

[338] B. Widrow et al. Adaptive noise cancelling: Principles and applications. *Proc. IEEE 63 (1975) 1692-1716*, 1975.

[339] J. Winkler and S. J. Elliott. Adaptive control of broadband sound in ducts using a pair of loudspeakers. *Acustica 81 (1995) 475-488*, 1995.

[340] R. Wood. *Development of neuro-adaptive active noise control systems*. PhD thesis, Department of Automatic Control and Systems Engineering, The University of Sheffield, UK, 1997.

[341] H. Xia and S. M. Veres. Improved efficiency of adaptive robust control by model unfalsification. *Automatica*, 36, 1999.

[342] L. Yao, W.A. Sethares, and D.C. Kammer. Sensor placement for on-orbit modal identification via a genetic algorithm. *AIAA Journal*, 31(10):1922–1928, 1993.

[343] D. S. Yoo and M. J. Chung. A variable structure control with simple adaptation laws for upper bounds on the norm of the uncertainties. *IEEE Transaction on Automatic Control*, 37(6):860–864, 1992.

[344] K. D. Young and U. Ozguner. Frequency shaping compensator design for sliding mode. *International Journal of Control*, 57:1005–1019, 1993.

[345] J. Yuh. Application of discrete-time model reference adaptive control to a flexible single-link robot. *Journal of Robotics Systems*, 4(5):621–630, 1987.

[346] A.C. Zander. *Active control of aircraft interior noise*. Ph.D. Dissertation, The University of Adelaide, 1994.

[347] T. Zhou and H. Kimura. Time domain identification for robust control. *Systems & Control Letters*, 20:167–178, 1993.

[348] O. C. Zienkiewicz. *The Finite Element Method*. McGraw-Hill, London, 1977.

[349] D.C. Zimmerman. A darwinian approach to the actuator number and placement problem with non-negligible actuator mass. *Mechanical Systems and Signal Processing*, 7(4):363–374, 1993.

# Index

acoustic box 333–4
acoustic echo cancellation 14
acoustic feedback 7–8, 26
acoustical short-circuit 11
active absorber 8
active feedforward cancellation of sound 7
active flow control 22–3
active mounts 19–20, 148–54
   transfer function model 148–9
active noise control 70–2
   adaptive prediction 224–6
   electric locomotive 319–40
   feedforward 59–65
   human head 223–39
   identification-based 83–5
   multichannel 75–95
   neural networks 167–82
   overview 4–16
   road booming noise 345–54
   three-dimensional propagation 25–55
active noise control structure 27–30
active noise control system
   performance hierarchy 186
active optics 20–1
active seats 334–7
active structural acoustic control 21–2
active vibration control 17–22
   beam 380–1, 388
   cantilever beam system 359–62
   feedforward 360–1
   flexible manipulators 275–318
   H-infinity design 135–58
   microvibrations 241–74
   neural networks 159
   overview 17–22
   passive 136
actuator noise attenuation 151, 158
adaptation algorithm 58–9, 69, 71
adaptive ANC system 27, 85–6
   neural networks 160
adaptive digital filter 7–8
adaptive feedback control 69–70
   stability 70
adaptive feedback controller 72

adaptive feedforward control 9, 75
   sound 7–8, 71, 75–6
adaptive harmonic control 97–116
   frequency selective LMS 107–10
   frequency selective RLS 103–7
adaptive inverse control 312, 314–17
   stability 315
adaptive methods 58
adaptive noise cancelling 9–10
adaptive optics 21
adaptive prediction active noise control
   system 224–6
   single-channel 224–5
   multiple-channel 225–6
adaptronics 19
aeroacoustic instability 23
aerodynamic noise 320, 326
air conditioner noise 326
aircraft 15–16, 19, 22–3, 188
   fly-over noise 16
   helicopters 19, 22, 64, 68
   passengers 14, 64–5, 71
   pilots 13
   propeller 71, 64–5
   safety 23
   skin vibrations 17
algorithms
   adaptation 58–9, 69, 71
   DFL scheme 172–3, 179
   direct neuro-modelling and control 174
   Feintuch 9
   filtered-reference LMS 57–8, 60–4, 70–1
   filtered-u 189
   filtered-x LMS 7–9, 76, 81, 87, 93–4, 98,
      103, 227, 334, 346–51, 353
   FRM scheme 169–70, 177
   genetic 185–220
   gradient descent 187, 189
   heterogeneous 358–9
   identification-based 83–5, 90
   IFL scheme 173, 179
   indirect adaptive 87–8, 93–4
   LMS 9
   orthogonal forward regression 163–4